DYNAMIC ASPECTS OF STRUCTURAL CHANGE IN LIQUIDS AND GLASSES

ANNALS OF THE NEW YORK ACADEMY OF SCIENCES
Volume 484

DYNAMIC ASPECTS OF STRUCTURAL CHANGE IN LIQUIDS AND GLASSES

Edited by C. Austen Angell and Martin Goldstein

The New York Academy of Sciences
New York, New York
1986

Cover: *Egyptian glass, probably first century B.C.–first century A.D. The piece, which measures 19 × 23 mm, is a fused mosaic plaque and shows an ancient Egyptian motif: the figure of an ibis, the sacred animal of Thoth, the moon deity and god of science. (Paperback volume only.)*

Courtesy of the Freer Gallery of Art, Smithsonian Institution, Washington, DC (Acc. no. 08.66).

Library of Congress Cataloging-in-Publication Data

Dynamic aspects of structural change in liquids and glasses.

(Annals of the New York Academy of Sciences, ISSN 0077-8923; vol. 484)
Results of a conference held in New York, N.Y. on December 1–3, 1985, by the New York Academy of Sciences.
Bibliography: p.
Includes index.
1. Glass—Congresses. 2. Liquids—Congresses.
3. Phase rule and equilibrium—Congresses. 4. Chemistry, physical and theoretical—Congresses. I. Angell, C. Austen, 1933– . II. Goldstein, Martin, 1919 Nov. 18– . III. New York Academy of Sciences. IV. Series: Annals of the New York Academy of Sciences; v. 484.
Q11.N5 vol. 484 500 s 86-33306
[QC176.8.A44] [530.4′2]
ISBN 0-89766-363-2
ISBN 0-89766-364-0 (pbk.)

SP
Printed in the United States of America
ISBN 0-89766-363-2 (cloth)
ISBN 0-89766-364-0 (paper)
ISSN 0077-8923

ANNALS OF THE NEW YORK ACADEMY OF SCIENCES

Volume 484
December 31, 1986

DYNAMIC ASPECTS OF STRUCTURAL CHANGE IN LIQUIDS AND GLASSES[a]

Editors and Conference Organizers
C. AUSTEN ANGELL and MARTIN GOLDSTEIN

CONTENTS

[a]This volume is the result of a conference entitled Dynamic Aspects of Structural Change in Liquids and Glasses, which was held in New York, New York on December 1–3, 1985 by the New York Academy of Sciences.

Part IV. Relaxation toward What?

Part V. New Perspectives on Structure

Poster Papers

Financial assistance was received from:

- ALLIED CORPORATION
- EXXON RESEARCH AND ENGINEERING COMPANY
- GENERAL ELECTRIC
- NATIONAL SCIENCE FOUNDATION
- OFFICE OF NAVAL RESEARCH/DEPARTMENT OF NAVY GRANT N00014-86-G-0027

Preface

This volume comprises the proceedings of the third in a series of conferences held under the auspices of the New York Academy of Sciences on problems of the glassy state. The perspective in all three conferences has been one that stresses the common features of glasses rather than features that are limited to particular chemical classes of glass-forming substances. It is striking how much the field has matured since the first conference a decade ago and how many new powerful and sophisticated approaches are being taken; it is equally striking how most of the problems focused on then remain unsolved.

We are grateful to the speakers, session chairs, poster presenters, and other participants for their contributions, which enable us to take great pride in the scientific quality of these proceedings. We are also grateful to the staff of the New York Academy of Sciences for, as usual, taking from our shoulders so much of the burden of preparing the conference and seeing that it ran smoothly, as well as for the care taken in the production of this volume. In particular, we wish to thank Ms. Ellen Marks and Mr. Albert Capozzelli of the Conference Department and Mr. Kevin M. Mayer of the Editorial Department. M. G. is grateful to Yeshiva University for releasing him from a portion of his teaching load to enable him to work on the organization of the conference.

C. AUSTEN ANGELL
MARTIN GOLDSTEIN

PART I. VITRIFICATION AND DEVITRIFICATION: THE INS AND OUTS OF IT

Tiling, Prime Numbers, and the Glass Transition

FRANK H. STILLINGER AND THOMAS A. WEBER

AT&T Bell Laboratories
Murray Hill, New Jersey 07974

BACKGROUND

Thermal motions in high-temperature liquids permit rapid and effective exploration of alternative molecular packings. But such exploration becomes increasingly sluggish if the liquid is supercooled, and it largely ceases if further cooling passes through a glass transition. This paper proposes and studies a new type of statistical mechanical model that is intended to explain such phenomena. These models possess several conceptual ingredients that seem not to have been considered before in the extensive theoretical literature devoted to supercooling and glass formation.[1-6] Although many of their features have yet to be worked out in quantitative detail, the early results reported here indicate success in describing qualitatively both static structure and relaxation behavior in real glasses. Furthermore, as discussed below, these models can in principle also be used to discuss the kinetics of crystal-phase nucleation from the supercooled liquid or vitreous solid states.

The inverse of the shear viscosity $\eta(T)$ provides a rough but convenient measure at absolute temperature T of the mean rate at which a liquid undergoes spontaneous structural change at the molecular level. For many substances the Tammann–Vogel–Fulcher (TVF) equation[7]

$$\eta(T) = A \exp\,[B/(T - T_0)] \tag{1}$$

provides a satisfactory fit to viscosity measurements, with positive constants A, B, and T_0. The apparent activation enthalpy $\Delta H^\ddagger$ and entropy $\Delta S^\ddagger$ can be formally extracted from the TVF equation:

$$\frac{\Delta H^\ddagger}{k_B T} = \frac{BT}{(T - T_0)^2} \tag{2}$$

$$\frac{\Delta S^\ddagger}{k_B} = \frac{BT}{(T - T_0)^2} - \frac{B}{T - T_0} - \ln A \tag{3}$$

Both of these quantities diverge as T decreases toward T_0. This suggests in turn that as T decreases, geometrically larger and larger zones of activation must be involved to permit macroscopic shear flow to take place. Evidently particles constituting the system manage to become well packed in domains of increasing size, so that flow and other relaxation processes require "unbundling" of these larger domains in order to permit repacking in an alternative way.

This point of view focuses attention on the temperature-variable mean size of well-packed domains as a relevant physical parameter, an attribute not explicitly clear in other models for glasses.

From this perspective we can now state several fundamental concepts upon which our new family of models will be based: 1) Any molecular configuration of the material system of interest can be resolved into particle domains that are well packed but spatially nonrepetitive. These domains have variable size, and in principle could encompass the entire system. 2) The set of possible domains may include a crystalline fragment, but in view of the nonrepetitiveness constraint only a few unit cells could be involved. Large well-packed domains of necessity then are amorphous. 3) Walls between contiguous domains usually will have positive misfit energy; that is, they are locations of poor packing. The exception to this rule would be the properly registered contacts between pairs of the small crystalline domains, which together then would form a larger crystal fragment. The perfect macroscopic crystal formally will be regarded as a periodic array consisting just of the special crystal fragment domains. 4) Transitions occur as thermal motion spontaneously unbundles and repacks particles into different domain patterns.

TILING MODELS

We will now convert the four informal statements written above into a precise statistical mechanical model. For simplicity of exposition and of visualization we will confine attention to two dimensions, although the extension to the more realistic case of three dimensions is quite straightforward.

Let the region occupied by the system be covered by a square lattice of N sites with unit spacing. We will suppose that the "well-packed but spatially nonrepetitive" particle domains are squares with integer sides ($1 \times 1, 2 \times 2, \ldots$), and that these must tile the lattice, that is, cover it without gaps or overlaps. For some positive integer a we will also suppose that any $a \times a$ square in the tiling represents a crystalline fragment but that all other square sizes represent well-packed amorphous domains. The allowed system configurations are all the distinct tilings of the given lattice. Periodic boundary conditions will apply.

The potential energy for any configuration will be assigned the form ($\lambda, \mu \geq 0$)

$$\Phi = 2\lambda \sum_{j \geq 1} j n_j - a\mu N_{aa} \tag{4}$$

Here n_j is the number of $j \times j$ squares, and N_{aa} is the number of pairs of $a \times a$ squares that are in full contact along a common side. The coupling constant λ represents mismatch energy per unit length of boundary between pairs of amorphous domains, between an amorphous and a crystalline domain, or between a pair of misregistered crystalline domains. The other coupling constant μ is present to favor crystallization, and indeed if

$$\mu > \lambda > 0 \tag{5}$$

the ground-state configuration for Φ consists of a periodic registered array just of $a \times a$ squares filling the entire macroscopic lattice. FIGURES 1A & 1B show the crystalline and typical amorphous domain patterns, respectively.

Static thermodynamic properties of the tiling model can be obtained from the canonical partition function ($\beta = 1/k_B T$). Thus

$$Z(\beta) = \sum_{\text{tilings}} \exp(-\beta\Phi) \tag{6}$$

provided that the system is ergodic and that sufficient time is available for equilibration to have occurred at the prevailing temperature. If the inequality shown in equation 5 is satisfied, Z would reflect the existence of a melting transition at some temperature T_m.

In view of primary interest in the glass transition, we elect to suppress crystallization altogether by setting μ equal to zero. Consequently

$$Z(\beta) = \sum_{\{n_j\}} \Omega(n_1, n_2 \ldots) \exp[-\beta\Phi(n_1, n_2 \ldots)] \tag{7}$$

where Ω is the number of tilings possible with the given numbers $n_1, n_2 \ldots$ of squares of various sizes.

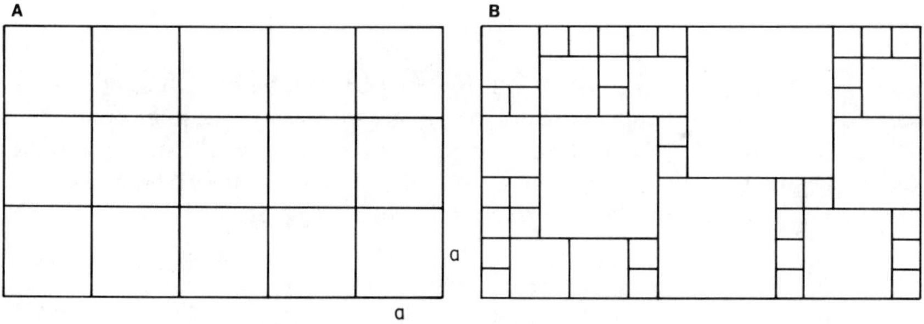

FIGURE 1. (**A**) A registered periodic arrangement of $a \times a$ squares representing the perfect macroscopic crystal. (**B**) A typical pattern of domains in the amorphous state.

Next we must specify the transitions allowed between configurations, and their respective rates. These will be selected according to a principle of parsimony, namely that they be a minimal set of transitions that assures ergodicity (any configuration can be reached from any other). Squares will be permitted both to fragment into smaller squares, and to aggregate into larger squares. The specific rules are given below.

Fragmentation: $(pq) \times (pq)$ can fragment into p^2 $q \times q$'s if and only if p is a smallest prime factor of pq. (The rate for this transition will be denoted by $r_f(pq)$.) Thus a 6×6 can transform into four 3×3's by a single fragmentation transition, but not into nine 2×2's. Squares of prime-number size such as 7×7 can only fragment fully into 1×1's.

Aggregation: p^2 $q \times q$'s in a square arrangement can eliminate a common boundary to form $(pq) \times (pq)$ if and only if p is a smallest prime factor of pq. (The rate will be denoted by $r_a(pq)$.) This is obviously the reverse of the above fragmentation process.

Any tiling can be converted to the "vacuum state" consisting entirely of 1×1's simply by implementing a series of fragmentations. Likewise, any other tiling can be created out of the vacuum by a suitable sequence of aggregations. This guarantees ergodicity.

The aggregation rates will be assigned the generic form

$$r_a(pq) = \nu_0 \alpha^{2pq(p-1)} \tag{8}$$

where

$$0 < \alpha < 1 \tag{9}$$

and ν_0 is a fundamental attempt frequency. The parameter α represents the probability that any given unit length of boundary is suitable for aggregation, and its exponent in equation 8 is the total length of boundary that would be annihilated by the aggregation.

Rates must satisfy detailed balance, of course. Consequently we have

$$r_f(pq) = \nu_0 \alpha^{2pq(p-1)} \exp(-\beta \Delta\Phi) \tag{10}$$

where $\Delta\Phi$ is the potential energy increment incurred by the fragmentation transition

$$\Delta\Phi = 2\lambda pq(p - 1) \tag{11}$$

At high temperature the system will typically be tiled by a mixture of squares mostly of small size. As temperature declines, larger squares become more prominent, but equations 8 and 10 show that transition rates are then substantially slowed. It is the blocking of structural change by large squares, particularly those with prime-number sizes, that make the present model attractive for describing glass transitions.

PARTITION FUNCTION EVALUATION

The combinatorial factor Ω required for evaluation of the partition function can be obtained by a sequential filling process for the lattice, placing squares in the system one at a time in random order. Suppose that a fraction ξ of the system's area has been covered at some intermediate stage, with use of the same fraction of each of the available numbers n_j of squares of various sizes. Suppose the next square to be added has size $l \times l$, and let $E_l(\xi)$ denote the expected number of ways it could be fitted onto the partially filled lattice without overlap. Then Ω can be obtained as the product of such factors for the entire filling sequence.

The Flory or mean field approximation[9] sets

$$E_l(\xi) \simeq (1 - \xi)^{l^2} N \tag{12}$$

which assumes that a common attrition factor $1 - \xi$ is applicable for each unit of area to be covered by the $l \times l$ square. This leads to the estimate

$$\ln \Omega \simeq N \left\{ \sum_{j \geq 1} (n_j/N) \, [\ln(N/n_j) + 1] - 1 \right\} \quad \text{(mean field)} \tag{13}$$

In the spirit of the scaled particle theory[10] it would seem reasonable that the $l \times l$

square's boundary length (linear in l) as well as its area (quadratic in l) should appear as exponents in an accurate $E_l(\xi)$ expression. Consequently we consider the following generalization of equation 12

$$E_l(\xi) \simeq [F(\xi)]^{l^2}\,[G(\xi)]^{l}N \qquad \textbf{(14)}$$

Two relations are available to determine F and G. The first is that equation 12 is actually an *exact* expression for $l = 1$. The second emerges in the large-l limit, where successful placement requires that all other squares be confined to a reduced macroscopic area, and thereby compressed to a higher effective filling fraction. One finds

$$F(\xi) = (1-\xi)\,[1-(1-\mu_1)\xi]^{-(1-\mu_0)/(1-\mu_1)} \qquad \textbf{(15)}$$

and

$$G(\xi) = [1-(1-\mu_1)\xi]^{(1-\mu_0)/(1-\mu_1)} \qquad \textbf{(16)}$$

TABLE 1. Predicted Values of $N^{-1}\ln\Omega$ for Uniform Tiling with $l \times l$ Squares

l	Mean Field[a]	Improved[b]
1	0.000	0.000
2	−0.403	−0.173
3	−0.645	−0.244
4	−0.764	−0.260
5	−0.831	−0.258
6	−0.873	−0.249
7	−0.900	−0.238
8	−0.919	−0.227
9	−0.933	−0.217
10	−0.944	−0.207
100	−0.999	−0.046
1000	−1.000	−0.007

[a] $N^{-1}\ln\Omega = (2/l^2)\ln l + 1/l^2 - 1.$
[b] $N^{-1}\ln\Omega = [(1-l)/l^2]\ln l.$

where the μ_k are moments of the size distribution for squares

$$\mu_k = \sum_{l\geq 1} l^k(n_l/N) \qquad \textbf{(17)}$$

This improved combinatorial argument subsequently yields

$$\ln\Omega \simeq N\left\{\sum_{j\geq 1}(n_j/N)\ln(N/n_j) + [(1-\mu_0)/(1-\mu_1)]\mu_1\ln\mu_1\right\} \quad \text{(improved)} \qquad \textbf{(18)}$$

That use of the improved combinatorics versus the mean field approximation may be important for our tiling model is illustrated by TABLE 1. It shows the respective mean field and improved values for $N^{-1}\ln\Omega$ in the event that the system were being tiled uniformly with $l \times l$ squares. In the large-system limit that is of primary concern, this quantity should be identically zero for all l. Although both approximations give zero for $l = 1$, both also drift to negative values for larger l. This imprecision is clearly worse for the mean field case, especially for large squares that ought to be significant for the tiling model at low temperature. The improved combinatorics at least has the feature that it approaches the correct result in the large-l limit, in contrast to the mean field approximation.

When the Ω expression equation 18 is placed in the partition function equation 7, the equilibrium state at given β may be identified by locating the maximum term. This results in the following expression for the equilibrium mean concentrations of the squares

$$\langle n_l(\beta)\rangle / N = \exp\,[\Lambda_2(\beta)l^2 + \Lambda_1\,(\beta)l + \Lambda_0(\beta)] \tag{19}$$

Consequently

$$\Lambda_0(\beta) = -\frac{\mu_1 \ln \mu_1}{1 - \mu_1} - 1 \tag{20}$$

and

$$\Lambda_1(\beta) = \left(\frac{1 - \mu_0}{1 - \mu_1}\right)\left(\frac{\ln \mu_1}{1 - \mu_1} + 1\right) - 2\beta\lambda \tag{21}$$

and $\Lambda_2(\beta)$ must be determined by the fixed total area condition $\mu_2 = 1$.

In view of the fact that the moments μ_0 and μ_1 depend on the concentrations, and that at the same time the concentrations depend on these moments, self-consistent numerical solutions must be sought at each temperature. TABLE 2 provides some results so obtained.

In the high-temperature regime for which

$$\beta\lambda < \beta_c\lambda \equiv 0.323\,106\,608\,8\ldots. \tag{22}$$

each of Λ_0, Λ_1, and Λ_2 is negative. Equation 19 then shows that the concentrations decrease monotonically with increasing l. TABLE 2, however, demonstrates that as β increases from zero to β_c, $\Lambda_2(\beta)$ drifts continuously upward to zero. Continuation past β_c would cause Λ_2 to become positive; the concentrations in equation 19 would no longer be monotonically decreasing with l, but would manifest a sudden appearance of squares near the maximum possible size (edge length proportional to $N^{1/2}$). Hence β_c is a condensation point at which the system in principle becomes macroscopically unstable with respect to the elimination of misfit boundary.

TABLE 2. Self-consistent Numerical Solutions

$\beta\lambda$	Λ_2	Λ_1	Λ_0	μ_1	μ_0
0.00	−0.37501	−0.27591	−0.16875	0.69843	0.56246
0.05	−0.31065	−0.39962	−0.18259	0.67668	0.53417
0.10	−0.24844	−0.52538	−0.19763	0.65357	0.50491
0.15	−0.18848	−0.65374	−0.21415	0.62874	0.47447
0.20	−0.13090	−0.78553	−0.23264	0.60163	0.44248
0.25	−0.07582	−0.92229	−0.25399	0.57122	0.40826
0.30	−0.02337	−1.06728	−0.28015	0.53527	0.37026
0.323 1066	0.00000	−1.13941	−0.29526	0.51514	0.35029

TABLE 3. Equilibrium Concentrations $\langle n_l(\beta)\rangle/N$ for Squares with $l \leq 10$

l	$\langle n_l(0)\rangle/N$	$\langle n_l(\beta_c)\rangle/N$
1	4.406×10^{-1}	2.382×10^{-1}
2	1.085×10^{-1}	7.622×10^{-2}
3	1.263×10^{-2}	2.439×10^{-2}
4	6.943×10^{-4}	7.806×10^{-3}
5	1.803×10^{-5}	2.498×10^{-3}
6	2.211×10^{-7}	7.994×10^{-4}
7	1.281×10^{-9}	2.558×10^{-4}
8	3.506×10^{-12}	8.186×10^{-5}
9	4.532×10^{-15}	2.620×10^{-5}
10	2.767×10^{-18}	8.383×10^{-6}

TABLE 3 shows equilibrium concentrations of squares computed from equation 19, with $l \leq 10$, at $\beta = 0$ and β_c. The effect of temperature reduction on large squares is particularly noticeable.

It should be mentioned in passing that the Flory or mean field approximation also leads to a condensation point, but at a substantially lower temperature

$$\beta_c\lambda = 0.628\ 028\ 3600 \text{ (mean field)} \tag{23}$$

To the extent that TABLE 1 suggests that the improved combinatorics has not entirely rectified the error in the mean field approximation, we might expect an exact solution for the model to display a condensation point at a temperature even higher (smaller $\beta\lambda$) than that given by equation 22.

The moment quantity μ_1 substantially gives the mean energy, and heat capacity follows from β differentiation. FIGURE 2 shows a plot of this heat capacity for temperatures above the condensation point ($0 \leq \beta \leq \beta_c$).

CONDENSED STATE

Although the preceding analysis appears to predict correctly the occurrence of the transition point, it is not reliable for indicating the structure of the condensed state itself. The reason is that one macroscopic square (or a small number of order unity of such squares) occupying a substantial fraction of the entire system area cannot be part of an essentially continuous filling operation. Yet precisely such continuous filling forms the basis of both the mean field and the improved estimates for the combinatorial quantity Ω. It is necessary to place macroscopic squares in the system first, then use the preceding continuous filling procedure as before to tile the remaining multiply connected region.

The end result of this generalization is that if the system shape is itself square, the condensed state consists just of the single tile of maximum size $N^{1/2} \times N^{1/2}$. With periodic boundary conditions this has degeneracy N because the corner of the tile can be located at any one of the N lattice sites. As the temperature decreases through the condensation point the system would (if equilibrium were obtained) discontinuously transform into a single well-packed amorphous domain. The mean energy on a per-site basis would drop discontinuously from 1.0303 λ just above the transition temperature

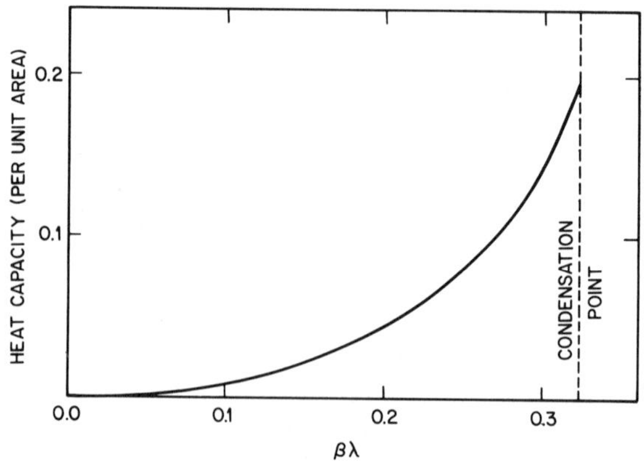

FIGURE 2. Heat capacity per unit area for $\beta \leq \beta_c$.

to zero below. This transition is unequivocally first order, with an entropy change $\Delta S/Nk_B = 0.3329$.

In the event that the system shape is not square, the condensed state may exhibit two or more macroscopic tiles, and the degeneracy of the corresponding ground state may be larger than N. A particularly interesting case arises when the system is rectangular, with a rational side ratio close to the irrational "golden mean," $\theta = (5^{1/2} + 1)/2$. The ground-state tiling then requires a hierarchy of square sizes, descending from the largest possible size down to the microscopic regime. One possible arrangement is shown in FIGURE 3. There are N places for the largest square, $(N/\theta)^{1/2}$ for the next largest, and two choices for each succeeding level in the size descent. The overall degeneracy of the ground state will be proportional to N^q, where

$$q = \frac{3}{2} + \frac{\ln 2}{2 \ln \theta}$$

$$= 2.2202\ldots. \qquad \textbf{(24)}$$

Although this degeneracy is substantially larger than that for the square-shaped system, it does not alter the conclusion that in the infinite system limit the configurational entropy per particle of the condensed state is zero.

GLASS TRANSITION

Although the preceding has assumed thermal equilibration, the tiling model is useful precisely because it cannot equilibrate below the condensation temperature. A high-temperature configuration containing only microscopic squares could in principle reach the ground (condensed) state by allowed transitions, and this might be possible via many different sequences of intermediate configurations. Equations 8 and 10, however, show that impossibly small rates would be involved, with the pure number $0 < \alpha < 1$ raised at least to a power comparable to $N^{1/2}$.

Therefore, as temperature declines toward and through the condensation point the texture of the tiling will coarsen, but kinetic arrest will certainly prevent the appearance of macroscopically large squares. The condensation point would never be observed as such, but would be inferred (for example) from positions of maxima in apparent heat capacities for slower and slower cooling rates. It is natural to expect that with a slow, finite cooling rate, the properties of the model would manifest a reasonably sharp "glass transition" in the neighborhood of the true condensation point.

A typical glassy configuration would be kinetically strangled by large squares, but should still contain a few small inclusions of the type shown in FIGURE 4. The 2×2 square at the left in this structure can fragment into 1×1's, then another 2×2 can form at the right, and then this sequence could reverse, and so forth. Such local oscillation would still have a respectably high rate below the condensation temperature, but would be totally ineffective in producing overall structural relaxation. Instead, it represents for the tiling model a simple analogue to the "two-level" systems, which appear almost universally in low-temperature amorphous solids.[11,12]

Side-by-side coexistence of rapidly oscillating inclusions and of larger, more sluggish features in the tiling model glassy state suggests that a broad spectrum of relaxation times should be present. This spectral breadth may lead to a "stretched exponential," or "KWW," form for relaxation functions.[13]

COMPUTER SIMULATION

In the near future, the most reliable source of information about kinetic properties of the tiling model is likely to be computer simulation. Consequently a Monte Carlo

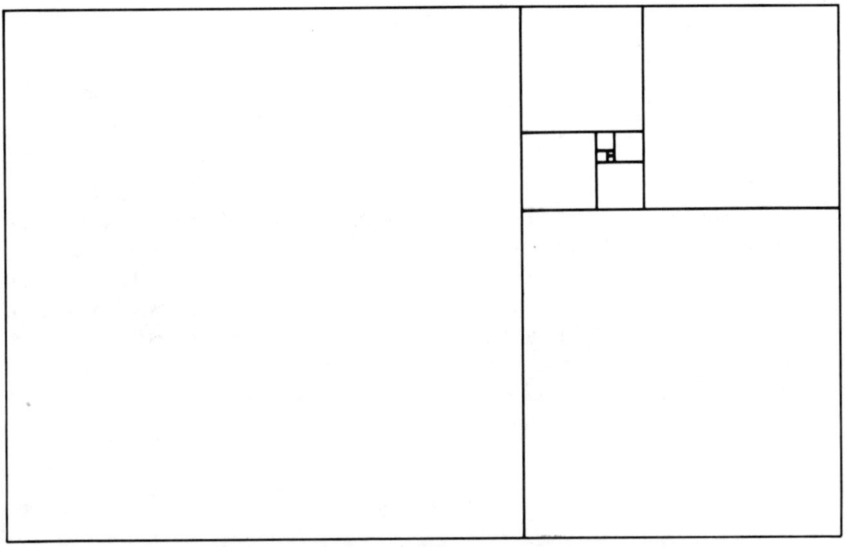

FIGURE 3. Condensed state tiling when the system shape is rectangular with the side ratio near the golden mean.

program for this purpose has been constructed and successfully tested. It embodies the transition rates shown in equations 8 and 10, and generates a stochastic sequence of tiling configurations appropriate to given initial conditions and the temperature of interest. At each Monte Carlo step it keeps full account of all possible transitions, and focuses on selecting the next configuration, while maintaining a record of the elapsed time.

Test runs with the Monte Carlo program (with $\alpha = 0.98$) have demonstrated the feasibility of simulating a two-dimensional system of size $N = 10^4$. This should be large enough to infer accurately properties for the infinite system limit. In particular we have carried out extended runs at $\beta\lambda = 0.15$ to assess the validity of the improved combinatorial method discussed above, and to verify that indeed it is superior to the mean field approximation.

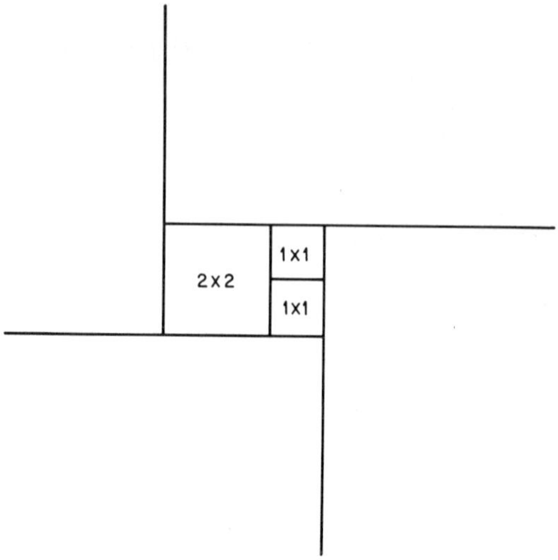

FIGURE 4. Randomly oscillating "two-state" inclusions expected in the low-temperature glass.

Further simulation calculations will be directed to establishing the following: 1) equilibrium square concentrations and the heat capacity over the entire high-temperature range; 2) the location of the underlying condensation point; 3) the cooling rate dependence of the glass transition and of the properties of the glass state itself; and 4) the time dependence of various linear and nonlinear relaxation functions, and how these vary with temperature near the condensation point.

Finally, it appears feasible both to include the crystallization-inducing interactions shown in equation 4 and to simulate three-dimensional models.

SUMMMARY

As a liquid is supercooled toward and through a glass transition, the constituent molecules collectively experience increasing difficulty in executing rearrangements that lower the overall interaction energy. A conceptually simple lattice model is introduced to represent this phenomenon. In its two-dimensional version the plane is tiled by squares with integer sides. Each tile represents a well-packed but nonrepetitive domain of molecules, and domain walls have positive mismatch energy. Subdivisions and aggregations of squares are kinetically allowed if and only if the smallest prime factors of square sides are involved. At equilibrium, cooling coarsens the texture of squares until a first-order transition point is reached, at which macroscopic squares appear. This transition, however, is totally inhibited kinetically, yielding instead an amorphous "glass" of microscopic but relatively large squares. Exploratory calculations have been carried out that demonstrate the feasibility of using computer simulation to study kinetic properties of the model.

REFERENCES

1. O'Reilly, J. M. & M. Goldstein, Eds. 1981. Ann. N.Y. Acad. Sci. **371.**
2. MacKenzie, J. D. 1960. Modern Aspects of the Vitreous State. Butterworths. Washington, DC.
3. Phillips, J. C. 1979. J. Non-Cryst. Solids **34:** 153.
4. Nelson, D. R. 1983. Phys. Rev. B **28:** 5515.
5. Palmer, R. G., D. L. Stein, E. Abrahams & P. W. Anderson. 1984. Phys. Rev. Lett. **53:** 958.
6. Leutheusser, E. 1984. Phys. Rev. A. **29:** 2765.
7. Ubbelohde, A. R. 1978. The Molten State of Matter: 414. John Wiley & Sons. NY.
8. Hill, T. L. 1956. Statistical Mechanics. McGraw-Hill. New York, NY.
9. Flory, P. J. 1953. Principles of Polymer Chemistry: 497–503. Cornell University Press. Ithaca, NY.
10. Reiss, H., H. L. Frisch & J. L. Lebowitz. 1959. J. Chem. Phys. **31:** 369.
11. Anderson, P. W., B. I. Halperin & C. M. Varma. 1972. Philos. Mag. **25:** 1.
12. Phillips, W. A. 1972. J. Low Temp. Phys. **7:** 351.
13. Ngai, K. L. 1979. Comments Solid State Phys. **9:** 127, 141.

DISCUSSION OF THE PAPER

R. G. Palmer (*Duke University, Durham, NC*): My guess is that the glassy behavior of your model comes from the limited transitions allowed: the system is highly contrained dynamically. Would you see the thermodynamic phase transition if you allowed more types of transition?

Stillinger: Glassy behavior in the tiling model arises from a combination of three attributes: the sparse set of allowed transitions, the intrinsic slowness of those transitions that create or destroy large domains, and the restriction to domains with only square shapes. By relaxing the first two of these the opportunity for observing a vestige of the thermodynamic phase transition (at least on cooling from high temperature) would certainly be enhanced. In particular if the parameter α in equations 8 and 10 were set equal to unity, aggregation processes would be greatly facilitated. The system, however, still has the problem that most tilings with large squares will have those squares juxtaposed in ways inconsistent with further aggregation. Unless a new set of transitions is allowed that involves restructuring domains inside nonsquare boundaries of arbitrary size, the system will still be kinetically unable to achieve the fully condensed state.

J. T. BENDLER (*General Electric Research and Development Center, Schenectady, NY*): Dr. Stillinger, can you say how the entropy factor, α, was assigned in the simulations and how sensitive the stretched exponential is to it?

STILLINGER: The parameter α was set equal to 0.98 in the simulation studies thus far carried out. It is not obvious how the stretched exponential relaxation behavior should depend on α, so further simulation studies are warranted.

P. C. HOHENBERG (*AT&T Bell Laboratories, Murray Hill, NJ*): You indicated that there is a static transition in your model, from an approximate analytical calculation. How is this transition temperature related to the dynamic slowing-down transition?

STILLINGER: Computer simulation reproducibly achieves thermal equilibrium at temperatures greater than or equal to the condensation temperature T_c. Cooling below T_c eventually causes glassy behavior to emerge, the temperature range for which depends on the cooling rate. To the extent that a glass transition temperature T_g can be identified, $T_c > T_g$.

HONENBERG: I understood your answer to be $T_{\text{dynamic}} < T_{\text{static}}$. Is this right?

STILLINGER: Yes.

J. C. PHILLIPS (*AT&T Bell Laboratories, Murray Hill, NJ*): Can you expect that this model contains a T_g as well as T_0? Will the specific heat exhibit the asymmetry usually observed in DTA experiments: S-shaped on cooling, peaked on heating?

STILLINGER: Under the condition of a constant cooling rate, the tiling model exhibits a smooth continuation of mean energy versus temperature through the transition point T_c. But especially with very slow cooling, this metastable continuation eventually halts when the prevailing arrangement of squares is blocked from further aggregation at any perceptible rate. At this point ("T_g") the mean energy remains nearly constant as further cooling proceeds.

Simulation on the tiling model has not yet been turned to the task of examining asymmetry in apparent heat capacity under cooling and heating conditions, respectively. There are some basic elements present, however, that might create such asymmetry. Upon cooling, aggregation requires neighboring squares to be in registry (in full contact along a shared side), whereas fragmentation has no analogous precondition. In addition, unfavorable Boltzmann factors are present only in fragmentation rates, not in aggregation rates. There is no reason to suppose that the time-dependent populations of squares of different sizes that appear on cooling would be identically mirrored upon heating at the same rate. Any difference would affect the corresponding heat capacities.

M. WEINBERG (*Jet Propulsion Laboratory, Pasadena, CA*): If different geometrical units were used in the model, would there be qualitative changes in the results?

STILLINGER: Some changes in the allowed shapes of domains should have little effect on qualitative properties. For example, replacing the underlying square grid with a triangular lattice would permit tiling with equilateral triangles, and if least prime factors of side lengths were again used to define permissible transitions, no substantial change in either equilibrium behavior or kinetic behavior would ensue. But if very complicated domain shapes were allowed (such as elongated, nonconvex, and multiply connected shapes), with some selection of transitions to assure ergodicity, drastic changes would no doubt be possible, including elimination of the first-order condensation point.

Slipping into the Glassy State via the Crystalline Solid State

W. L. JOHNSON

W. M. Keck Laboratory of Engineering Materials
California Institute of Technology
Pasadena, California 91125

INTRODUCTION

Although a glass is often defined as a configurationally frozen liquid that failed to crystallize upon cooling, common experience suggests that the glassy state be defined by its atomic scale structure and by its thermodynamic properties rather than by the method by which it is prepared. The former definition has led many researchers to apply the term amorphous (and not glassy) to materials whose origin is other than the liquid state. Ignoring this question, the term glass will be used in the present context to refer to noncrystalline solids of any type, irrespective of their origin. The well-known methods of synthesizing such materials include quenching of vapors onto substrates held at low temperature, sputtering, electrodeposition, and quenching from the liquid state. Under certain circumstances, crystalline matter is also observed to transform to the glassy state. For example, it is known that covalent solids such as crystalline Si[1,2] and metallic crystals such as TiNi[3] become glassy when irradiated by energetic neutrons or ions. The application of hydrostatic pressure to crystalline H_2O at temperatures below 130 K to produce glassy H_2O is described in this volume.[4] Finally, it is noteworthy that solid state chemical reactions often lead to a transformation from a crystalline reactant to a glassy product. The thermal oxidation of Si crystals in fact results in the formation of a glassy oxide layer.[5] Recently, it was found that a crystalline metallic alloy can become glassy upon reaction with hydrogen gas to form a metallic hydride.[6,7] More surprisingly, it has been observed that diffusion couples of two crystalline metals can react by solid state interdiffusion to form and grow a binary metallic glass. The relative ease with which crystalline phases nucleate and grow in metallic melts as well as the observations that most metallic glasses crystallize in short times at temperatures near the glass transition temperature, T_g,[8] make this observation especially intriguing. It is the purpose of the present article to examine the kinetic and thermodynamic factors that conspire to permit these solid state crystal-to-glass reactions in metallic alloys.

REVIEW OF EXPERIMENTAL RESULTS

As mentioned above, it has been known for some time that glassy products are formed by oxidation and other types of solid state chemical reactions of covalent materials. In experiments on thin films, it has also been observed that a polycrystalline silver layer deposited onto a glassy covalent solid, such as amorphous Te, can dissolve by interdiffusion into the glass.[9] The first example of a crystal to glass transformation in metallic solids in which one of the reactants was not already glassy was reported by Yeh *et al.*[6] FIGURE 1 shows a series of X-ray diffraction patterns taken at various stages of isothermal reaction at 175° C of a polycrystalline Zr_3Rh alloy (Cu_3Au-type structure) with hydrogen gas at 1 atm. The final product (at $t = 46$ hr) is an amorphous metallic hydride of composition $Zr_3RhH_{5.5}$. One sees, as the reaction proceeds, a gradual reduction in intensity of the crystalline Bragg peaks, a noticeable broadening of the peak widths, and a slight shift in the peak positions. Concurrently, there is a gradual increase of the broad diffuse bands characteristic of the amorphous hydride. This suggests a transformation of the crystalline material to the amorphous state. The slight shift of the crystalline lattice parameter suggests that some hydrogen does in fact enter the interstitial sites of the crystal. Electron microscopy was carried out at various stages of the transformation in an attempt to better understand the reaction mechanism.[7] FIGURE 2 shows electron micrographs taken after thinning unreacted and partially reacted samples. One sees the crystalline grains of the unreacted sample. The accompanying diffraction pattern is that expected for an appropriately oriented Cu_3Au-type crystallite. In the partially reacted sample one sees the remaining crystals surrounded by a featureless gray zone. Diffraction reveals that the intergranular gray zone is indeed an amorphous phase. One notes that the remaining crystalline grains are separated from the gray zone by a sharp interface. This reflects the first-order nature of the phase transition from the crystalline state to the glassy state. By following the

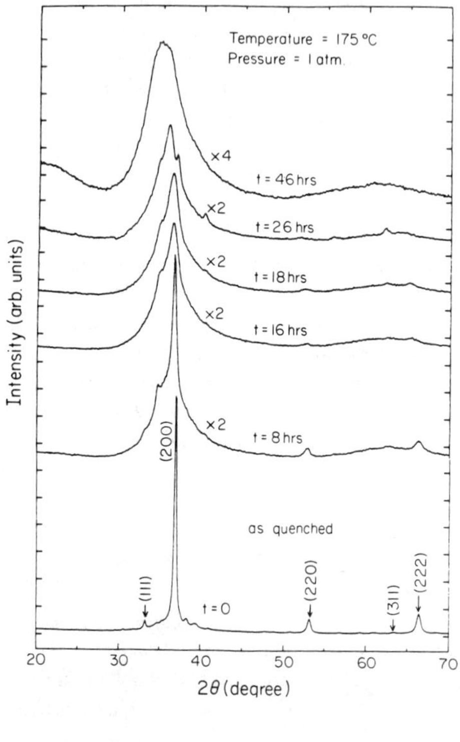

FIGURE 1. A series of X-ray diffraction patterns taken at various stages of the reaction of polycrystalline Zr_3Rh with hydrogen gas (CuKα radiation).

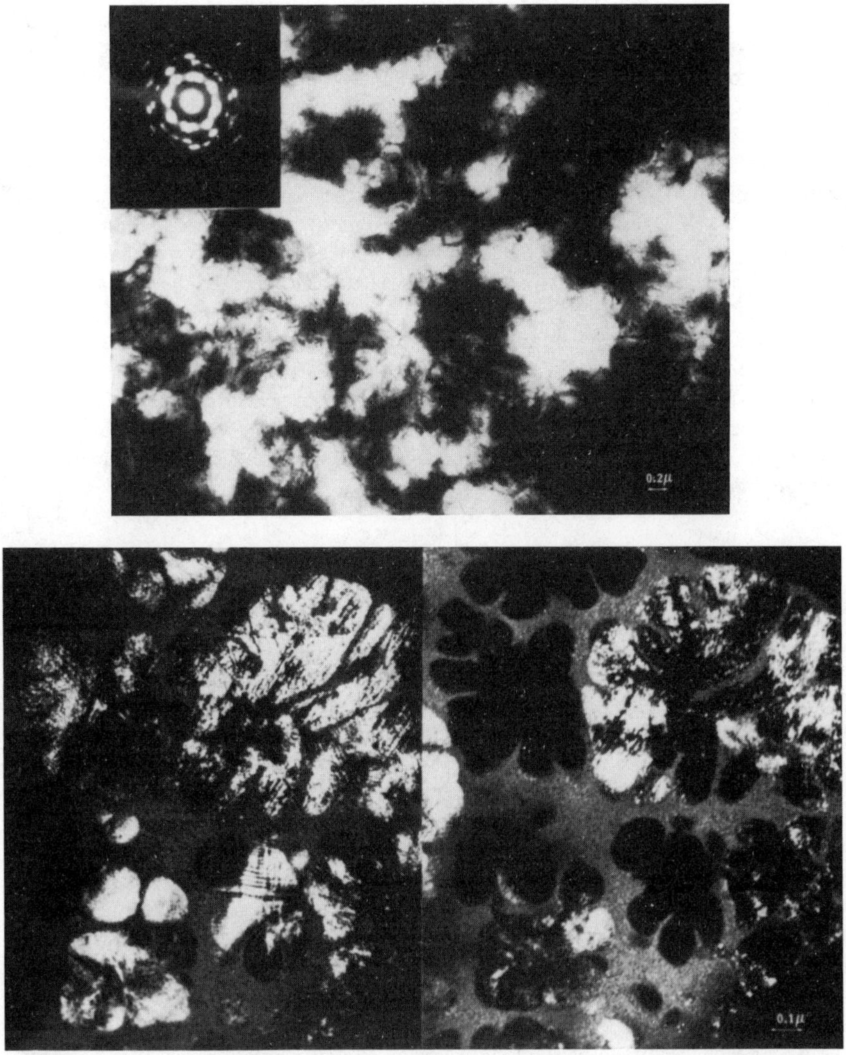

FIGURE 2. An electron micrograph of and a diffraction pattern from an unreacted polycrystalline Zr_3Rh sample (upper panel) and electron micrographs of a partially reacted sample, in bright and in dark fields (lower panels).

progress of the reaction using TEM, it is clear that the transformation proceeds by the motion of this interface into the crystals until no crystalline material remains.

The overall reaction can be compared to melting. In a polycrystalline metal, melting begins heterogenously at surfaces and grain boundaries and then proceeds by the advance of the liquid front to crystal interiors. In effect, the crystals in the above experiment have "melted" to a liquid below T_g. The process of solid state glass transformation is thus morphologically quite similar to ordinary melting. For ordinary melting to occur, a crystal must be (at least slightly) superheated. One may inquire as to how this comes about in the present case. The answer, as discussed by Yeh *et al.*,[6] lies in the fact that the free energy of amorphous Zr_3RhH_x lies below that of coherent crystalline Zr_3RhH_x when x exceeds some critical value $x_c < 5.5$. Thus, although hydrogen dissolution into crystalline Zr_3RhH_x system is thermodynamically possible, it results in a less rapid lowering of the free energy of the system than does the hydrogen gas–amorphous Zr_3RhH_x reaction. This results in a crystal that is superheated with respect to the glass, and provides the driving force for the crystal-to-glass transformation.[7,10]

One may inquire as to why the glassy hydride does not transform further to form equilibrium crystalline phases. In fact this does occur when the isothermal reaction is carried out at only slightly higher temperatures. At temperatures above 225° C, one observes the nucleation and growth of fcc-ZrH_2 crystals during reaction. Above 250° C the final product of the reaction is a two-phase mixture of fcc-ZrH_2 and Rh metal. We can summarize by the following reactions equations

$$H_2\text{ (gas)} + \text{crystalline } Zr_3Rh \xrightarrow{<225^\circ\text{ C}} \text{amorphous } Zr_3RhH_{5.5} \quad (1)$$

$$H_2\text{ (gas)} + \text{crystalline } Zr_3Rh \xrightarrow{>250^\circ\text{ C}} \text{crystalline } ZrH_2 + \text{crystalline Rh} \quad (2)$$

To explain the observation of the metastable amorphous intermediate at temperatures below 225° C, one can appeal to the schematic diagram in FIGURE 3. The formation of crystalline ZrH_2 and Rh metal at a rate V along interface A in reaction 2 requires the interdiffusion of Zr and Rh metal ($J_{Rh\text{-}Zr}$) in addition to H transport in the crystalline Zr_3Rh (J_H). In fact, this diffusion must occur over a distance at least as large as the dimension of a critical nucleus of crystalline fcc-ZrH_2 and Rh metal (d_1 and d_2). Owing to surface energy σ_{12}, this dimension would have to be on the order of tens of angstroms. Below 225° C, this is apparently impossible. On the other hand, although reaction 1 requires transport of H from the gas phase across the sample surface (B) and through the specimen (J_H), reaction 1 requires no interdiffusion of metal atoms. It is the fast diffusion of H in the polycrystalline sample at relatively low temperatures, contrasted with the much slower interdiffusion of metal atoms ($J_{Rh\text{-}Zr}$), which allows one to carry out reaction 1 without having reaction 2 take place at the same time.

As a second example of solid state crystal-to-glass reactions, we turn to recent experiments on thin-film diffusion couples of polycrystalline metals. Schwarz and the author first reported the reaction of polycrystalline multilayers of two metals to form a glass.[11] It is noteworthy that Herd *et al.*[12] independently reported that thin films of Rh metal can be thermally reacted with amorphous Si to form an amorphous silicide. To

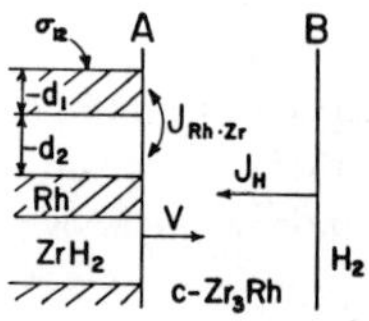

FIGURE 3. A schematic illustration of the reaction of polycrystalline Zr_3Rh with H_2 gas to form an equilibrium mixture of crystalline ZrH_2 and Rh metal.

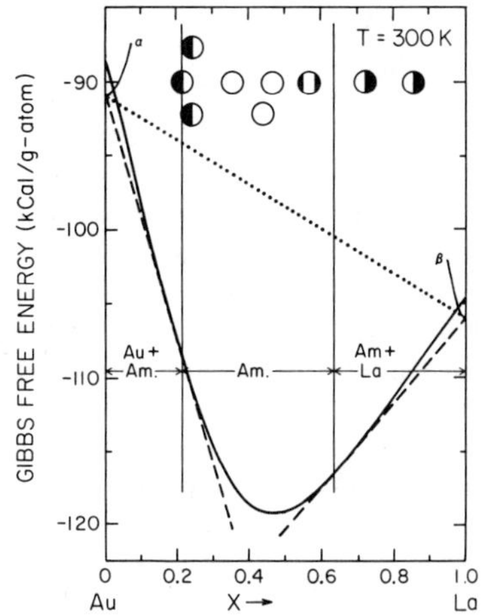

FIGURE 4. A free energy diagram of the binary Au-La system at 300 K. Circles represent the results of the reactions: ◐: Au + amorphous alloy; ○: amorphous alloy; ◑: La + amorphous alloy. The diagram was taken from reference 11.

understand these reactions, it is essential to examine the thermodynamic and kinetic aspects of binary diffusion couples. FIGURE 4 shows the free energy diagram given in reference 11 for the binary Au-La system. The diagram applies for the system at room temperature and ambient pressure. Under these conditions, one notes that the two metals exhibit restricted terminal crystalline fcc solutions. This is because of the large difference in Goldschmidt radii (1.87 Å for La and 1.44 Å for Au) of the two atoms. This is manifested in the free energy diagram by the sharply upturned free energy curves of the terminal solutions (α and β curves). On the other hand, several intermetallic compounds having crystal structures different from that of the parent phases are known to form exothermically with a very large negative free energy of mixing. The free energy of the amorphous (undercooled liquid) phase is depicted in the figure. The amorphous phase is also characterized by a large negative free energy of mixing, although it must in fact lie everywhere higher in free energy than a one- or two-phase mixture of intermetallic compounds (compounds not shown). This follows from the fact that the lowest free energy state is not the glassy state.

When one brings together a layer of crystalline Au with a layer of crystalline La, the free energy lies along the dotted line shown on the diagram. Formation of a glassy alloy can clearly lower the free energy. Formation of intermetallic compounds (not shown) could lower it still further. Both reactions are thermodynamically allowed: the one that occurs depends on kinetic factors. Binary Au-La glasses have been prepared by rapid cooling of the liquid alloy and are known to crystallize rapidly at temperatures between 150° and 200° C.[13] As such, one may surmise that if glass is to form by isothermal interdiffusion of the metals, then this must occur below the known crystallization temperature (in this case, ~150° C) of the glass. Because metallic glasses are observed in general to crystallize near their respective T_g values, it follows that this is a general requirement for glass growth in the solid state. In contrast to a required drop in free energy, a thermodynamic requirement, the upper temperature limit of T_g is a kinetic requirement. A second kinetic question is whether a glass-forming reaction is possible at such low temperatures. Anomalous "fast" diffusion of Au in crystalline La at low temperature was reported earlier.[14] It was this observation that provided Schwarz and the author the motivation to carry out the thin-film diffusion couple experiment. FIGURE 5 shows the X-ray diffraction patterns of a deposited polycrystalline multilayer of Au and La together with those of several multilayers of differing overall composition reacted for several hours at 75° C. The individual layers are several hundred angstroms in thickness. The data show clearly that an amorphous alloy can be formed by interdiffusion of the layers. The amorphous nature of the product has since been confirmed by TEM studies.[15,16] Similar multilayer reactions have been observed in several binary systems including Au-Y,[17] Ni-Zr,[18,19] and Co-Zr.[20,21] In a revealing study, Schroeder *et al.*[21] have used the technique of cross-sectional TEM analysis to directly observe the progress of the reaction. FIGURE 6 shows a micrograph taken from their work. A partially transformed Co-Zr mutilayer structure is shown. One sees well-defined polycrystalline layers of Co (thin textured layers) and Zr (thick textured layers) as well as a characteristic thin gray zone that

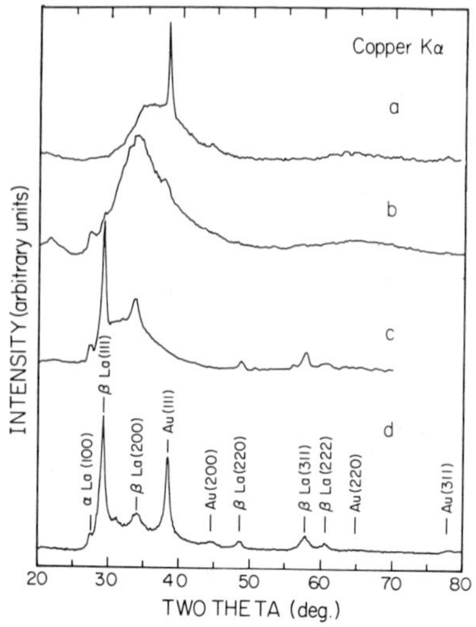

FIGURE 5. X-ray diffraction patterns of an as-reacted Au-rich multilayer sample (a), a reacted equiatomic multilayer (b), a reacted La-rich multilayer (c), and an as-deposited equiatomic multilayer (d). The samples were reacted at 70° C.

separates the crystalline multilayers. Schroeder *et al.*[21] demonstrated that this gray zone is indeed a growing interlayer of an amorphous Co-Zr alloy. Again one notes the resemblance of the growing glass interlayer to a growing zone of melt. Such experiments show clearly that the glassy phase grows more or less uniformly as a planar interlayer between the polycrystalline multilayer.

To study the kinetic aspects of the glass interlayer growth, Van Rossum *et al.*[22] employed the technique of Rutherford backscattering spectroscopy to Ni-Hf bilayers and multilayers.[15,16] They were able to demonstrate the growth of a well-defined amorphous alloy phase during isothermal reactions carried out at temperatures between 300° and 380° C. A typical backscattering spectra is shown in FIGURE 7 for a bilayer reacted at 340° C for various times as indicated. The growth of an amorphous interlayer can be clearly followed. Using such data, it was possible to obtain the time dependence of the amorphous interlayer thickness. Dolgin *et al.*,[15,16] following the

FIGURE 6. ATEM micrograph of a partially reacted Co-Zr multilayer (200 Å Co layers and 340 Å Zr layers). The micrograph was obtained from H. Schroeder (personal communication). See reference 21.

earlier work of Deal and Grove,[23] Guezin *et al.*,[24] and Gosele and Tu,[25] developed a model for the interlayer growth. The model is based on the interdiffusion constant, $\tilde{D}$, of the two metals in the growing amorphous layer together with an interface response parameter, κ, which measures the response of the amorphous–crystalline interface to a change in the chemical potential of the diffusing species as it crosses the interface. By fitting the backscattering spectra, Dolgin *et al.*[16] deduced values of $\tilde{D}$ and κ at the various temperatures of the isothermal reaction of Ni-Hf bilayers. FIGURE 8 shows the values of ln $\tilde{D}$ ($\tilde{D}$ in Å^2/sec) plotted against T^{-1}. An Arrhenius behavior is seen with an activation energy Q of about 1.4 eV. This was compared with data obtained by tracer diffusion methods by Cantor and Cahn[26] on similar binary Ni-Zr glasses. The activation energy of 1.4 eV was found to be reasonable. Actual values of $\tilde{D}$ range from $\sim 10^{-16}$ to $\sim 10^{-15}$ cm^2/sec. It is of considerable interest that interdiffusion with such

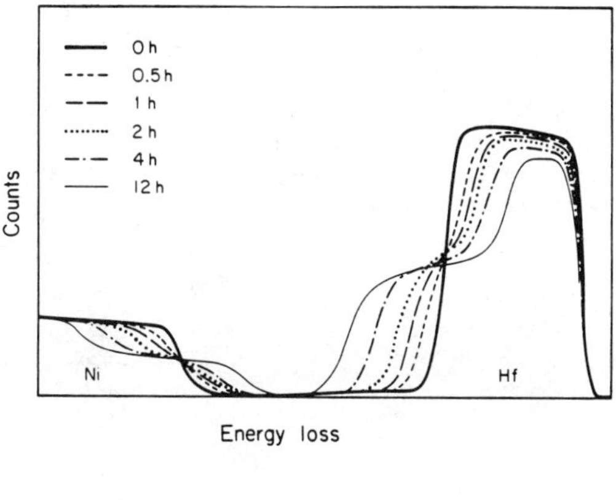

FIGURE 7. Rutherford backscattering spectra of an Ni-Hf bilayer reacted for the indicated times at 340° C. The Ni layer was 2500 Å thick; the Hf layer was 1500 Å thick. The spectra were taken from reference 16.

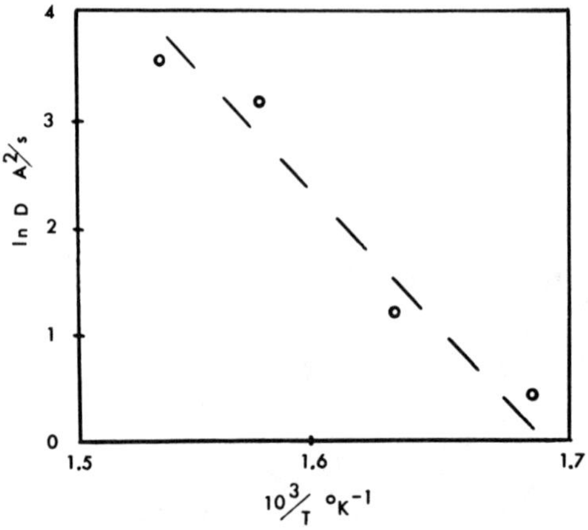

FIGURE 8. Arrhenius plot of interdiffusion constant $\tilde{D}$ for the glassy Ni-Hf interlayer. The plot was taken from references 15 and 16.

large values of $\tilde{D}$ over time scales of several hours can occur within the growing amorphous layer without resulting in the crystallization of the glass. A model for diffusion and crystallization behavior of metallic glasses proposed by Buschow and Beekmans[27] suggests that crystallization of the glassy interlayer should have occurred under these conditions.

To understand the apparent problem posed by the above observations, Cheng *et al.*[28] performed a Kirkendall marker-type experiment using the Rutherford backscattering technique. The result is illustrated in FIGURE 9, which shows the relative motion of the thin W-marker (4 Å of W) signal with respect to the signals of the Ni and Zr layers between which it was placed. Comparing the energy shift in the lower edge of the Zr signal, ΔE, with energy shift of the W-marker, ΔE_m, shows that Ni atoms move from the lower Ni layer through the W-marker. Many spectra were taken as the

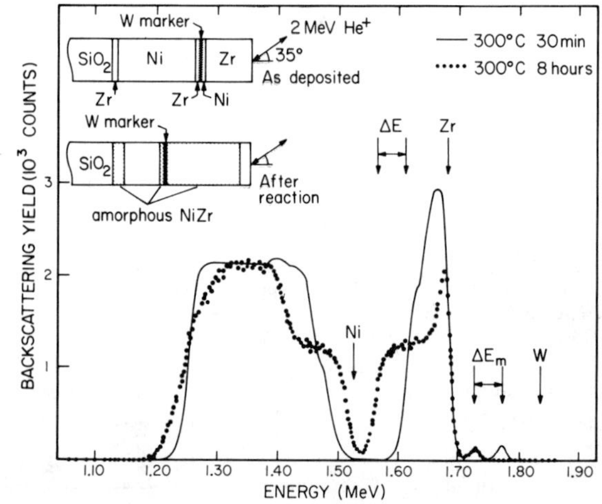

FIGURE 9. Rutherford backscattering spectra of a W-marker between an Ni layer and a Zr layer following partial reaction. The spectra were taken from reference 28.

reaction progressed and have clearly demonstrated that Ni is the only moving species (within the resolution of the experiment) throughout the reaction.[28] Several such experiments using different marker atoms have been performed on Ni-Zr bilayers, and all these experiments support the same conclusion. This provides a clue as to why the amorphous layer does not crystallize in spite of substantial atomic interdiffusion across the layer. Apparently, crystallization of an intermetallic Ni-Zr compound requires not only diffusion of one atomic species, as suggested by Buschow and Beekmans,[27] but the correlated diffusive motion of both atomic species (in this case, Ni atoms and Zr atoms). As such, the nucleation of the intermetallic compounds of Ni-Zr is apparently suppressed by a lack of mobility of the Zr atoms at the temperature at which the amorphous interlayer has been shown to grow (that is, $T \sim 300°$ C). The crystallization

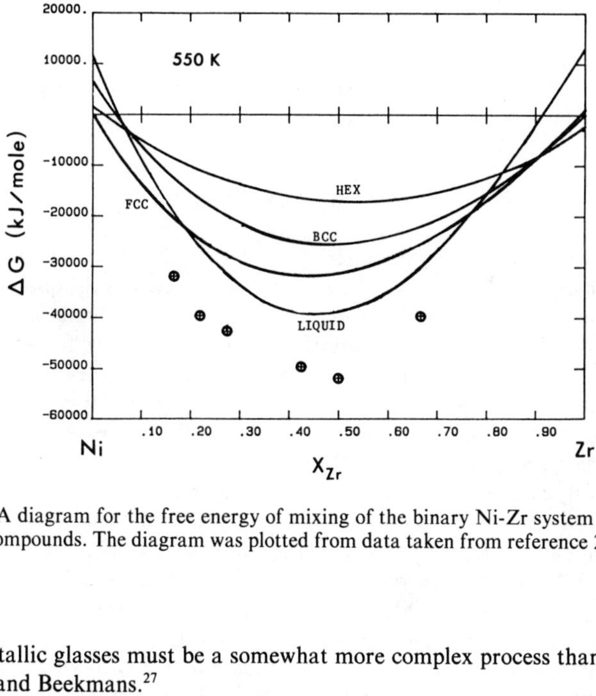

FIGURE 10. A diagram for the free energy of mixing of the binary Ni-Zr system at 250° C. ●: Equilibrium compounds. The diagram was plotted from data taken from reference 29.

of binary metallic glasses must be a somewhat more complex process than envisioned by Buschow and Beekmans.[27]

A free energy diagram devised by Saunders and Miodownik[29] of the Ni-Zr system that applies at 250° C is shown in FIGURE 10. Intermetallic compounds are indicated by crosses. The other curves represent the bcc, fcc, and hcp solid solution phases as well as the undercooled liquid phase at the same temperature. Indeed, one sees that there are at least six equilibrium intermetallic compounds, the nucleation and growth of which would reduce the free energy below that of the undercooled liquid phase. Saunders and Miodownik[29] in fact point out that the actual amorphous curve lies somewhat below their extrapolated undercooled liquid curve. This does not, however, alter the conclusion that the intermetallic compounds represent lower free energy states than the glass. It is very surprising that the constraint imposed by the lack of diffusive mobility of Zr atoms during glassy interlayer growth is sufficient to prevent the nucleation and growth of all of these intermetallic compounds. This is perhaps one of the most striking features of this crystal-to-glass reaction.

As a final point, we note that there is a qualitative difference between the free energy diagram of Au-La (FIG. 4) and that of Ni-Zr (FIG. 10). For the former case, the terminal solutions are characterized by a positive free energy of mixing whereas in the latter case, the hcp-, bcc-, and fcc-terminal solutions are characterized by a negative free energy of mixing. In the latter case, the diffusion of an Ni layer into an existing hcp-Zr layer lowers the free energy of the system. Such diffusion will thus occur spontaneously. This leads, as pointed out by Saunders and Miodownik,[29] to the possibility of forming terminal solid solutions that are superheated with respect to the glass. In the Au-La case, this is not possible because of the positive free energy of mixing. As such, the glass must necessarily form first at the interface before significant interdiffusion of Au and La can occur. Thus the mechanism of glass formation must be through a local interfacial instability in the case of in Au-La, as opposed to instability of a superheated solution, as in the case of Ni-Zr. More work clearly needs to be done to clarify the detailed microscopic mechanism by which the glassy phase forms.

FUTURE PROSPECTS

The growth of metallic glass interlayers in thin-film diffusion couples might be viewed as an academic curiosity. Of more practical interest is the question of growing bulk metallic glass structures. Recently, Atzmon *et al.*[30] and Schultz[31] have extended solid state glass-forming-reaction principles demonstrated by the thin-film experiments to three-dimensional-composite structures produced by mechanical deformation of physical mixtures of metal powders and foils. By co-deformation, bulk composites of Ni-Zr,[30,31] and more recently, of Cu-Er and Ni-Er,[32] have been produced with submicron filamentary and lamellar domains of the two metals. These composites can be isothermally reacted to form a bulk amorphous product.

In 1983, Koch *et al.*[33] reported that powder mixtures of Ni and Nb could be alloyed by energetic ball milling to produce a metallic glass powder. Schwarz *et al.*[34] and Politis and Johnson[35,36] have reported similar results on Ni-Zr, Ni-Ti, and Cu-Ti powder mixtures. At this point, the detailed microscopic mechanism by which the glassy phase is formed in these experiments has not been determined, although it has been proposed[34] that a solid state interdiffusion reaction occurs during ball milling. Detailed explantation of these results must await further study. Nevertheless, these experiments point toward a variety of new synthesis techniques for the production of metallic glasses in forms that are of potential technological interest. Together with the scientific issues raised by the thin-film experiments, the results of these other studies provide both a promise and a challenge for future research in this relatively new field.

REFERENCES

1. MAYER, J. W. 1971. *In* Radiation Effects in Semiconductors. J. W. Corbett & G. D. WATKINS, Eds.: 367. Gordon Breach. New York, NY.
2. BROWN, L. M., & R. M. MAYER. 1969. Philos. Mag. **19:** 721.
3. BRIMHALL, J. L., H. E. KISSINGER & L. A. CHARLOT. 1983. Radiat. Eff. **77:** 273.

4. WHALLEY, E., O. MISHIMA, J. P. HANDA & D. D. KLUG. 1986. Ann. N.Y. Acad. Sci. This volume.
5. DEAL, B. E. & M. SKLAR. 1965. J. Electrochem. Soc. **112:** 430.
6. YEH, X. L., K. SAMWER & W. L. JOHNSON. 1983. Appl. Phys. Lett. **42:** 242.
7. YEH, X. L. & W. L. JOHNSON. 1986. J. Appl. Phys. **59:** in press.
8. SPAEPEN, F. & D. TURNBULL. 1976. *In* Rapidly Quenched Metals. B. C. Giessen & N. J. Grant, Eds. Vol. **2:** 205. M.I.T. Press. Boston, MA.
9. HAUSER, J. J. 1981. J. Phys. Colloq. (Paris) **42:** C4–C943.
10. YEH, X. L. & W. L. JOHNSON. 1986. *In* Proceedings of the Materials Research Society Symposium on Rapidly Quenched Metals. A. Taub & B. C. Giessen, Eds. Materials Research Society. Pittsburgh, PA.
11. SCHWARZ, R. & W. L. JOHNSON. 1983. Phys. Rev. Lett. **51:** 415.
12. HERD, S. R., K. N. TU & K. Y. Ahn. 1983. Appl. Phys. Lett. **42:** 597.
13. JOHNSON, W. L., S. J. POON & P. DUWEZ. 1975. Phys. Rev. B **11:** 159.
14. LE CLAIR, A. D. 1978. J. Nucl. Inst. Methods **69:** 70.
15. DOLGIN, B. 1984. Ph.D. dissertation. California Institute of Technology. Pasadena, CA.
16. DOLGIN, B., M. VAN ROSSUM & W. L. JOHNSON. Submitted for publication.
17. SCHWARZ, R. B., K. L. WONG & W. L. JOHNSON. 1984. J. Non-Cryst. Solids **61** & **62:** 129.
18. CLEMENS, B. M., R. B. SCHWARZ & W. L. JOHNSON. 1984. J. Non-Cryst. Solids **61** & **62:** 817.
19. CLEMENS, B. M. 1985. *In* Proceedings of the Materials Research Society Symposium on Lavered Structures Epitaxy and Interfaces. J. M. Gibson & L. R. Dawson, Eds. Vol. **37:** 559. Pittsburgh, PA.
20. SAMWER, K. 1984. *In* Amorphous Metals and Non-Equilibrium Processing. M. Von Allmen, Ed.: 123. Materials Research Society of Europe. Les Editions de Physique. Courteabeouf.
21. SCHROEDER, H., K. SAMWER & U. KÖSTER. 1985. Phys. Rev. Lett. **54:** 197.
22. VAN ROSSUM, M., M.-A. NICOLET & W. L. JOHNSON. 1984. Phys. Rev. B **29:** 5498.
23. DEAL, B. E. & A. S. GROVE. 1965. J. Appl. Phys. **36:** 377.
24. GEUZIN, Y. Y., Y. S. KAGANOUSKIY, L. M. PARITSKAYA & V. I. SOLUNSKIY. 1980. Phys. Met. Metallogr. (USSR) **47:** 127.
25. GOSELE, U. & K. N. TU. 1982. J. Appl. Phys. **53:** 3252.
26. CANTOR, B. & R. CAHN. 1983. Amorphous Metallic Alloys. F. Luborsky, Ed.: 487. Butterworths. London.
27. BUSCHOW, H. J. & N. M. BEEKMANS. 1980. Solid State Commun. **35:** 233.
28. CHENG, Y. T., M.-A. NICOLET & W. L. JOHNSON. 1985. Appl. Phys. Lett. **47:** 800.
29. SAUNDERS, N. & A. P. MIODOWNIK. 1986. J. Mater. Res. **1:** 38.
30. ATZMON, M., E. GIBSON, J. D. VERHOEVEN & W. L. JOHNSON. 1984. Appl. Phys. Lett. **45:** 1052.
31. SCHULTZ, L. 1985. *In* Rapidly Quenched Metals. S. Steeb & H. Warlimont, Eds.: 1585. Elsevier/North-Holland. Amsterdam.
32. ATZMON, M., K. UNRUH & W. L. JOHNSON. 1985. J. Appl. Phys. **58:** 3865.
33. KOCH, C. C., O. B. CAVIN, C. G. MCKAMEY & J. O. SCARBROUGH. 1983. Appl. Phys. Lett. **43:** 1017.
34. SCHWARZ, R. B., R. R. PETRICH & C. K. SAW. 1985. J. Non-Cryst. Solids **76:** 281.
35. POLITIS, C. 1985. Physica **135B:** 286.
36. POLITIS, C. & W. L. JOHNSON. 1986. J. Appl. Phys. **59:** in press.

DISCUSSION OF THE PAPER

A. NAVROTSKY (*Princeton University, Princeton, NJ*): Regarding the questions raised whether glasses obtained via solid state reaction are identical to those produced by quenching melts, we have calorimetric evidence that vapor-deposited amorphous SiO_2 can be up to 40 kJ/mol more metastable in terms of enthalpy than bulk glass (Huffman *et al.*, *J. Electrochem. Soc.* In press). This energy is released in annealing at 400–900° C before the onset of crystallization. Its magnitude depends on preparation conditions.

JOHNSON: It is certainly true that the metallurgical state of amorphous samples, including metallic glasses, depends on sample preparation and history. What I am claiming here is that many key features (such as crystallization temperature and pair correlation function) of glasses formed in the solid state are very similar to those of melt-quenched glasses. There are differences even among melt-quenched samples, and surely our solid state synthesis process will result in such differences.

G. M. FAHY (*American Red Cross, Bethesda, MD*): 1) Can you envision the crystal-to-glass transition in partially frozen aqueous–organic solute systems, for example, in H_2O–glycerol systems? Are there or could there be substances capable of invading and solubilizing ice below −135° C in the presence of a vitreous matrix? 2) How resistant are your metallic glasses to devitrification upon warming? Do you think stable glasses (slowly devitrifying above T_g) of any sort could be formed by your technique?

JOHNSON: 1) Such reactions should in principle be possible. I don't see any reason why they shouldn't be. 2) As I have pointed out, the crystallization behavior of metallic glass formed in the solid state is similar to that of melt-quenched glasses. These glasses have a stability to devitrification similar to that of metallic glasses prepared by more conventional methods. In particular, they tend to crystallize rapidly near T_g.

U. KÖSTER (*University of Dortmund, Dortmund*): Can you form metal–metalloid glasses by the solid state reaction?

JOHNSON: Our group has not made an effort to systematically explore this possibility. I know of a few unsuccessful attempts to do this in transition metal–boron systems. I am not sure as to why this does not work. The crystallization behavior of metal–metalloid glasses may be rather different than that of metal–metal glasses. There may be no practical temperature interval where glasses can grow while crystallization remains suppressed. More work needs to be done in this area.

H. C. ANDERSEN (*Stanford University, Stanford, CA*): If only one of the two species is diffusing, does stress build up or do cracks form in the glassy layer?

JOHNSON: There is evidence of stress build-up and of crack or void formation. Crack formation occurs in relatively thick multilayer samples during the latter stages of reaction. X-ray diffraction studies show evidence that stress builds up as the reaction proceeds.

NAVROTSKY: Another way of slipping into the glassy state is through high-pressure phases, which, when quenched to ambience, are higher in free energy than the glass. For silicates with Si in octahedral coordination, such as SiO_2 (stishovite-rutile structures) and $MgSiO_3$ (ilmenite and perovskite polymorphs), heating at atmospheric pressure to several hundred degrees centigrade results in formation of glass rather than of the stable crystaline assemblage (Ito and Novrotsky, *Am. Mineral.* **70:** 1020, 1985). Is this route to the glassy state feasible for your metallic systems?

JOHNSON: It should be possible to use pressure as a thermodynamic variable to invert the free energy relationship between the crystalline state and the glassy state in metals. For example, certain intermetallic compounds are known to expand on freezing. I believe that $CoSi_2$ is one example. Applying sufficiently high pressure at low temperature (below the T_g of the glassy alloy) could produce a situation where the crystal is superheated with respect to the glassy state. This could produce vitrification. The idea is similar to the one you've described for the case of the covalent glasses. In my example, increasing pressure induces vitrification whereas you referred to a pressure-quench case.

Nucleation of Crystals from the Melt[a]

DAVID W. OXTOBY

Department of Chemistry
and
James Franck Institute
University of Chicago
Chicago, Illinois 60637

INTRODUCTION

The formation of a glass as a liquid is cooled is the result of the avoidance of another type of event: the crystallization of the melt. Although supercooled liquids and glasses have some properties that are inherent to those phases and independent of underlying crystalline states, other important properties (such as "glass-forming tendency") are directly related to the competition between glass formation and crystallization, and thus depend directly on the rate of crystallization. Although the traditional view is one of glasses as arising from dynamic entanglements in the liquid phase, it may be more useful instead to examine the nature of the crystalline solid and the ease with which molecules can be packed into a periodic lattice. The differing glass-forming tendencies of benzene and of toluene, or of the xylenes, arise more from packing differences in the crystal than from different interactions in the liquid. This view has recently been stressed by Angell *et al.*[1]

An important component of the study of glass-forming liquids is therefore the rate at which crystals nucleate from the melt, because over a wide range of thermodynamic states the nucleation step is rate determining and subsequent growth of crystals to macroscopic dimensions is rapid. The classical theory for the nucleation of crystals from the melt was developed by Turnbull and co-workers.[2] It assumes that the free energy of a crystallite (assumed spherical with radius R) is the sum of a bulk and a surface contribution

$$\Delta G(R) = -\frac{4}{3}\pi R^3|\Delta G_v| + 4\pi R^2\sigma_{SL} \tag{1}$$

Here $\Delta G_v = \Delta H_f - T\Delta S_f = (T_f - T)\Delta S_f$ is the bulk free energy charge per unit volume on freezing, and σ_{SL} is the interfacial solid-liquid surface free energy; ΔH_f, ΔS_f, and σ_{SL} are assumed to be only weakly dependent on temperature in the supercooled liquid. $\Delta G(R)$ has a maximum at

$$R^* = \frac{2\sigma_{SL}}{|\Delta G_v|} = \frac{2\sigma_{SL}}{(T_f - T)\Delta S_f} \tag{2}$$

[a]This work was supported by Grant CHE82-19440 from the National Science Foundation and by the National Science Foundation Materials Research Laboratory at the University of Chicago.

This defines the radius of the critical nucleus, which forms the barrier to nucleation. Crystallites with radii larger than R^* will tend to grow rapidly. The height of the barrier is

$$\Delta G^* = \Delta G(R^*) = \frac{16\,\pi\sigma_{SL}^3}{3(T_f - T)^2(\Delta S_f)^2} \tag{3}$$

As the supercooling increases ($T_f - T$ becomes larger), the barrier height drops and the nucleation rate increases.

The nucleation rate then has the form

$$J = A(T)\exp(-\Delta G^*/k_B T) \tag{4}$$

where $A(T)$, the preexponential factor, is an attempt frequency for the addition of particles to the crystal across the crystal-melt interface. In most cases, the exponential factor dominates the temperature dependence, and changes so rapidly as the liquid is cooled that the nucleation rate goes from very slow to very fast over a small temperature range. The temperature dependence of $A(T)$ does become significant near the glass transition because it scales roughly with the diffusion constant. This fact has been used in the recent work of Angell and co-workers[3] to trace out the effect of the cooling rate on nucleation.

Three steps are necessary to achieve a quantitative theory of nucleation rates that goes beyond equations 1–4. They are as follows: 1) locate the equilibrium freezing transition accurately; 2) describe inhomogeneous crystal-liquid systems and the interface between the two phases; 3) calculate the rate at which nucleation occurs. The first two steps are a necessary prerequisite to the third. In this paper I shall describe the use of a density functional theory to carry out this program.[4–7] I shall then present some new results on the effect of pressure on the nucleation of liquids.

DENSITY FUNCTIONAL THEORY OF CRYSTALLIZATION

An order parameter theory of crystallization was proposed by Ramakrishnan and Yussouff[4] and extended and reformulated by Haymet and Oxtoby.[5] The order parameters are the Fourier components of the average density

$$\rho(\mathbf{r}) = \rho_o(1 + \eta) + \rho_o \sum_i \mu_i\, e^{i\mathbf{k}_i\cdot\mathbf{r}} \tag{5}$$

Here ρ_o is the average density of the liquid, η is the fractional density change on freezing, and the infinite set μ comprises structural order parameters corresponding to reciprocal lattice vectors $\mathbf{k}_i$. This one particle level set of order parameters is sufficient for many purposes, but it cannot account for thermodynamic phases with only bond orientational order.[8]

The free energy $\mathcal{F}$ is a functional of the one-particle density $\rho(\mathbf{r})$. One form in which it can be written is as the exactly calculable free energy of a noninteracting system of particles with density $\rho(\mathbf{r})$, $\mathcal{F}_{id}(\rho(\mathbf{r}))$, plus a term due to interactions, which is expanded in a functional series expansion about the liquid density ρ_o. This gives

$$\mathcal{F}(\rho(\mathbf{r})) = \mathcal{F}_{id}(\rho(\mathbf{r})) - \frac{kT}{2}\int d\mathbf{r}d\mathbf{r}'c(\mathbf{r},\mathbf{r}')\,[\rho(\mathbf{r}) - \rho_o]\,[\rho(\mathbf{r}') - \rho_o] + \ldots \tag{6}$$

where $c(\mathbf{r}, \mathbf{r}') = c(|\mathbf{r} - \mathbf{r}'|)$ is the two-particle direct correlation function of the liquid and the higher order terms (omitted in equation 6) come from third and higher order direct correlation functions that are functional derivatives of $c(\mathbf{r}, \mathbf{r}')$. The two-particle function is directly obtainable from X-ray or neutron scattering data because its Fourier transform $c(k)$ is related to the structure factor through

$$S(k) = (1 - \rho c(k))^{-1} \tag{7}$$

The higher order direct correlation functions are less accessible to measurement or calculation, and have generally been omitted.

The location of the equilibrium freezing transition in this approach uses equations 5 and 6 to find the dependence of the free energy on the order parameters, truncated to a finite set. The solid must be locally stable (that is, it must represent a local minimum of $\mathcal{F}$), and at the transition the chemical potentials and pressures of solid and liquid must be equal. Some low-order calculations (using two to three order parameters) have been carried out for a variety of systems,[4] and an exhaustive and successful calculation has been reported[9] for the freezing of hard spheres.

The next step is to extend our considerations from uniform systems (solids or liquids) to nonuniform systems in which interfaces are present. For the simplest case of an infinite planar interface we can generalize equation 5 to

$$\rho(\mathbf{r}) = \rho_o(1 + \eta(z)) + \rho_o \sum_i \mu_i(z)\, e^{i\mathbf{k}_i \cdot \mathbf{r}} \tag{8}$$

where z is the coordinate perpendicular to the interface and the order parameters change continuously from zero (in the liquid phase) to their bulk solid values. The free energy is now a nonlocal functional of the order parameters. If this nonlocal dependence is expanded in a series of gradients and only the lowest nontrivial order retained, the result is a "square gradient" theory in which the order parameters are assumed to vary slowly over interatomic distances. A variational principle for the interface order parameter profiles then leads to a set of coupled second-order nonlinear differential equations[5,6] that can be solved numerically.

A low-order calculation of the solid-liquid surface free energy was carried out for the cases of sodium and potassium, retaining only the first set of reciprocal lattice vectors in equation 8.[6] Those results suggested that the interface was relatively broad (about 10 lattice spacings, or 15–20 Å in width) and that the surface free energy was only weakly dependent on the orientation of the crystal surface. The result calculated for σ_{SL} was

$$\sigma_{SL} = \rho_o kT\, [-c''(k_1)]^{1/2}\, \lambda \tag{9}$$

where $c''(k_1)$ is the second derivative of the Fourier transform of $c(r)$, evaluated at the first reciprocal lattice vector of the solid, k_1, and λ is a dimensionless constant characteristic of the crystal orientation. Because λ is given by a sum of terms from each order parameter, it should be approximately proportional to the entropy change on freezing. If the length $[-c''(k_1)]^{1/2}$ scales as $\rho_o^{-1/3}$, this can then be written as

$$\sigma_{SL} \propto \rho_o^{2/3}\, T\Delta S_f = \rho_o^{2/3}\, \Delta H_f \tag{10}$$

a relationship proposed first on empirical grounds by Turnbull (see reference 10).

Because the surface free energy for the alkali metals depends only weakly on crystal orientation, it should be a good approximation to take the critical nucleus to be spherical. Its density can then be written

$$\rho(\mathbf{r}) = \rho_o(1 + \eta(r)) + \rho_o \sum_i \mu_i(r)\, e^{i\mathbf{k}_i \cdot \mathbf{r}} \tag{11}$$

where r is the distance from the center of the nucleus. Nucleation involves supercooled

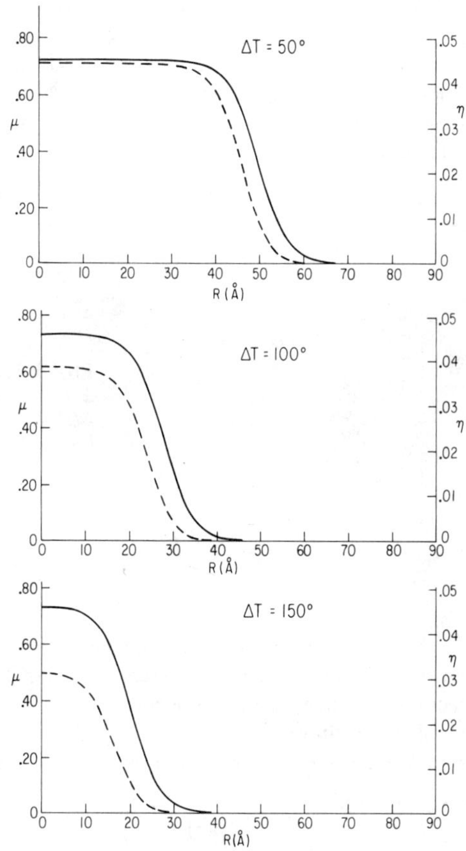

FIGURE 1. Order parameter profiles $\mu(R)$ (solid line) and $\eta(R)$ (dashed line) for several supercoolings ΔT. From reference 7.

liquids, so the liquid structure factor must be extrapolated to below the freezing point. Instead of looking for a simple maximum as a function of nucleus radius R, we are now searching for a saddle point in functional space. The numerical methods for finding this saddle point are described in more detail in reference 7.

The same low-order (single set of reciprocal lattice vectors) theory used for calculating the planar interface for liquid sodium was then applied to the nucleation problem. The results for the shape of the critical nucleus at several undercoolings are shown in FIGURE 1, and the dependence of the excess free energy ΔG^* on undercooling is compared with that predicted by classical theory (equation 3) in FIGURE 2. These results illustrate the qualitative, though not quantitative, success of the simple nucleation theory embodied in equation 3. Even for relatively large undercoolings the center of the nucleus resembles the bulk solid in its order parameters, whereas the interfacial width stays close to that for the infinite planar interface. Any singularity associated with the local instability of the liquid (such as is seen at the spinodal for binary mixture phase separation[11,12]) is far enough away that it does not qualitatively affect the observed results. Reference 7 contains further discussion of these questions.

We have considered only the free energy portion ΔG^* of equation 4. The dynamical preexponential factor $A(T)$ has been studied by Grant and Gunton,[13] who used Langer's field theoretical methods[14,15] with the density functional theory of freezing. Their result is a generalization of the classical theory, and explicitly incorporates the thermal conductivity of the liquid. As they point out, the two approaches are closely related.

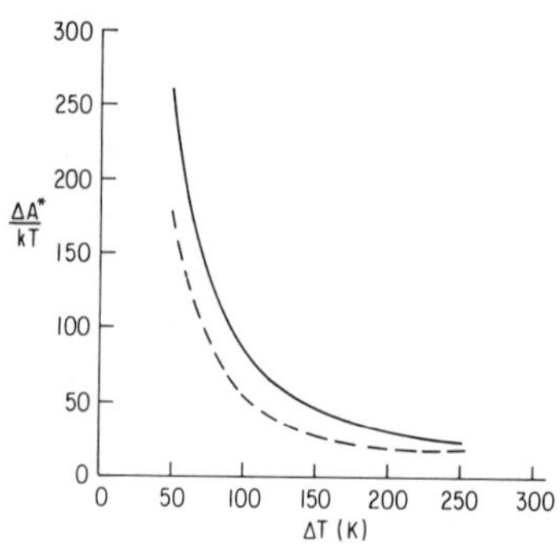

FIGURE 2. Calculated free energy of the nucleus $\Delta A^*/kT$ (solid line) compared with the result for $\Delta G^*/kT$ from the capillarity approximation (dashed line).

The next step to be taken in nucleation theory is to extend previous work to incorporate more order parameters. This can be done by retaining more terms in the reciprocal lattice expansions of equations 5, 8, and 11, at the cost of coupling increasing numbers of differential equations in calculated interfacial and nucleus properties. A simple, but approximate, alternative is to use instead the functional form

$$\rho(\mathbf{r}) = \left(\frac{\alpha}{\pi}\right)^{3/2} \sum_{\mathbf{R}} \exp\left[-\alpha^2|\mathbf{r} - \mathbf{R}|^2\right] \tag{12}$$

Such a "real space" ansatz for $\rho(\mathbf{r})$ was first used by Tarazona[16] for uniform solids and includes the quasi-harmonic approximation of a Gaussian distribution of density about each lattice site $\mathbf{R}$. Its advantage is that with a single order parameter α it incorporates the effects of many terms in the summation of equation 5, albeit approximately, and thus can be expected to be more successful than a low-order truncation of equation 5. If α is generalized to become z- or r-dependent, the properties of equilibrium interfaces and nucleation can then be studied within the square gradient approximation.

EFFECT OF PRESSURE ON NUCLEATION

Nucleation experiments on undercooled liquids[17,18] are typically carried out by suspending small (micron-sized) droplets of the liquid in an oil emulsion and measuring the heat evolved from the sample as it is cooled. Those droplets that contain nucleating impurities give thermal signals as they freeze at a range of relatively low undercoolings, whereas the other droplets freeze together at a reproducible lower temperature. It is this latter temperature that is identified as the nucleation temperature T_n. Although our theoretical approach has so far only considered homogeneous nucleation (in the bulk of a pure liquid) we need to recognize that even in the absence of impurities many of the experimentally measured undercooling limits are set by heterogeneous nucleation on the interface between the liquid and the emulsion. This is demonstrated by the fact that a different emulsion preparation can lead to a greater undercooling than previously reported, as shown especially in the work of Perepezko and co-workers.[18]

Experiments on pure liquids at atmospheric pressure typically can measure only a single parameter: the undercooling temperature T_n. Because the surface free energy σ is in most cases not measurable by independent techniques, this one experimental measurement is often used to fit σ using the classical theory. In such cases no test of this theory is possible. If a single liquid could be studied at a variety of pressures, however, additional information could be obtained and used to test nucleation theories further. Such experiments have recently been begun by Perepezko and co-workers (personal communication). In this section I shall outline a theoretical analysis of the problem.

If we consider liquids sufficiently far from their glass transition that the preexponential factor $A(T)$ in equation 4 changes only slowly with pressure, then nucleation is characterized by a constant value of the exponential factor $\Delta G^*/k_B T$ as a function of pressure

$$\frac{\Delta G^*}{k_B T} = \frac{16\pi\sigma_{SL}^3}{3k_B(T_f - T_n)^2 T_n (\Delta S_f)^2} = \text{constant} \tag{13}$$

where T_n is the temperature at which nucleation takes place. If we take the derivative of this expression with respect to the pressure along the coexistence curve, we find

$$\frac{3}{\sigma_{SL}}\frac{\partial\,\sigma_{SL}}{\partial P} - \frac{2}{T_f - T_n}\left(\frac{\partial T_f}{\partial P} - \frac{\partial T_n}{\partial P}\right) - \frac{1}{T_n}\frac{\partial T_n}{\partial P} - \frac{2}{\Delta S_f}\frac{\partial \Delta S_f}{\partial P} = 0 \tag{14}$$

which can be rearranged to give

$$\frac{\partial T_n}{\partial P} = \frac{2-2\theta}{2-3\theta}\frac{\partial T_f}{\partial P} + \frac{2-2\theta}{2-3\theta}\theta\frac{T_f}{\Delta S_f}\frac{\partial \Delta S_f}{\partial P} - \left(\frac{2-2\theta}{2-3\theta}\right)\frac{3}{2}\theta\frac{T_f}{\sigma_{SL}}\frac{\partial\sigma_{SL}}{\partial P} \tag{15}$$

where $\theta = (T_f - T_n)/T_f$ is the relative undercooling.

For very small undercoolings ($\theta \rightarrow 0$) we of course find

$$\frac{\partial T_n}{\partial P} = \frac{\partial T_f}{\partial P} \tag{16}$$

and the slope of the nucleation curve approaches that for the liquidus. For larger values of θ (experimentally values of 0.35 are obtainable for tin, and 0.5 for gallium) the two slopes differ. The first term on the right side of equation 15 leads to a slope $\partial T_n/\partial P$ that is larger in magnitude than $\partial T_f/\partial P$. Note that the latter is given by the Clapeyron equation as

$$\left(\frac{\partial T_f}{\partial P}\right)_{\text{coex}} = \frac{\Delta V}{\Delta S} \tag{17}$$

where ΔV and ΔS are the volume and entropy changes, respectively, on fusion. For most substances $\Delta V > 0$ whereas ΔS is always positive, so usually T_f increases with pressure.

The contribution of the second term in equation 15 depends on the sign of the pressure derivative of ΔS_f. Experimentally, entropy changes usually decrease slightly in magnitude with increasing pressure. The entropy of fusion of argon decreases from 1.686 R at 0.7 bar to 1.047 R at 15.7 kbar, and an extrapolation to infinite pressure yields 0.9 R.[19] For sodium, ΔS_f decreases from 0.846 R at 35 bar to 0.771 R at 22.1 kbar, and extrapolates to 0.7 R at infinite pressure.[19] These changes are relatively small, however. For sodium we estimate that

$$\frac{\partial T_f}{\partial P} = 8.8 \times 10^{-8}\ \text{K} \cdot \text{Pa}^{-1} = 9.1\ \text{K} \cdot \text{kbar}^{-1}$$

whereas

$$\frac{T_f}{\Delta S_f}\frac{\partial \Delta S_f}{\partial P} = 1.4 \times 10^{-8}\ \text{K Pa}^{-1} = 1.5\ \text{K} \cdot \text{kbar}^{-1} = 0.16\,\frac{\partial T_f}{\partial P}$$

The third term involves the pressure derivative of σ_{SL}, a quantity that is not well studied experimentally. Perepezko (personal communication) has measured the pressure dependence of T_n for tin, and finds it necessary to assume that $(\partial \sigma_{SL}/\partial P) < 0$ in order to fit his data to the nucleation theory of Spaepen.[20] As we shall show below, the simple expectation for the sign of this derivative is positive, based on the density functional approach.

Let us consider initially soft repulsive spherical potentials of the form

$$V(r) = \epsilon\left(\frac{\sigma}{r}\right)^n \tag{18}$$

The usual expectation is that repulsive forces dominate the structure of a liquid near its triple point, and this dominance should increase as the temperature and pressure increase. Thus such a potential should provide a good reference system for many liquids. Such potentials have a very useful scaling property. If the volume is reduced by a factor of α^3, then all distances decrease by a factor of α. If the temperature is simultaneously increased by a factor of α^n, then all thermodynamic properties remain unchanged. This arises because of the presence of the combination

$$\frac{V(r)}{kT} = \frac{\epsilon}{kT}\left(\frac{\sigma}{r}\right)^n \tag{19}$$

As a result, the thermodynamic properties depend only on the combination $\rho T^{-3/n}$ and not on ρ and T separately. In particular, the liquidus curve will be traced out with $\rho T^{-3/n}$ = constant, or

$$T \propto \rho^{n/3} \propto V^{-n/3} \tag{20}$$

The direct correlation function $c(r)$ on the liquidus will be a scaled function of $r\rho^{1/3}$

$$c(r) = f(r\rho^{1/3}) \qquad \text{for } \rho\, T^{3/n} = \text{constant}$$

so that

$$\begin{aligned} \rho c''(k_i) &= \rho^{-2/3} \int \rho d\mathbf{r}\, e^{i\mathbf{k}_i\rho^{-1/3}\cdot\mathbf{r}\rho^{1/3}} f(r\rho^{1/3}) \\ &= \rho^{-2/3} f_i \end{aligned} \tag{21}$$

The parameter λ in equation 9 should, like the entropy of fusion, be constant along the liquidus for such a potential. The overall result is then

$$\begin{aligned} \sigma_{SL} &= \rho_o k_B T\, [-c''(k_1)]^{1/2}\lambda \\ &\propto \rho_o^{1/6}\, T \\ &\propto T^{1+1/2n} \end{aligned} \tag{22}$$

σ_{SL} is thus an increasing function of T and therefore of P. Moreover, we can write

$$\frac{T_f}{\sigma_{SL}} \frac{\partial \sigma_{SL}}{\partial P} = \left(1 + \frac{1}{2n}\right) \frac{\partial T_f}{\partial P} \tag{23}$$

so that $T_f/\sigma_{SL}\, \partial\sigma_{SL}/\partial P$ ranges from $\partial T_f/\partial P$ to $\frac{3}{2}\, \partial T_f/\partial P$ as n ranges from infinity to zero. These conclusions are not limited to the single order parameter calculation of references 6 and 7; they still hold true if equation 9 is generalized to a sum of contributions from order parameters.

How well does a potential of this form represent the properties of liquid sodium? Singh and Holz have shown that the model of charged hard spheres in the mean spherical approximation provides an excellent fit to the experimental structure factor,[21,22] one significantly better than that obtained for uncharged hard spheres or for the one-component plasma (point charge) model. In their work, a dielectric screening function is used that affects only the long-wavelength behavior of $S(k)$ and therefore of $c(k)$; in particular, $c(k)$ evaluated at the first and subsequent peaks of the structure factor is unchanged by this screening. The quality of the agreement with experimental data is illustrated in FIGURE 3. The model of charged hard spheres is not a scaled potential of the form of equation 18 because it includes contributions both with $n = \infty$ and $n = 1$. For states near the freezing point, however, Singh has shown[25] that $c(r)$ is very close to the scaled function

$$c(r) = \text{constant} \times \Gamma \times f(r\rho^{1/3}) \tag{24}$$

where $\Gamma = e^2/kTA$ and $a^3 = 3/4\pi\rho$. As a result, the predictions outlined above with $n = 1$ should be rather well satisfied for sodium, and a positive $\partial\sigma/\partial P$ expected. This scaling cannot hold exactly, of course, because it would predict the entropy of fusion of sodium to be independent of pressure. We have seen above that its experimental pressure derivative, while quite small, is not zero. Nevertheless, we regard this prediction of a positive sign for $\partial\sigma/\partial P$ for simple metals like sodium to be worthy of testing.

Why then does tin apparently have a negative pressure derivative? There are three possible reasons for this difference: 1) The nucleation process is very sensitive to small changes in the liquid structure factor. In other words, the very small differences between the scaled form and the experimental data for $S(k)$ may have a disproportionately large effect on the nucleation. This can be tested by studying what happens when $S(k)$ is changed by small amounts. 2) Even at the largest undercoolings achieved, heterogeneous (surface) nucleation still most likely dominates the experimental results. It is possible that the discrepancy between theory and experiment would disappear if true homogeneous nucleation could be studied. Alternatively, it would be desirable to extend the theory to heterogeneous nucleation. 3) Tin differs fundamentally from sodium in that it shows important attractive ion core polarization effects (covalent interactions between atoms). One effect of this is the appearance of a second small peak in $S(k)$ beyond the large first peak,[26] as seen in FIGURE 4. The effect of these attractive forces can be studied in perturbation theory. Suppose we write

$$u(r) = u_0(r) + u'(r) \tag{25}$$

where $u_0(r)$ is a repulsive reference potential (such as charged hard spheres with dielectric screening) and $u'(r)$ is an attractive potential. Then using perturbation

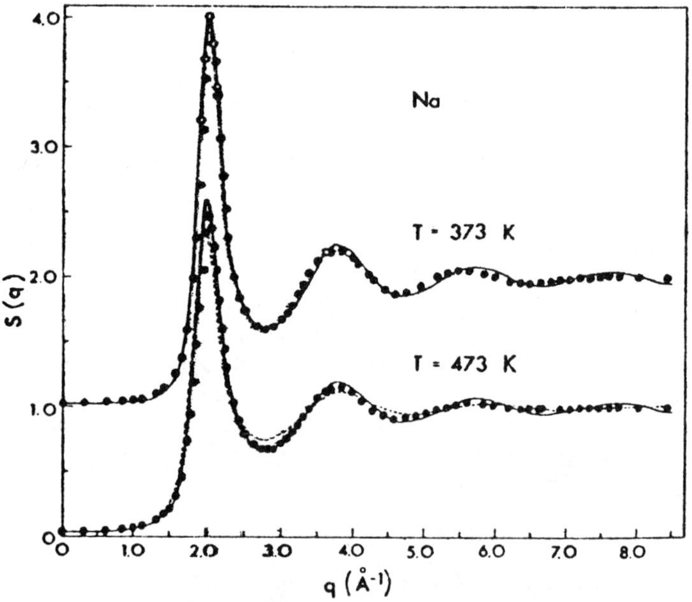

FIGURE 3. Calculated structure factor $S(q)$ of liquid sodium (solid line) compared with experimental measurements from reference 23, (filled circles) and reference 24 (open circles). From reference 21.

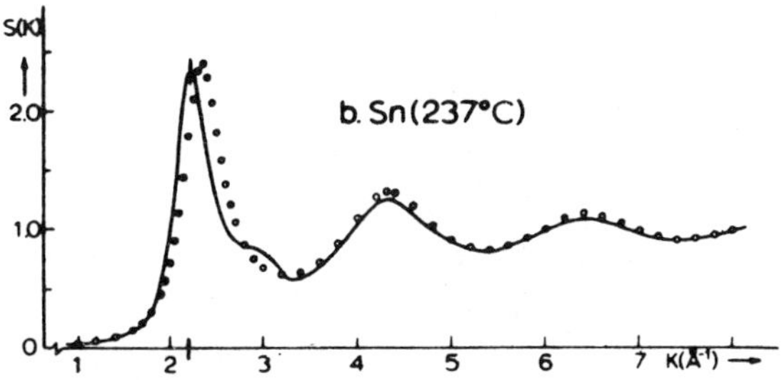

FIGURE 4. Experimental structure factor of liquid tin (solid line) compared with the hard-sphere model (circles). From reference 26.

theory we can take

$$c(r) = c_0(r) - \frac{1}{k_B T} u'(r) \tag{26}$$

where $c_0(r)$ is the direct correlation function of the reference liquid with potential $u_0(r)$. The attractive forces may cause different behavior for partially covalent metals like tin, compared with alkali metals. These three possibilities are presently under further investigation.

SUMMARY

A theoretical study of homogeneous crystal nucleation from the melt was carried out, and a density functional theory was used to test the assumptions of classical nucleation theory. The results show that the simple classical approach is qualitatively reasonable, although for quantitative purposes it may lead to errors of 10–20% in the estimation of the surface free energy. The effect of pressure on nucleation rates was discussed, and the predictions were compared with experimental data on the crystallization of liquid metals.

ACKNOWLEDGMENTS

I would like to thank John H. Perepezko for his valuable contributions to our discussions.

REFERENCES

1. Angell, C. A., L. E. Busse, E. I. Copper, R. K. Kadiyala, A. Dworkin, M. Ghelfenstein, H. Szwarc & A. Vassal. 1985. J. Chim. Phys. **82:** 267.
2. Turnbull, D. & J. C. Fisher. 1949. J. Chem. Phys. **17:** 71.
3. MacFarlane, D. R., R. K. Kadiyala & C. A. Angell. 1983. J. Chem. Phys. **79:** 3921.
4. Ramakrishnan, T. V. & M. Yussouff. 1979. Phys. Rev. B **19:** 2775.
5. Haymet, A. D. J. & D. W. Oxtoby. 1981. J. Chem. Phys. **74:** 2559.
6. Oxtoby, D. W. & A. D. J. Haymet. 1982. J. Chem. Phys. **76:** 6262.
7. Harrowell, P. & D. W. Oxtoby. 1984. J. Chem. Phys. **80:** 1639.
8. Steinhardt, P. J., D. R. Nelson & M. Ronchetti. 1983. Phys. Rev. B **28:** 784.
9. Haymet, A. D. J. & D. W. Oxtoby. 1986. J. Chem. Phys. **84:** 1769.

10. WOODRUFF, D. P. 1973. The Solid-Liquid Interface. Cambridge University Press. Cambridge. England.
11. CAHN, J. W. & J. E. HILLIARD. 1958. J. Chem. Phys. **28:** 258.
12. CAHN, J. W. & J. E. HILLIARD. 1959. J. Chem. Phys. **31:** 688.
13. GRANT, M. & J. D. GUNTON. 1985. Phys. Rev. B **32:** 7299.
14. LANGER, J. S. 1967. Ann. Phys. (N.Y.) **41:** 108.
15. LANGER, J. S. 1969. Ann. Phys. (N.Y.) **54:** 258.
16. TARAZONA, P. 1984. Mol. Phys. **52:** 81.
17. TURNBULL, D. 1956. *In* Solid State Physics. F. Seitz & D. Turnbull, Eds. Vol. **3:** 225. Academic Press. New York, NY.
18. PEREPEZKO, J. H. & J. S. PAIK. 1982. *In* Proceedings of the Materials Research Society Symposium: Rapidly Solidified Amorphous and Crystalline Alloys. B. H. Kear & B. C. Giessen, Eds. Elsevier/North-Holland. New York, NY.
19. STISHOV, S. M. 1975. Sov. Phys. Usp. **17:** 625.
20. SPAEPEN, F. 1975. Acta Metall. **23:** 729.
21. SINGH, H. B. & A. HOLZ. 1983. Phys. Rev. A **28:** 1108.
22. SINGH, H. B. & A. HOLZ. 1984. Phys. Rev. A **29:** 1554.
23. GREENFIELD, A. J., J. WELLENDORF & N. WISER. 1971. Phys. Rev. A 4: 1607.
24. HUIJBEN, M. J. & W. VAN DER LUGT. 1979. Acta Crystallogr. Sect. A **35:** 431.
25. SINGH, H. B. 1983. J. Stat. Phys. **33:** 371.
26. VAHVASELKÄ, K. S. 1978. Phys. Scr. **18:** 266.

DISCUSSION OF THE PAPER

F. H. STILLINGER (*AT&T Bell Laboratories, Murray Hill, NJ*): I would like to raise a question about your way of mathematically representing the spatially varying density $\rho(\mathbf{r})$. For any such function the set of plane waves $e^{i\mathbf{k}_i\cdot\mathbf{r}}$ provides a complete set, and the resulting Fourier expansion has coefficients depending only on $\mathbf{k}_i$. In particular, such position-independent coefficients could have been used in your application: a crystalline nucleus embedded in an amorphous medium (where local stresses might have substantially perturbed lattice spacings).

You are certainly at liberty to extend the representation to incorporate position-dependent expansion coefficients. But then the representation becomes over-complete, that is, not unique. In spite of this you seem to have found a unique solution.

A simple one-dimensional example illustrates the underlying ambiguity. Suppose hypothetically that the local density variation through the crystallite were given by

$$\delta\rho(x) = A \exp(-\alpha x^2) \cos(k_o x)$$

where A, α, and k_o are suitable constants. We have the following identity

$$\delta\rho(x) \equiv (2\pi)^{-1} \int_{-\infty}^{\infty} R(x, k, \lambda) \exp(ikx) \mathrm{d}k$$

where

$$R(x, k, \lambda) = A(\pi\lambda)^{1/2} \exp\{-[\alpha - (4\lambda)^{-1}]x^2\}$$
$$\cdot (\exp[-\lambda(k - k_o)^2] + \exp[-\lambda(k + k_o)^2]$$

The conventional Fourier representation sets $\lambda = (4\alpha)^{-1}$, but any larger positive value of λ would serve just as well. The free energy as a functional of R would have to be rigorously invariant to such variations in λ.

Surely analogous ambiguity exists in three dimensions regarding Fourier coefficients, with corresponding free energy invariance. What determines the preferred choice?

OXTOBY: I would like to point out first that the set of plane waves $e^{i\mathbf{k}_i \cdot \mathbf{r}}$ provides a complete set only if $\mathbf{k}_i$ is allowed to range over a continuously infinite set of values. In our approach, $\mathbf{k}_i$ is limited to a discrete set of values corresponding to the reciprocal lattice vectors of the solid, so that we could not have used this expansion with position-independent coefficients to represent the shape of the crystalline nucleus. Only functions that are periodic with the lattice spacing throughout all space can be represented in such a way.

Nonetheless, I agree that there is some arbitrariness in the form of our expansion. It will be unique only if there is a strict separation of distance scales between the characteristic interatomic spacings and the distances over which the average density in the nucleus changes. We have assumed such a separation in our calculation, and significant deviations from this approximation would lead to nonunique solutions.

G. M. FAHY (*American Red Cross, Bethesda, MD*): Can you say anything about how high solute concentrations, for example, 60% glycerol in water, affect either homo- or heterogeneous nucleation rates (or nucleation temperatures)? Can you say anything about the combined effect of solutes and pressure on nucleation rates (or nucleation temperatures)?

OXTOBY: Even very small concentrations of a solute species can have a large effect, by providing sites for heterogeneous nucleation of the solvent. When the solute concentrations become high, still other factors come into play, such as the thermodynamics of binary mixtures and the dynamical processes that lead to phase separation or partitionless solidification. Experiments on liquid metal alloys show that in most cases the nucleation temperature roughly follows the liquidus curve in terms of dependence on composition. The reasons for this behavior are not understood at present.

M. WEINBERG (*Jet Propulsion Laboratory, Pasadena, CA*): The nucleation rate has been measured as a function of undercooling for inorganic oxide glasses. In those cases where classical theory has been compared with experimental results, it has been found that there is a large discrepancy (many orders of magnitude) between the experimental and theoretical values of the prefactor. Hence, I believe one must consider corrections (or reformulations) of the "kinetic part" of nucleation theory as well as the "thermodynamic part."

OXTOBY: The cases you are referring to are the ones in which a glass transition occurs close to the nucleation temperature. These are exactly the cases that I deliberately excluded in my treatment. I agree that there are interesting dynamical problems related to the magnitude of the prefactor in such cases, which require further study.

S. A. RICE (*University of Chicago, Chicago, IL*): I believe the effective electron densities in solid and liquid tin are very different: two electrons/atom and four electrons/atom, respectively. Then the effective ion–ion (or, better, pseudoatom–pseudoatom) potential will be very different in the two phases. Perhaps this is the reason that the pressure dependence of the interface tension is of opposite sign to that predicted for an r^{-n} potential. I note that your argument assumes the pair potential to be the same in the liquid and solid phases.

OXTOBY: The approach I have outlined assumes the local structure to be similar in the solid and liquid phases, in that it uses a direct correlation function from the liquid for the solid as well. When the structure and bonding in the two phases are very different, the approximations implicit in the theory will not hold. Thus, for example, the theory could not describe the covalent gray tin form that is stable at lower temperatures. If the electronic properties of liquid tin and solid white tin are as different as you suggest, this may well contribute to the fact that the behavior of the interfacial tension is opposite to that seen for the alkali metals.

M. LIPKIN (*Cornell University, Ithaca, NY*): If the number of order parameters is essentially chosen to be finite and small, would this not mean that a crystalline minimum is always selected, whereas another "global" minimum (essentially disordered) involving very many nonzero order parameters is excluded?

OXTOBY: I disagree that a state with many nonzero order parameters is a disordered state. It is rather an ordered state with perhaps a more complex unit cell. I agree that one must be cautious in assuming that the state that one finds by gradually increasing the number of order parameters is the globally stable one. There could be another crystalline phase that would be difficult to locate by the methods we use.

C. A. ANGELL (*Purdue University, Lafayette, IN*): It's interesting that you have shown that the end of the supercooled liquid state must always be caused by a nucleation and growth process for the case of the alkali metals because, in view of their soft repulsions, these would be expected to be the most likely liquids to encounter a true mechanical instability on extended supercooling. It seems, in this case, that the possibility of a mechanical instability in liquids is eliminated (the case of water which has a reentrant spinodal (R. J. Speedy, *J. Phys. Chem.* **86:** 982, 1982) would not be included in this generalization). Can you tell us what changes in the structure factor in the supercooled region would be necessary to change this prediction?

OXTOBY: A mechanical instability in the liquid phase would be signaled by an apparent divergence, or at least dramatic increase, in the structure factor of the liquid for some value of the wave vector **k**. No sign of such an anomaly is seen for the alkali metals, nor as far as I know for any other liquid. The magnitude of the first peak in $S(k)$, for example, increases with supercooling but in a smooth and nonsingular fashion.

Micromechanisms and Kinetics of Crystallization below the Glass Transition Temperature[a]

UWE KÖSTER

Department of Chemical Engineering
University of Dortmund
Dortmund, Federal Republic of Germany

INTRODUCTION

In the last few years the micromechanisms and kinetics of the crystallization of metallic glasses have become subjects of an increasing research effort and have been reviewed by several authors.[1-3] Metallic glasses have been observed to crystallize by nucleation and by growth processes. Nucleation may be homogeneous, or, heterogeneous in the bulk or at the surface; nucleation rates may be transient or in steady state. Growth may be primary, eutectic, or polymorphic. Unlike most silicate glasses, metallic glasses have been found to crystallize in reasonable experimental times even far below the glass transition temperature T_g. As mentioned recently by Uhlmann,[4] the occurrence of nucleation and crystal growth in this temperature range indicates a decoupling between the kinetic coefficient for transport at the crystal-glass interface and the bulk viscosity, and relatively high diffusivity of the constituents of the glass.

The aim of this paper is to verify the crystallization behavior of several metallic glasses at temperatures below T_g in some detail in order to better understand the physical basis of this phenomenon.

THERMODYNAMICS OF CRYSTALLIZATION

Metallic glasses have been found to crystallize by nucleation and by growth processes. The driving force is the difference in free energy between the glass and the appropriate crystalline phase(s). In order to gain an overall picture of all possible crystallization reactions, a (hypothetical) diagram of the free energy for the various phases versus concentration has been found to be very useful.[5] In FIGURE 1 such a hypothetical diagram is shown for Fe-B alloys. Depending on their composition, metallic glasses have been observed to crystallize by 1) polymorphic, 2) primary, and 3) eutectic crystallization.

Polymorphic crystallization of one phase with the same composition as the glass can occur only in concentration ranges near those of pure elements or compounds and needs only single jumps of atoms across the crystallization front.

During primary crystallization of one crystal phase with a composition different

[a]The financial support of the Deutsche Forschungsgemeinschaft is gratefully acknowledged.

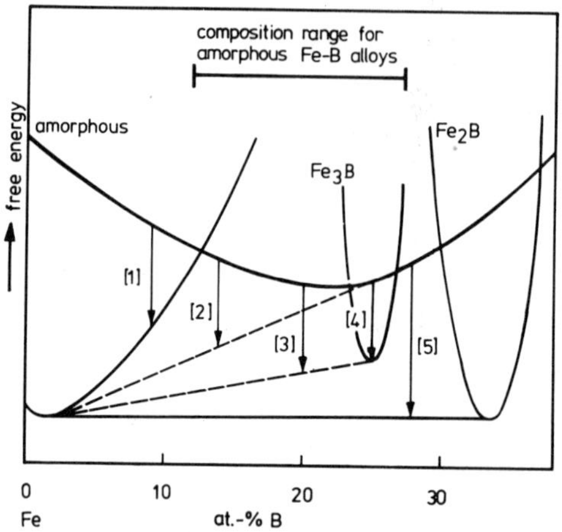

FIGURE 1. Hypothetical diagram of the free energy for the various phases in Fe-B alloys versus concentration.[1] The following crystallization reactions can occur due to the difference in free energy: 1) polymorphic crystallization of α-Fe; 2) primary crystallization of α-Fe; 3) eutectic crystallization of α-Fe + Fe_3B; 4) polymorphic crystallization of Fe_3B; and 5) eutectic crystallization of α-Fe + Fe_2B.

from the glassy matrix (for example, α-Fe) a concentration gradient is built up ahead of the crystallization front. The amorphous phase becomes enriched in B until crystallization is stopped after reaching the metastable equilibrium; subsequently, the B-enriched matrix transforms by one of the other mechanisms discussed here.

During eutectic crystallization two crystalline phases grow cooperatively by a discontinuous reaction. There is no difference in the overall concentration across the reaction front; diffusion takes place parallel to the reaction front, and the two components have to separate into the two phases, thus reducing the growth rate of this reaction compared with that during polymorphic crystallization.

Appropriate free energy-versus-composition diagrams can be constructed from available thermodynamical data of the melt and the crystalline phases by considering the amorphous alloy an extension of the liquid phase to temperatures in the undercooled region.[6,7] If not enough thermodynamical data are available, regular solution behavior might be assumed and the heat of mixing of the liquid phase could be estimated using the method of Miedema.[8,9] The differences in free energy between the amorphous and the appropriate crystalline phases can be calculated from the observed enthalpy of fusion ΔH_f and an approximation given by the following equation[10]

$$\Delta G = \frac{\Delta H_f \cdot \Delta T}{T_f} \cdot \frac{2T}{T_f + T} \tag{1}$$

The driving force (difference in free energy) is such that most metallic glasses can crystallize by two or even more different reactions. Which crystallization reaction does occur depends not only on its thermodynamic driving force, but on its kinetics.

CRYSTAL GROWTH

Crystal growth may be primary, eutectic, or polymorphic, and has been observed to be always thermally activated. Growth rates have been calculated from the diameter of the largest crystal found after each heat treatment. *In situ* observations cannot be used because surfaces may change the reaction mode or the kinetics. The crystallization mode that applies determines whether linear or parabolic growth rates will be observed.[1]

Parabolic Growth

Primary crystallization is generally assumed to be controlled by volume diffusion. If we may assume the diffusion rate D is concentration independent, the radius r of a spherical particle will be proportional to the time t as shown below

$$r = \alpha \cdot \sqrt{D \cdot t} \tag{2A}$$

and the growth rate u_p will be given by

$$u_p = \frac{1}{2} \cdot \alpha^2 \cdot D \cdot r^{-1} \tag{2B}$$

where α is a dimensionless parameter evaluated from the composition at the particle interface and the composition of the sample. This time-dependent growth rate will be reduced drastically as soon as the diffusion fields overlap and the regime of Ostwald ripening is achieved.

On the other hand, this relation can be used to calculate diffusivities from primary crystallization data.[11,12] The values of D in FIGURES 2 & 3 were calculated using the equation below

$$D = D_0 \cdot \exp\left(\frac{-Q_D}{R \cdot T}\right) \tag{3}$$

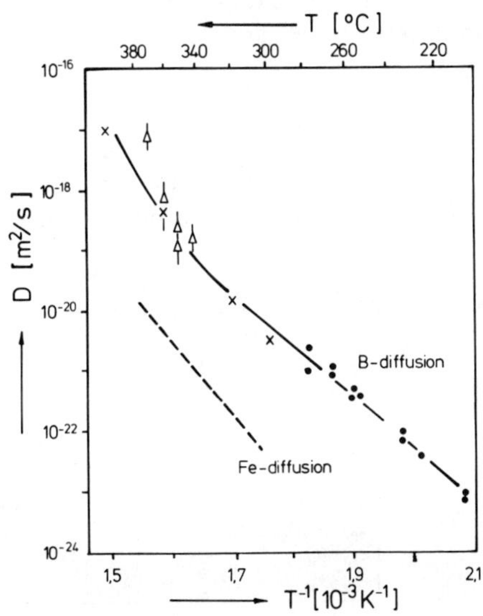

FIGURE 2. Diffusion in $Fe_{40}Ni_{40}B_{20}$ metallic glasses. × = Data from primary crystallization; Δ: data from Kijek *et al.*[13]; ●: data from Brodowski and Sagunski[14]; ---: data from Horvath *et al.*[15]

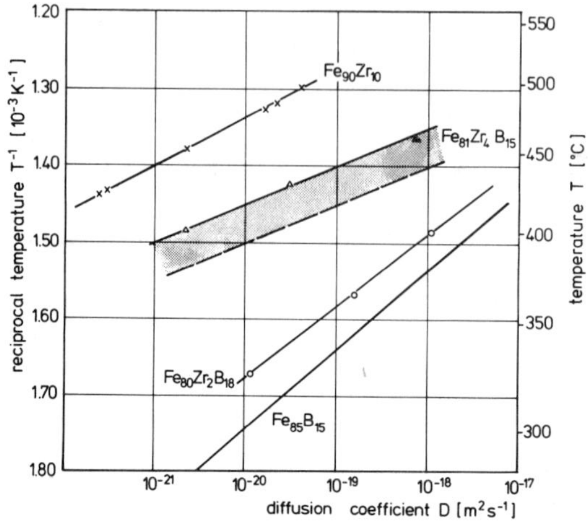

FIGURE 3. Diffusion in Fe-Zr-B glasses.

Prefactors D_0 and activation energies for diffusion Q_D for several glasses are listed in TABLE 1.

In $Fe_{40}Ni_{40}B_{20}$ glasses primary crystallization is controlled by B diffusion and may deviate from Arrhenius-type behavior above T_g. The Fe diffusion was measured by Horvath *et al.*[12] by means of the radiotracer technique in combination with ion-beam sputtering and is shown in FIGURE 2 for comparison. The rate of the Fe diffusion was about two orders of magnitude slower than that for the B diffusion.

Of special interest are Zr-based metallic glasses where a tremendous difference in the diffusivities of the late transition metal and the Zr has been realized. Assuming diffusivities independent of concentration at 250° C, diffusion coefficients of $2 \cdot 10^{-20}$ and $7 \cdot 10^{-30}$ m^2/sec may be calculated for Fe and Zr, respectively. This enormous difference might be responsible for the ability of glass formation by solid state reaction in several Zr-based systems.

Linear Growth

Linear growth has been found to occur in crystallization reactions that do not involve any compositional change (polymorphic or partitionless crystallization) or in a

TABLE 1. Diffusion Data Estimated from Primary Crystallization

	D_0 (m^2/sec)	Q_D (kJ/mol)
$Fe_{86}B_{14}$	$2 \cdot 10^{-4}$	180
$Fe_{84}C_8B_8$	$1.1 \cdot 10^{-3}$	180
$Fe_{78}Zr_2B_{20}$	$1.8 \cdot 10^{-3}$	195
$Fe_{78}Cu_2B_{20}$	1500	270
$Fe_{42}Ni_{42}B_{16}$	1100	260
$Fe_{90}Zr_{10}$	20	305
$Fe_{20}Zr_{80}$	2	200

cooperative transformation into two crystalline phases involving long-range diffusion in such a way that the mean composition of the crystallized region is equal to that of the glassy matrix (eutectic crystallization). FIGURE 4 shows the typical microstructure of a partially crystallized metal-metalloid as well as that of an intertransition-metal glass.

Once a stable nucleus has formed, it will continue to grow during isothermal annealing at a rate u_1, which is given by

$$u_1 = u_0 \cdot \exp\left(\frac{-Q_g}{R \cdot T}\right) \tag{4}$$

where Q_g is the activation energy for growth. The preexponential u_0 may be expressed as

$$u_0 = \delta \cdot \nu \tag{5}$$

where ν is a characteristic frequency on the order of the Debye frequency and δ is the distance across the crystallization front. With an interface thickness δ of about 0.5 nm the prefactor will be on the order of $5 \cdot 10^3$ m/sec. In a eutectic crystallization the preexponential exhibits the interlamellar spacing, which itself depends on the temperature.

FIGURE 5 shows the temperature dependence of crystal growth for several glasses

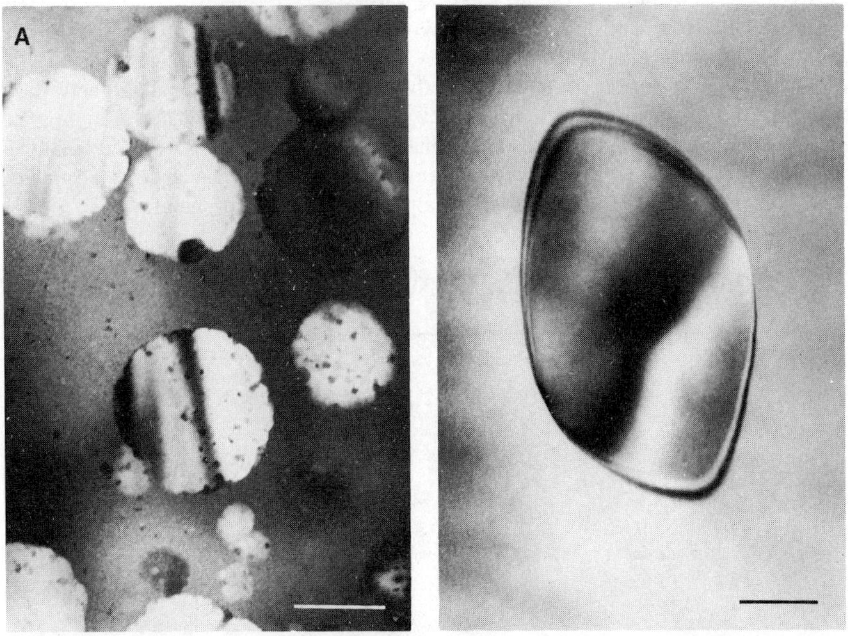

FIGURE 4. Typical microstructure of partially crystallized metallic glasses. (**A**) $Co_{33}Zr_{67}$: 23 hr at 346° C. (**B**) $Fe_{76}B_{24}$: 10 min at 400° C.

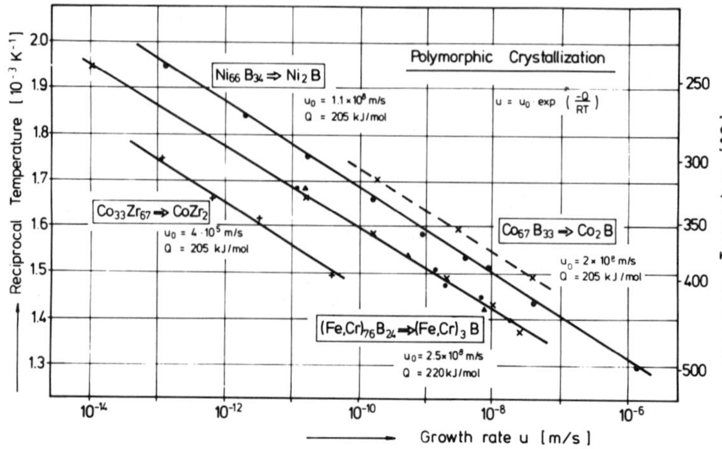

FIGURE 5. Temperature dependence of crystal growth during polymorphic crystallization of metallic glasses.

crystallizing by the polymorphic reaction. The prefactors u_0 as well as the activation energies for growth during polymorphic crystallization in several metallic glasses are shown in TABLE 2. The range of annealing temperatures investigated as well as the glass transition temperatures are indicated; for systems whose glass transition temperature is hidden by initial crystallization, T_x is used as an estimate of T_g. The observed preexponentials u_0 are normally several orders of magnitude larger than those predicted using the above theory, especially in ternary or more complicated glasses.

Extremely large preexponentials are always correlated with very high activation energies and might therefore indicate growth by simultaneous transfer of a group of atoms from the glassy to the crystalline phase rather than singly. Other explanations assume that impurities are present and that the temperature dependence of the growth rate does not only give the activation enthalpy for boundary movement.

TABLE 2. Crystallization Parameters for Polymorphic Crystallization of Metallic Glasses

	Crystallization Product	u_0 (m/sec)	Q_g (kJ/mol)	Temperature Range (°C)	T_g or T_x (°C)
$Fe_{76}B_{24}$	Fe_3B	$7 \cdot 10^6$	200	240–440	460[16]
$Fe_{66}Ni_{10}B_{24}$	$(Fe,Ni)_3$-B	$9 \cdot 10^{13}$	300	360–460	—
$Fe_{38}Ni_{38}B_{24}$	$(Fe,Ni)_3$-B	$3 \cdot 10^{14}$	290	320–390	380[17]
$Co_{75}B_{25}$	Co_3B	$2 \cdot 10^9$	220	320–400	460[18]
$Ni_{67}B_{33}$	Ni_2B	$1.1 \cdot 10^8$	205	240–500	460[19]
$Fe_{40}Zr_{60}$[a]	$FeZr_2$	$1 \cdot 10^3$	170	300–375	—
$Fe_{30}Zr_{70}$[a]	$FeZr_2$	$5 \cdot 10^3$	175	300–350	—
$Fe_{24}Zr_{76}$[a]	$FeZr_3$	$1.2 \cdot 10^5$	170	290–340	—
$Co_{33}Zr_{67}$	$CoZr_2$	$4 \cdot 10^5$	205	300–400	475[20]
$Ni_{33}Zr_{67}$[21]	$NiZr_2$	$5 \cdot 10^6$	200	340–375	395

[a]The T_g for $Fe_{33}Zr_{67}$ is 416°C.[20]

Because the large prefactors have only been observed associated with anisotropic growth, growth probably occurs by the lateral motion of steps. If the fraction of these active sites is assumed to be x_a, the growth rate in equation 3 has simply to be multiplied by x_a. Because x_a increases with the annealing temperature as indicated by a decrease of the length-to-thickness ratio of the prolate crystals, this will lead to an effective higher activation energy and a larger prefactor.

NUCLEATION

Nucleation phenomena can only be investigated indirectly, for example, by analyzing the microstructure after some amount of crystallization by extrapolating growth reactions to shorter times.[22] Complementary techniques such as differential scanning calorimetry (DSC) provide data on reaction kinetics in a larger sample volume. With growth rates from transmission electron microscopy (TEM) and assuming an isokinetic reaction with a constant number of nuclei, Greer[23] developed a simple model for crystallization in $Fe_{80}B_{20}$ glasses and was able to predict the rate of reaction in good agreement with experimental DSC results.

Nucleations for primary and polymorphic crystallizations are single-step reactions, but eutectic crystallization has been observed to be a process with two nucleation steps. Because the analysis of primary crystallization is complicated due to the radius dependence of the growth rate, we will concentrate on nucleation for polymorphic crystallization.

In classic nucleation theory the steady state homogeneous nucleation rate I_{st} is given by

$$I_{st} = I_0 \cdot \exp\left(\frac{L \cdot \Delta G_c}{R \cdot T}\right) \cdot \exp\left(\frac{Q_N}{R \cdot T}\right) \tag{6}$$

where I_0 is a constant factor; L, the Loschmidt number; Q_N, the activation energy for the transfer of atoms across the liquid-nucleus interface; and ΔG_c, the free energy required to form a nucleus of critical size. For heterogeneous nucleation at a limited number Z of nucleation sites, the activation energy ΔG_c is reduced due to the gain in surface energy, thus leading to a significant higher nucleation rate I^*_{st}.

At the very beginning of annealing a finite period is expected during which the steady state distribution of clusters assumed in classical nucleation theory is established. This involves a transient or time-dependent nucleation rate $I(t)$ given by the following equation[24]

$$I(t) = I_{st} \cdot \left[1 + 2\sum_{n=1}^{\infty}(-1)^n \cdot \exp\left(-n^2 \cdot \frac{t}{\tau}\right)\right] \tag{7}$$

where τ is the time lag, which is expected to increase significantly with decreasing temperature. Kelton *et al.*[25] have shown recently that this equation is a very good approximation of the nonstationary nucleation problem.

Nucleation can be studied in some detail by crystallization statistics[26,27] of partially crystallized glasses, that is, from the fit of calculated and measured crystal diameter distributions. Let us assume a polymorphic crystallization of spherical crystals and a constant growth rate u, and let us apply the calculations below only for the initial states of crystallization, when noninterfering crystals are formed. The operating nucleation mode that applies determines how the number of crystals ΔN_i nucleated during the time interval Δt can be calculated (FIG. 6). In cases where the homogeneous mode applies, such a calculation may be carried out with the equation

$$\Delta N_i = I(t) \cdot (1 - X_{i-1}) \cdot \Delta t, \quad i = 1, \ldots, t/\Delta t \tag{8}$$

with

$$X_i = \frac{4\pi}{3} \cdot u^3 \cdot \sum_{j=1}^{i} \Delta N_j \cdot [\Delta t \cdot (i + 1 - j)]^3$$

In cases where the heterogeneous mode applies, such a calculation may be carried out with the equation

$$\Delta N_i = \overset{*}{I}(t) \cdot (1 - X_{i-1}) \cdot \left(1 - Z^{-1} \cdot \sum_{j=1}^{i} \Delta N_j\right) \cdot \Delta t \tag{9}$$

for i where $\Sigma_{j=1}^{i} \Delta N_j = Z$; $\Delta N_i = 0$ for all other i.

If there is a limited number Z of preexisting postcritical nuclei, they will start growing as soon as the annealing temperature is reached. This will lead to a very narrow diameter distribution as shown in FIGURE 6.

The mode of nucleation observed during crystallization of metallic glasses depends strongly on the family of the glass as well as on the annealing temperature. In $Co_{33}Zr_{67}$ glasses (intertransition-metal glass), polymorphic crystallization of spherical $CoZr_2$ crystals has been observed (FIG. 5). The size distribution of these crystals after isothermal annealing was determined using a modified Schwartz–Saltykov method (reference 24 and M. Blank-Bewersdorff, personal communication) and is plotted in FIGURE 7A. A good agreement with the calculated crystal distribution can be obtained only when assuming a transient-type homogeneous nucleation. From such fits of experimental measured and calculated crystal diameter distributions the time lag τ as well as the steady state nucleation rate $\overset{*}{I}_{st}$ can be determined for the different annealing temperatures (Q_N = 280 kJ/mol; Q_τ = 340 kJ/mol).[28] These activation energies are in good agreement with the activation energy for Zr diffusion in similar Zr-based glasses. Nucleation studies above T_g are under way.

For polymorphic crystallization of metal-metalloid glasses below T_g the number of crystals per unit volume does not change with annealing temperature, and, as shown elsewhere,[26,27] only heterogeneous nucleation at a limited number Z of quenched-in nucleation sites (fluctuations) can lead to the narrow crystal diameter distribution that is typically observed. The number N_v of these quenched-in nucleation sites has been found to depend on the alloy system and the composition, but decreases in any case with increasing cooling rates, that is, with higher surface velocities of the quenching wheel.

FIGURE 8A shows a typical micrograph of partially crystallized $Fe_{65}Ni_{10}B_{25}$ glass with several elliptical sections of prolate $(Fe, Ni)_3B$ crystals. The size distributions of these crystals were determined using the method of van't Hoff[29] and are plotted in FIGURE 7B. The agreement with the diameter distributions calculated by crystallization statistics assuming transient-type heterogeneous nucleation is very good. From

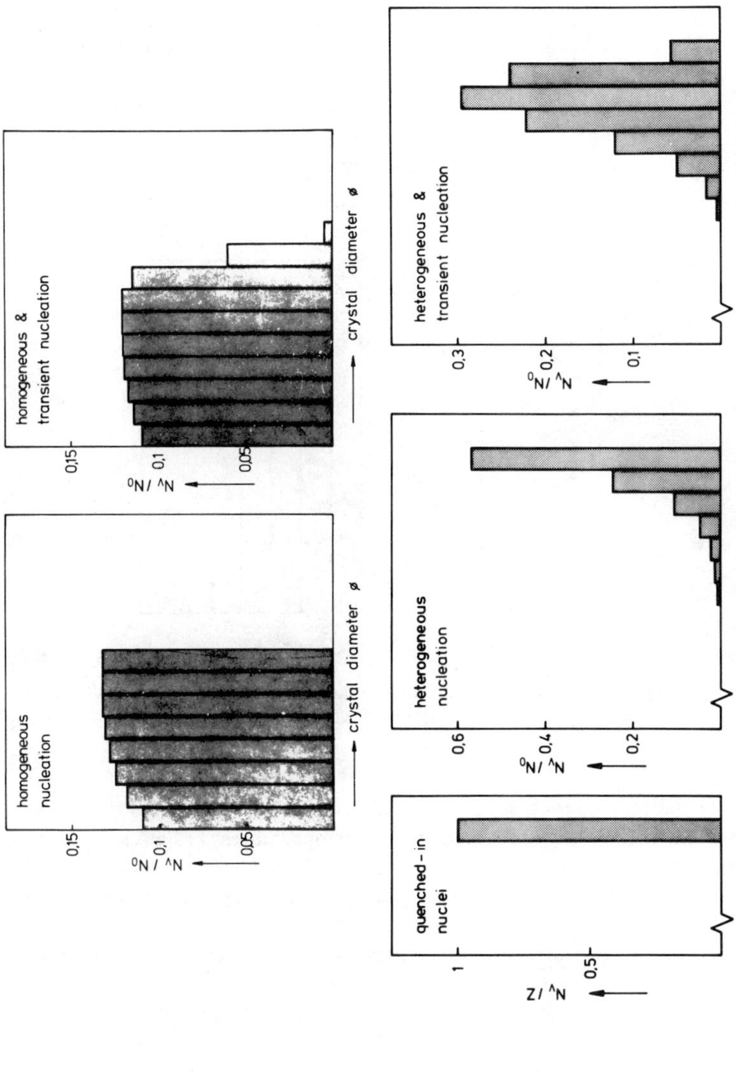

FIGURE 6. Schematic histograms for the crystal diameter distribution depending on the nucleation mode.

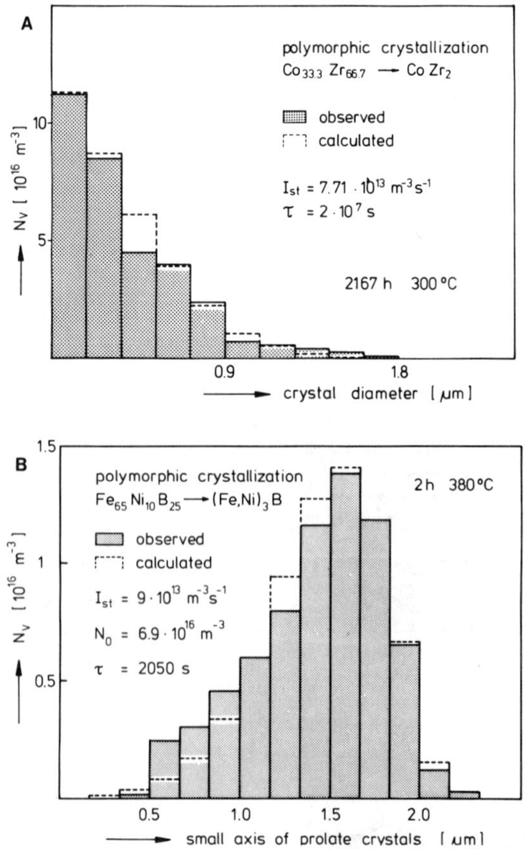

FIGURE 7. Crystal size distribution in partially crystallized metallic glasses. **(A)** $Co_{33}Zr_{67}$ (M. Blank-Bewersdorff, personal communication). **(B)** $Fe_{65}Ni_{10}B_{25}$.[27]

these fits of experimental measured and calculated crystal diameter distributions τ and I_{st} can be determined for the different annealing temperatures (FIG. 9A). As in the case of the $Co_{33}Zr_{67}$ glass, the activation energies ($Q_N = 300$ kJ/mol; $Q_\tau = 283$ kJ/mol) are in good agreement with the activation energy for diffusion in similar (Fe, Ni)-B glasses.

The number of crystals per unit volume after polymorphic crystallization of metal-metalloid glasses does not change with the annealing temperature up to T_g. Above T_g, however, their number is increased by orders of magnitude, an increase that is probably due to homogeneous nucleation (FIG. 8B). Even in this temperature range transient effects with a significant τ have been observed. As shown by Herold,[31] the large crystal in FIGURE 8B, which is assumed to be formed by heterogeneous nucleation at the very beginning of annealing, can be used for calculating τ for the homogeneous nucleation mode. In $Fe_{65}Ni_{10}B_{25}$ glasses τ ranges between 4 sec at 480° C and 10^{-2} sec at 550° C.

The steady state nucleation rate and the time lags for the nucleation above T_g are shown in FIGURE 9A in comparison with those below T_g. The observed activation energies of more than 1000 kJ/mol are not well understood. But there exists a striking similarity between the temperature dependence of the nucleation rate and that of the viscosity (compare FIG. 9A with FIG. 9B). This similarity may indicate that nucleation above T_g is controlled by viscous flow rather than diffusion of single atoms. If one assumes nucleation is controlled in this way, diffusivity in equation 6 has to be replaced by viscosity, which is usually given by the Fulcher–Vogel equation

$$\eta = A \cdot \exp\left(\frac{B}{T - T_0}\right) \qquad \mathbf{(10)}$$

and the temperature dependence of I_{st} should be given by

$$I_{st} \propto \exp\left(\frac{-B}{T - T_0}\right) \cdot \exp\left(-\frac{L \cdot \Delta G_c}{R \cdot T}\right) \qquad \mathbf{(11)}$$

In the temperature range between 460° and 530° C, where $L \cdot \Delta G_c$ is assumed to still be small compared to $R \cdot T$, the parameters B and T_0 can be estimated from our experimental data to be 4.700 and 631 K, respectively, which is close to the values predicted for (Fe, Ni)-B glasses by Davies.[30] The Fulcher–Vogel temperature T_0

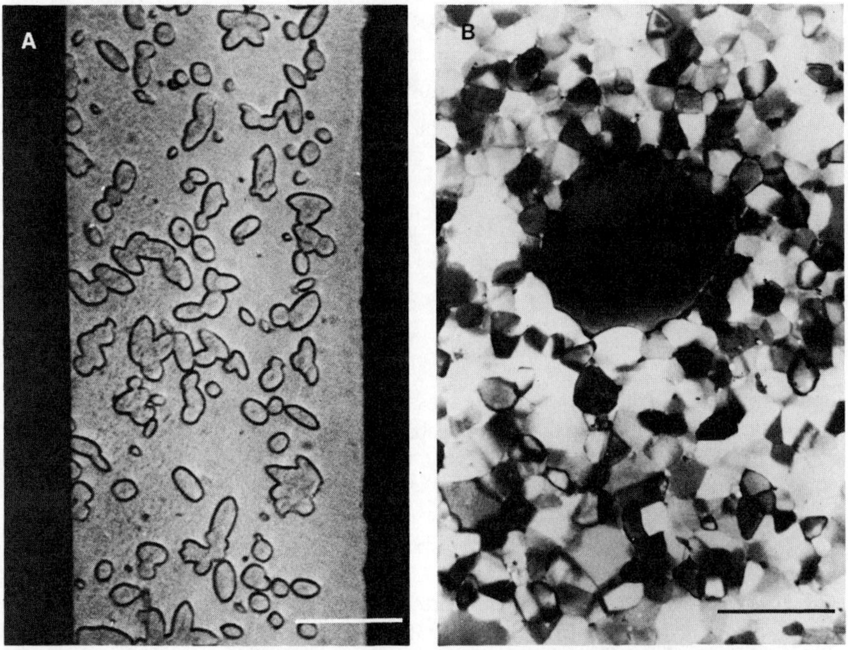

FIGURE 8. Typical microstructure after annealing of $Fe_{65}Ni_{10}B_{25}$ metallic glasses. (**A**) Below T_g for 350 min at 360° C (LM-Nomarski contrast). (**B**) Above T_g for 10 sec at 550° C (TEM). Bar: 0.3 μm.

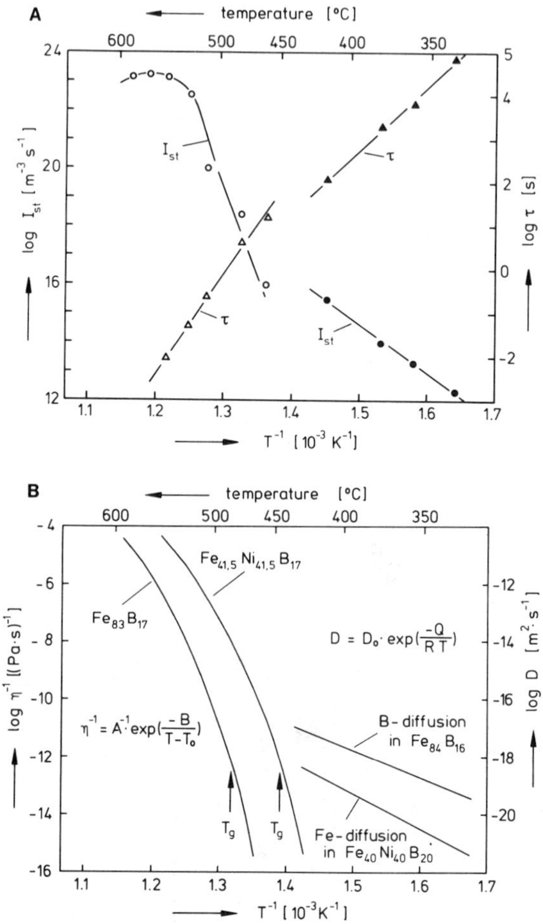

FIGURE 9. Temperature dependence for (**A**) τ and I_{st} in $Fe_{65}Ni_{10}B_{25}$ glasses and for (**B**) the viscosity[30] and diffusion in (Fe, Ni)-B glasses.

defines an ideal T_g where, in a silicate glass, all atomic motions should be perfectly "frozen-in." As mentioned above we have observed nucleation and growth in this glass even far below a temperature of 600 K.

DISCUSSION

Below T_g crystallization has been found to proceed by heterogeneous nucleation at several quenched-in nucleation sites (metal-metalloid glasses) or by homogeneous nucleation (Co-Zr glasses); in both cases the nucleation rates are transient. Above T_g there is evidence for a transient nucleation process controlled by viscous flow of the undercooled melt rather than diffusion of single atoms.

There is indeed a decoupling between the kinetic coefficient controlling nucleation and growth of crystals and the bulk viscosity. As mentioned in the study by Greer[32] in which creep measurements were made, it seems likely that the Stokes–Einstein relationship between viscosity and diffusion is not valid for metal-metalloid glasses well below T_g. On the contrary, there is some evidence for diffusion jumps that do not contribute to viscous flow.

Our results indicate that reasonable diffusion of at least one of the components of the glass as a prerequisite for any crystallization can occur far below T_g. In the case of primary crystallization due to the change in concentration, this can lead locally to a reduced T_g.

Crystallization of metallic glasses may proceed differently to that of silicate glasses because of the different structure and strength of their bonds. Whereas we observe, for example, strong Fe-Fe bonds in Fe-B glasses, there are no Si-Si bonds in glassy silica. The strongest bond is the Si-O bond. (The Si-O bond has an energy of 445 kJ/mol; the Si-Si bond, an energy of 180 kJ/mol.) Details of this behavior are not yet well understood: there is an urgent need for systematic investigations of nucleation and crystal growth in other metallic glass systems.

ACKNOWLEDGMENTS

I would like to thank M. Blank-Bewersdorff for her helpful comments. I am indebted to her for permission to quote some of her results prior to their publication.

REFERENCES

1. KÖSTER, U. & U. HEROLD. 1981. *In* Glassy Metals I. Topics in Applied Physics. H.-J. Güntherodt & H. Beck, Eds. Vol. **46:** 225. Springer–Verlag. Berlin.
2. SCOTT, M. 1983. *In* Amorphous Metallic Alloys. F. E. Luborsky, Ed.: 114. Butterworths. London.
3. KÖSTER, U. 1983. *In* Phase Transformations in Crystalline and Amorphous Alloys. B. L. Mordike, Ed.: 113. Deutsche Gesellschaft für Metallkunde. Oberursel.
4. UHLMANN, D. R. 1983. J. Non-Cryst. Solids **73:** 585.
5. HEROLD, U. & U. KÖSTER. 1978. *In* Rapidly Quenched Metals 3. B. Cantor, Ed. Vol. **1:** 281. Metals Society. London.
6. WEISS, P. 1975. Ph.D. dissertation. Department of Mechanical Engineering. Ruhr Universität. Bochum.
7. KÖSTER, U. & P. WEISS. 1975. J. Non-Cryst. Solids **17:** 359.
8. SCHWARZ, R. B., K. L. WONG & W. L. JOHNSON. 1984. J. Non-Cryst. Solids **61 & 62:** 129.
9. MIEDEMA, A. R. 1976. Philips Tech. Rev. **36:** 217.
10. THOMPSON, C. V. & F. SPAEPEN. 1979. Acta Metall. **27:** 1855.
11. KÖSTER, U., U. HEROLD, H.-G. HILLENBRAND & J. DENIS. 1980. J. Mater. Sci. **15:** 2125.
12. KÖSTER, U., U. HEROLD & A. BECKER. 1981. *In* Rapidly Quenched Metals 4. T. Masumoto & K. Suzuki, Eds. Vol. **1:** 587. Japan Institute of Metals. Sendai.
13. KIJEK, M., M. AHMADZADEH, B. CANTOR & R. W. CAHN. 1980. Scr. Metall. **14:** 1337.
14. BRODOWSKY, H. & H. SAGUNSKI. 1984. Z. Phys. Chem. N. F. **139:** 149.
15. HORVATH, J., K. FREITAG & H. MEHRER. 1984. *In* Rapidly Quenched Metals 5. S. Steeb & H. Warlimont, Eds. Vol. **1:** 751. Elsevier/North-Holland. New York, NY.
16. RAY, R., R. HASEGAWA, C.-P. CHOU & L. A. DAVIS. 1977. Scr. Metall. **11:** 973.
17. WALTER, J. L. 1979. Mater. Sci. Eng. **39:** 95.

18. TAKAHASHI, M., C. O. KIM, M. KASHIMURA & T. SUZUKI. 1978. Jpn. J. Appl. Phys. **17:** 1911.
19. DONALD, I. W. & H. A. DAVIES. 1980. J. Phys. C. **41:** 118.
20. ZIELINSKI, P. G., J. OSTATEK & M. KIJEK. 1978. *In* Rapidly Quenched Metals 3. B. Cantor, Ed. Vol. **1:** 337. Metals Society. London.
21. SCOTT, M. G., G. GREGAN & Y. D. DONG. 1981. *In* Rapidly Quenched Metals 4. T. Masumoto & K. Suzuki, Eds. Vol. **1:** 671. Japan Institute of Metals. Sendai.
22. KÖSTER, U. & U. HEROLD. 1981. *In* Rapidly Quenched Metals 4. T. Masumoto & K. Suzuki, Eds. Vol. **1:** 717. Japan Institute of Metals. Sendai.
23. GREER, A. L. 1982. Acta Metall. **30:** 171.
24. GUTZOW, I. & S. TOSCHEV. 1971. *In* Advances in Nucleation and Crystallization of Glasses. L. L. Hench, Ed.: 10. American Ceramic Society. Columbus, OH.
25. KELTON, K. F., A. L. GREER & C. V. THOMPSON. 1983. J. Chem. Phys. **79:** 6261.
26. KÖSTER, U. & H. BLANKE. 1983. Scr. Metall. **17:** 495.
27. BLANKE, H. & U. KÖSTER. *In* Rapidly Quenched Metals 5. S. Steeb & H. Warlimont, Eds. Vol. **1:** 227. Elsevier/North-Holland. New York, NY.
28. KÖSTER, U. & M. BLANK-BEWERSDORFF. 1985. Paper presented at the fall meeting of the Materials Research Society. Boston, MA.
29. UNDERWOOD, E. E. 1970. Quantitative Stereology. Addison–Wesley. New York, NY.
30. DAVIES, H. A. 1978. *In* Rapidly Quenched Metals 3. B. Cantor, Ed. Vol. **1:** 1. Metals Society. London.
31. HEROLD, U. 1982. Ph.D. dissertation. Department of Mechanical Engineering. Ruhr Universität. Bochum.
32. GREER, A. L. 1984. J. Non-Cryst. Solids **61** & **62:** 737.

DISCUSSION OF THE PAPER

D. MONROE (*AT&T Bell Laboratories, Murray Hill, NJ*): What is the origin of the delay to nucleation required by your analysis? Are recent ideas of nucleation of first order structural transitions of ferroelectrics relevant? In these experiments, by McWhan and Aeppli, and models, by Littlewood, the density difference between the phases means that the nucleus is strained by its embedding in the surrounding medium, giving a delay to nucleation.

KÖSTER: In classical nucleation theory at the very beginning of annealing a finite period is necessary during which the distribution of clusters or fluctuations approaches thermal equilibrium. This involves the time lag for the nucleation process. A numerical simulation technique that directly models the reactions by which clusters are produced gives similar results (Kelton *et al.*, *J. Chem. Phys.* **79:** 6261, 1983).

The time lag may be increased due to additional elastic strain. Such an influence is well known in nucleation for precipitation leading to preferred nucleation at dislocations.

C. A. ANGELL (*Purdue University, West Lafayette, IN*): Has anyone seen an oxide glass crystallize below T_g?

A. NAVROTSKY (*Princeton University, Princeton, NJ*): Yes. The $NaAlSi_3O_8$-$CaMgSi_2O_6$ system has glass-glass immiscibility predicted from thermodynamic properties, with a miscibility gap closing near 550° C. This immiscibility was seen with one of the exsolving glasses crystallizing (Henry *et al.*, *Geochim. Cosmochim. Acta* **47:** 277, 1983).

C. T. MOYNIHAN (*Rensselaer Polytechnic Institute, Troy, NY*): Shear viscosities (stress relaxation to zero stress) can and have been measured below T_g for glasses with constant structure. Below T_g the shear viscosity activation energy is much less than that above T_g, reflecting the effect of structure or fictive temperature on the viscosity. Have you or anyone else measured shear viscosity below T_g to compare with the diffusivities?

KÖSTER: Data on viscosity and diffusion in metallic glasses have been reviewed (Greer, *J. Non-Cryst. Solids* **61** & **62:** 737, 1984). Whereas activation energies for viscous flow and diffusivity below T_g are in the same order and much smaller than they are above T_g, diffusivities—at least in Pd-Si glasses—have been observed to be about 300 times the value that would be predicted from measured viscosities using the Stokes–Einstein relationship.

PART II. STRUCTURAL ASPECTS: NEUTRON SCATTERING AND ELECTRON MICROSCOPE STUDIES OF ALMOST GLASSY CRYSTALS AND ALMOST CRYSTALLINE GLASSES

How Disordered Can a Crystallized Glass Be?

A. F. WRIGHT, A. N. FITCH, AND B. E. F. FENDER

Institut Laue–Langevin
38042 Grenoble Cedex, France

INTRODUCTION

By controlled heat treatment, certain glasses can be transformed to a submicrocrystalline state with crystalline dimensions in the 70–100-Å range. The formation and development of the structures have been studied by small-angle neutron scattering (SANS), neutron diffraction, extended X-ray absorption fine structure (EXAFS), and spectroscopic techniques. For crystallites in the 20–30-Å range, changes in neutron diffraction and EXAFS from the quenched glass are barely perceptible, but small-angle neutron scattering is readily detected. Analysis of SANS data as a function of heat treatment suggests a continuous evolution from the nucleation stage to a fully developed state (80–100 Å diameter). Invariably, the first crystalline phase is either structurally or chemically disordered. Chemical disorder takes the form of solid-solutions with ready exchange of selected ions from the glass. The disorder is thought to play a role in reducing the excess interface energy of the nucleus. The fully developed yet metastable microstructure with characteristic dimensions <100 Å is remarkably resistant to further transformation under prolonged heat treatment.

CRYSTAL NUCLEATION

Glasses that retain a high nucleation rate at temperatures at which the crystal growth rate is low can be readily heat treated to give a fine microcrystalline solid.[1] This is the basis of the formation of glass-ceramic materials. Typically the first crystalline phase produced is a metastable phase that transforms to a thermodynamically stable phase or phases when the material is heated to high temperatures.[2,3] Although the bulk free energy change or driving force, ΔG, plays a role in crystallization, the process is controlled by the thermodynamic and kinetic free energy barriers to nucleation, ω^* and ΔG_D, respectively.[4,5] The nucleation rate is given by

$$I = A \exp\left[-(\omega^* + \Delta G_D)/KT\right]$$

where for a spherical nucleus

$$\omega^* = 16\pi\sigma^3 V_m^2/3\,\Delta G^2$$

and where σ is the crystal-liquid interterfacial energy, ΔG is the bulk free energy change, and ΔG_D is the free energy barrier for diffusion across the interface.

For a single-component glass, the crystallization process is purely structural in nature. The first step can only be a transformation directly to a crystalline phase via the route that presents the lowest barrier to rearrangement. In the case of SiO_2, the crystalline phase is always β-cristobalite, the most highly disordered phase of silica. Because reconstructive transformations between the various polymorphs of SiO_2 are extremely slow compared to this crystallization process, this may be evidence that silica glass and β-cristobalite are structurally similar. The only remaining parameters to describe the condition of the product of the transformation of a single-component amorphous solid are the critical size of the nucleus and the density of nuclei formed. The former is considerably influenced by the interfacial energy, σ, between the crystallite and the matrix. An increase in the interfacial energy increases the thermodynamic barrier for nucleation[5] and, therefore, the critical radius, r_c ($r_c = -2\sigma V_m/\Delta G$), that must be exceeded before the crystallite becomes stable relative to the parent glass.

The situation is more complicated for a nonstoichiometric multicomponent glass. Metastable states of different compositions and structures compete for the first phase transformation. A strong contender could be a glassy phase of slightly different composition from the host. We then have amorphous phase separation. But the first crystalline phase of glass-ceramics is often a solid-solution.[3,4] The extra compositional degree of freedom and chemical disorder of these crystalline phases offer alternative pathways on the free energy surface to a new crystalline phase. The critical radius for composition n, r_{cn}, is then a function of the interfacial energy, σ_n, and will vary with the composition at the surface of the nucleus, C_n, and with the degree of supersaturation, C_s.[6]

$$r_{cn} = 2\sigma_n V_m \Big/ \left(RT \ln \frac{C_n}{C_s} \right)$$

Once past the critical size, the solid-solution can then move to lower bulk free energies by continuous compositional change during growth. These extra degrees of freedom of the solid-solution phases, coupled with high supersaturations obtainable for certain nucleation promotors, such as TiO_2, ZrO_2, or Cr_2O_3, can lead to high-density submicrocrystalline structures with particle dimensions in the 20–100-Å range. In the smallest crystallites, the discontinuity at the interface means that long-range order may not be fully achieved, and it may be difficult to distinguish structurally between very small crystallites and local structures with similar short-range order. The considerable thermodynamic stability of these structures and their influence on the remaining glass matrix, however, puts them in a special class of microstructure intermediate between glasses and glass-ceramics.

For our investigation, by SANS[7] and neutron diffraction, we have chosen an extensively studied multicomponent glass: a cordierite system $2MgO \cdot 2Al_2O_3 \cdot 5SiO_2$ that contains 11.5% TiO_2 as a nucleating agent.[3] The same samples were also investigated by EXAFS analysis by the group of Jacqueline Petiau in Paris.[8] TiO_2 is a good nucleating agent at high concentrations (>10%). It precipitates as a solid-solution with Al_2O_3 and MgO, and thus reflects the bulk transformation of all the nonsilica content of the glass. The volume fraction is high, which makes structural changes sensitive to detection by scattering techniques. In addition, we have examined the same glass nucleated by Cr_2O_3.[9] Cr_2O_3 gives high crystal nucleation densities at low concentrations (<1%). It also forms a solid-solution with Al_2O_3 and MgO, and spectroscopic techniques such as fluorescence line narrowing and electron paramagnetic resonance are sensitive to changes in the local environment of the Cr^{3+} ion that may occur during nucleation.

SANS

The coherent scattering cross section per atom of a sample of N atoms containing particles of scattering-length density ρ_{bp} is given by

$$\frac{d\sigma}{d\Omega} = \frac{1}{N}(\rho_{bp} - \rho_{bm})^2 \left| \int_{V_t} \exp(iQ \cdot r) d^3r \right|^2$$

The integral includes all correlations within and between particles over the total sample volume, V_t. The scattering function for a single particle of volume V_p, normalized to give $|P(Q = 0)| = 1$, is written

$$P(Q) = \frac{1}{V_p^2} \left| \int_{V_p} \exp(iQ \cdot r) d^3r \right|^2$$

For the special case of a sample of identical spherical particles the scattered intensity

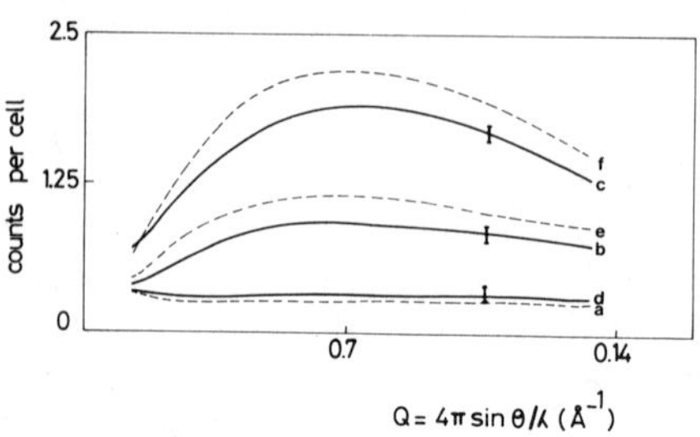

FIGURE 1. Differential SANS from cordierite-TiO_2 glass following intense nucleation heat treatments for 36, 72, and 120 hr (a, b, and c, respectively) at 720° C and for 48, 96, and 144 hr (d, e, and f, respectively) at 711° C.

$I(Q)$ is the product of $P(Q)$ and the interparticle interference function $S(Q)$, scaled by the contrast $(\Delta\rho)^2$ between the particle and the matrix and experimental constants combined in k. We then have

$$\frac{I(Q)}{k} = \frac{d\sigma}{d\Omega}(Q) = \frac{N_p}{N} V_p^2 (\rho_{bp} - \rho_{bm})^2 P(Q) S(Q)$$

This expression is a good approximation for the controlled nucleation and growth of crystallites from glasses where the nucleation and growth processes are so well separated by temperature that a long-established nucleus has little growth advantage over a newly formed one.[10,11] The particle size distribution is then sufficiently narrow that an average spherical particle function $P(Q)$ adequately describes the scattering due to isolated particles.

The interparticle interference function $S(Q)$ depresses the scattering at low Q, resulting in a broad peak in the observed scattering pattern. The calculation and simulation of $S(Q)$ has been treated elsewhere.[7,12,13] The particle pair correlation function $g(r)$, which describes the spatial distribution of particles, is derived from

TABLE 1.

Heat Treatment		Intensity	
Temperature (°C)	Time (hr)	(Q = O) (Arbitrary Units)	Dimensions (Å)
710	48	0.31	—
	96	1.07	19.5
	144	2.57	26.1
720	36	0.26	—
	72	1.38	21.1
	120	2.97	26.6

$S(Q)$. The full analysis then gives average particle size and spatial information, and from a logical series of measurements, one can also obtain information about growth processes and changes in scattering contrast, and so control the microstructure of the product.[11] SANS measurements on the cordierite-TiO_2 system have shown that the interesting temperature range for the study of nucleation is 710–720° C.[7] At 720° C an induction period of 5 hr is followed by rapid nucleation of crystallites leading to a submicrocrystalline state, whereas at higher temperatures concurrent crystal growth reduces the supersaturation and quickly brings the nucleation process to a halt. Here we report on extended heat treatments at 710° and 720° C intended to push the nucleation density to its limit. The scattering from such small particles is very weak, but is compensated partly by the many nuclei created. FIGURE 1 shows corrected SANS data for 36, 72, and 120 hr at 720° C, and for 48, 96, and 144 hr at 710° C. It is interesting to note that the scattering for the 36- and 48-hr samples is totally flat, showing no coherence effects at all. This is consistent with scattering from uncorrelated density fluctuations,[14] which could precede the appearance of crystallites. A short heat treatment at higher temperatures, however, shows that these glasses are very highly nucleated indeed, even though discrete particles are not detected at this stage. The glasses heated for longer times at 710° and 720° C scatter coherently. The results of data analyses are given in TABLE 1.

The intensity data closely obey the Ostwald ripening law $I(Q = 0) = kr^3$(FIG. 2),[11] which suggests that the total mass of precipitated material remains constant at this stage. From the uncorrelated scattering intensity we estimate a lower limit to the critical particle size of ~12 Å. Heat treatment at higher temperatures leads to limited

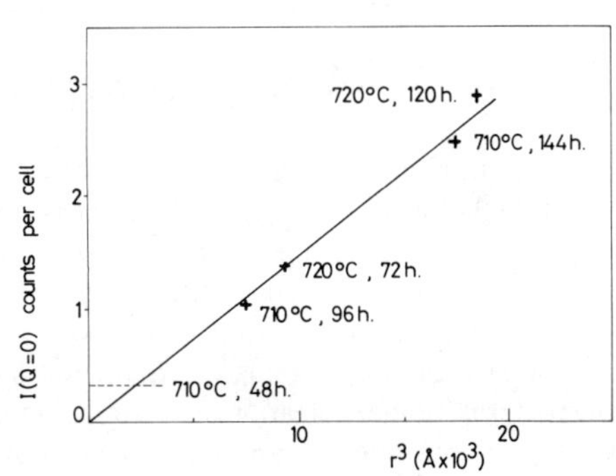

FIGURE 2. Intensity at $I(Q = 0)$ as a function of r^3 using the data from TABLE 1 (plus signs). $I(Q = 0)$ is defined in FIGURE 3. For an Ostwald ripening process $I(Q = 0) = kr^3$ (solid line).

growth of the crystallites. *In situ* SANS measurements show that the crystallite size of the heavily nucleated sample (720° C for 120 hr) reaches a diameter of only 66 Å after heating to 840° C over 8 hr. Previous measurements[7] on the same glass submitted to less extensive nucleation treatment have shown that the crystallite size can achieve 1000 Å after 1 hr at 835° C. The scattered intensity here is in accordance with a pure-growth law, $I(Q = 0) = k'R_s^6$ (there is no reduction in the number of crystallites by ripening), even though the size stabilizes after a few hours, presumably because of a scarcity of nutritive material. FIGURE 3 shows the scattering curve after development to 66 Å and the components of the simulation. Note that the weak maximum in $S(Q)$ (in units of Q_D) occurs near 2π, a value consistent with random close packing of hard spheres. The crystallites thus enter into contact with each other. The pair correlation function $g(r)$ (FIG. 4) obtained from the model parameters shows this more clearly: the probability of finding a neighboring particle at the contact distance (one diameter) is only slightly different from the mean particle density everywhere. This is in sharp contrast to glasses crystallized with a lower nucleation density where each crystallite grows in isolation, surrounded by a zone depleted in nutrient material.[7] This submicrocrystalline state is extremely stable. Under less severe nucleation conditions a second phase transition occurs at 860–900° C with the appearance of large (>10^4 Å) crystals

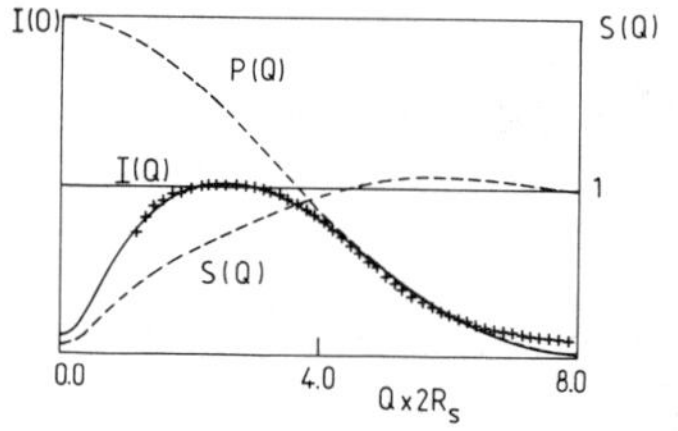

FIGURE 3. SANS after nucleation for 120 hr at 720° C and development for 8 hr at 840° C (plus signs). $I(Q)$ is the product of $S(Q)$ and $P(Q)$ in the simulation refined for spheres with diameters of 66 Å.

of a β-quartz structure (magnesium aluminium silicate). We have heated the present glass in its submicrocrystalline state to 960° C for some hours. The β-quartz structure is not produced and the particle size never exceeds 70 Å. The product remains perfectly transparent, although an opaque surface layer is found. The submicrocrystalline structure would appear to frustrate the crystallization of the remaining silica-rich matrix.

WIDE-ANGLE NEUTRON DIFFRACTION

The samples previously used for SANS were recut into rods 0.8 cm in diameter and 2 cm high for neutron diffraction measurements. A high-flux low-resolution D2 diffractometer operating at 0.93 Å was chosen because of the small differences expected during nucleation. FIGURE 5 shows the diffraction patterns obtained for the untreated glass compared to that treated at 711° C for 144 hr and known to be heavily nucleated with particles 26 Å in diameter. There are no obvious differences between the patterns. Some indication of crystallization was observed on further heating to 800° C for 30 min. SANS analysis shows particles 50 Å in diameter at this stage.

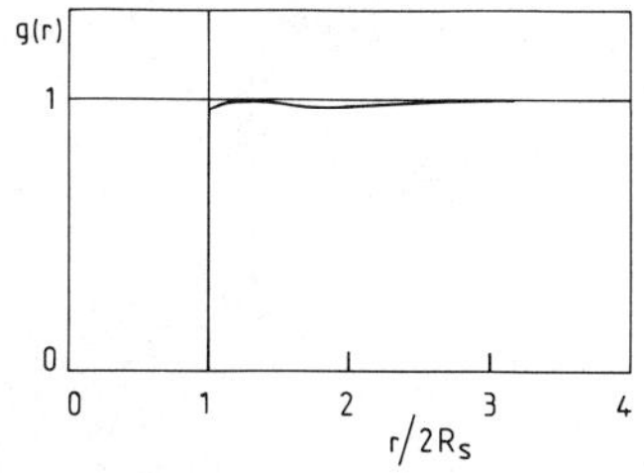

FIGURE 4. Pair correlation function $g(r)$ obtained from $S(Q)$ in the simulation of FIGURE 3 (r is in units of the particle diameter).

Subtracting the scattering from that of the untreated glass we obtain the differential diffraction patterns shown in FIGURE 6. It is then clear that the broadened but undeniably crystalline pattern obtainable after 30 min at 800° C also contributes in a weakened and more broadened form to the diffraction immediately following the nucleation treatment of 144 hr at 710° C. Further treatment at 900° C for 1 hr simply sharpens the diffraction pattern (FIG. 7). Previous X-ray diffraction studies have identified the first crystalline phase appearing in this glass as a pseudobrookite solid-solution $Al_2TiO_5 \cdot MgTi_2O_5$. The neutron diffraction patterns suggest the same solid-solution, but we cannot identify it with certainty because of line broadening that caused servere overlapping of peaks. (Note the progressive splitting of the group of peaks near 12° (2θ) as the heat treatment proceeds.) Direct heating of the glass to 900° C (FIG. 7) without nucleation treatment produces the β-quartz solid-solution phase with some contamination of pseudobrookite.

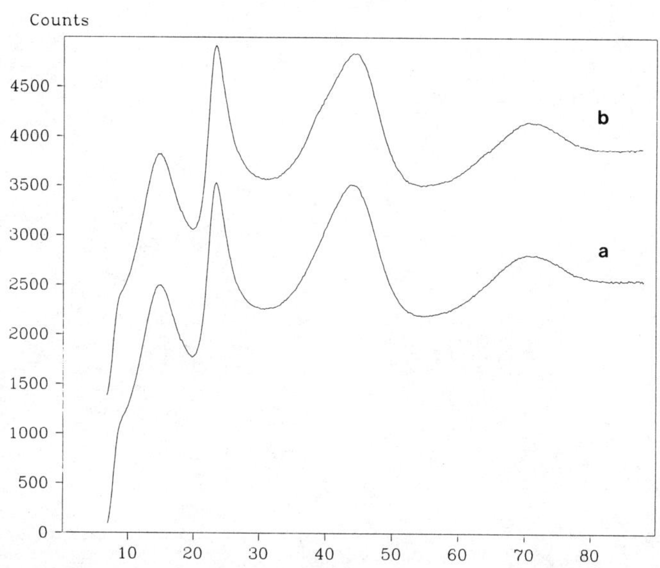

FIGURE 5. Wide-angle neutron diffraction patterns from (a) untreated and (b) nucleated glass heated for 144 hr at 711° C.

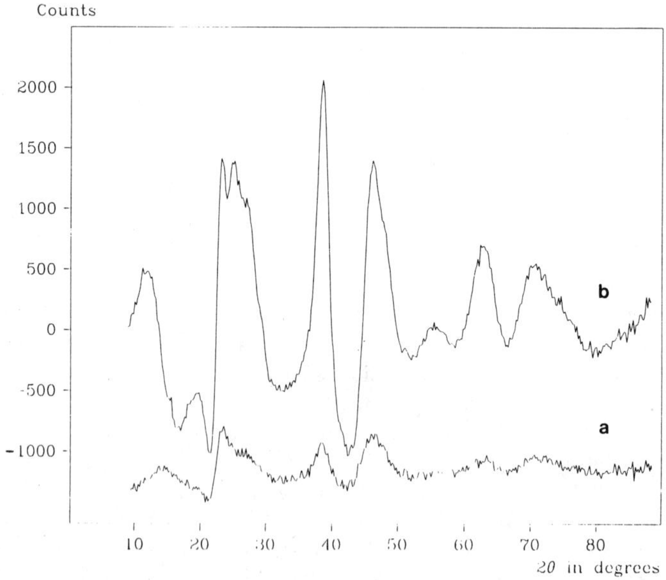

FIGURE 6. Differential wide-angle diffraction patterns from (a) nucleated glass heated for 144 hr at 711° C and (b) nucleated and developed glass heated for 144 hr at 711° C, then for 30 min at 800° C.

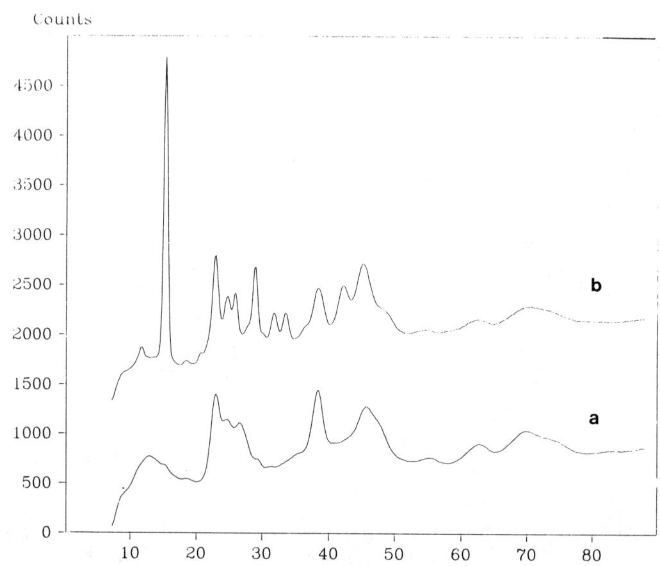

FIGURE 7. Wide-angle neutron diffraction patterns from (a) nucleated glass developed for 1 hr at 900° C and (b) nonnucleated glass heated directly for 1 hr at 900° C.

EXAFS ANALYSIS

EXAFS analysis gives structural information on the environment about a selected element, particularly the partial radial distribution function out to a distance of 4 to 5 Å. EXAFS on the titanium edge has been measured for the untreated, nucleated, and crystallized glass.[8] The pre-peaks situated about 10 eV before the absorption edge were also measured. The pre-peak is weak when the target ion is in a centrosymmetric (octahedral) site and is intense when the target ion is in a noncentrosymmetric (tetrahedral) site. For nucleation treatments of 48 and 96 hr at 710° C the pre-peak remains practically unchanged. Only after further heating at 900° C does the pre-peak broaden and weaken substantially. It must be concluded that during nucleation and the formation of 20-Å particles after 96 hr at 710° C that the vast majority of the Ti^{4+} ions remain in tetrahedral coordination. Transformation to octahedral coordination occurs during the growth stage at 900° C. The EXAFS spectra on the K edge of Ti during heat treatment up to 800° C failed to reveal any changes greater than the uncertainty of the measurement. Heat treatment at 835° C and 900° C resulted in an increase in

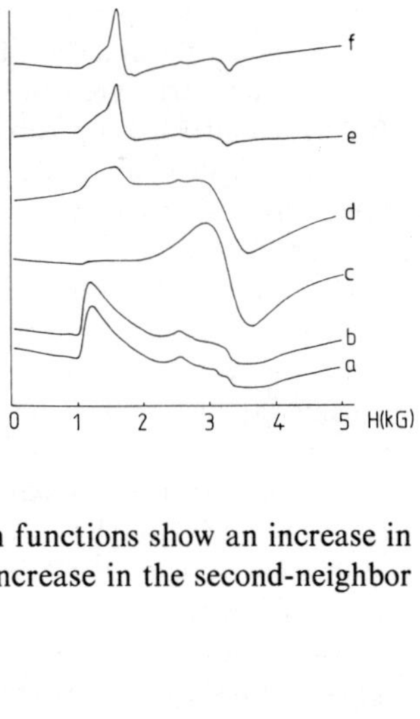

FIGURE 8. The Cr^{3+} electron paramagnetic resonance of cordierite (Cr^{3+}) glass during nucleation and crystallization: (a) not heated treated; (b) heat treated for 4 hr at 875° C; (c) for 4 hr at 875° C, then for 2 hr at 900° C; (d) for 4 hr at 875° C, then for 5 hr at 900° C; (e) for 10 min at 950° C; and (f) $MgAl_2O_4$ (Cr^{3+}) spinel powder.

the amplitude of the first peak. The radial distribution functions show an increase in the oxygen coordination from 3.7 ± 0.5 and a slight increase in the second-neighbor distance.

DISCUSSION

Crystal Disorder and the Stability of Small Particles

The correlation between submicrocrystallization and structural or chemical disorder has been observed in other glassy systems. Cordierite glass nucleated by Cr_2O_3 produces a spinel solid-solution ($MgAl_{2-x}Cr_xO_4$) rich in Cr^{3+}, which is identified by its fluorescence emission spectrum and by electron paramagnetic resonance. The Cr^{3+} content (initially 0.8% Cr_2O_3) of the glass is rapidly exhausted during the early stages of growth as the crystallite size approaches 100 Å. Further crystal growth then occurs by addition of Al^{3+} to the spinel structure, with a consequent dilution of the Cr^{3+}. This is most clearly shown by the electron paramagnetic resonance spectrum (FIG. 8). In this type of crystal the M^{3+} sites are normally randomly populated with Cr^{3+} or Al^{3+} ions. This disorder on the crystallographic lattice is the extra degree of freedom that allows the selection of the most stable microcrystal. The broad, intense signal at $H \simeq 3.25$ kG is associated with an exchange interaction due to a phase extremely rich in Cr^{3+}, possibly $MgCr_2O_4$ (Al^{3+}). This signal diminishes as that due to $MgAl_2O_4$ (Cr^{3+}) increases as the crystallization continues. Cr_2O_3 is very insoluble in glass, but pure Cr_2O_3 is not precipitated. By combining in a spinel solid-solution with Al^{3+} as the exchangeable ion, one can speculate that the Cr would prefer to populate the M^{3+} sites in the bulk of the particle, leaving Al to occupy those sites on the surface and make the link to the glassy matrix, itself still rich in Al^{3+} ions. Because the Al^{3+} and Cr^{3+} ions can interchange with little change in energy, it is conceivable that such a radially inhomogeneous particle would be more stable in this environment than in a random one. Similarly, the β-quartz solid-solutions precipitated from magnesium alumino-silicate glasses[3] and from lithium alumino-silicate glasses benefit from the exchange of Si for the combination of Al^{3+} and a charge-compensating ion, Li^+ or $1/2(Mg^{2+})$, situated in the C-axis channels. These exchanges introduce a small lattice expansion.[15] Could this also be a variable to help minimize the bonding mismatch at the crystal-matrix interface?

Whereas β-quartz has a twofold statistical disorder[16] of the O atoms, which could help to reduce the interfacial energy only slightly, β-cristobalite has a sixfold O disorder.[17] The structure can be described as a superposition of six orientations of a tetrahedral unit cell comparable to α-cristobalite. The Si atoms are almost ordered and show only a little strain release, whereas the O sites are only one-sixth occupied. The structure is interpreted in terms of ordered microdomains with a sixfold orientational disorder.

The transition from the glass to β-cristobalite is therefore only a partial ordering: translational ordering of the Si is achieved, but orientational ordering is not. The latter occurs at the β-to-α transition near 200° C, and is accompanied by microfracturing of the material due to the anisotropic contraction of the domains constituting the structure.

Controlled crystallization of the aqueous electrolyte $LiCl(H_2O)_9$ produces cubic ice,[18] the water isomorph of β-cristobalite. In cubic ice the O atoms are ordered, but the H atoms have orientational disorder in addition to the usual H-bond disorder (A. N. Fitch, A. F. Wright, and J. Dupuy, unpublished results). Cubic ice appears only to be stable when subdivided into small particles. Cubic ice precipitated from $LiCl(H_2O)_9$ begins to transform to hexagonal (normal) ice when the diameter exceeds ~200 Å. Above this size the extra stability of the bulk hexagonal ice outweighs the interfacial energy advantage of the disordered cubic phase. Cubic ice can only be formed by rapid quenching to the glass followed by extended annealing near the glass transition temperature, T_g.

Crystallization of the Residual Glassy Matrix

We have reported that the submicrocrystalline state appears particularly stable with respect to further transformations. In the normal course of cascade crystallization the formation of the first phase by elimination of solute shifts the composition of the residual glass to such an extent that secondary crystallization is possible. This second stage occurs at temperatures where crystal growth is rapid, and so a high nucleation density is not necessary to completely crystallize the remaining glass. In a microcrystalline state where the volume fraction of the first precipitate is already quite high (in the case of the cordierite-TiO_2 glass, this could be 30%), the residual glass is subdivided by crystallized particles every 70–150 Å. The area of the glass-crystal interface is determined by the crystalline phase, and even though the glass occupies more than twice the volume of the crystal, the glass volume is interstitial with concave reentrant surfaces. The unit size and shape of the residual glass is then not ideal for abundant nucleation, where spherical particles would give the lowest value of ω^*. The interstitial volumes of residual glass may simply not exceed the critical size for crystallization. They would then be more stable in the glassy state than in the crystalline state unless, but this is unlikely, any new crystal-crystal interface were energetically more favorable than the existing glass-crystal interface. High-density nucleation frustrates further crystallization of the residual glass whereas low-density nucleation promotes it.

The Submicrocrystalline State

In this paper we have presented experimental evidence for the formation of submicrocrystalline structures in glass. How do these structures compare with the submicrocrystalline models for glass structure, first proposed by Porai–Koshits[19] more than 20 years ago, and revived by Phillips[20] and Goodman?[21] In such models the basic argument is that quasi-crystalline clusters in different orientations[21] or of different polymorphs[21] form the bulk of the solid. The boundaries between these clusters are strained, but this strain is minimized by either the intervention of defect states such as double bonds or by favorably organized structural disorder. We emphasize particularly in this work the importance of the crystalline solid-solution as a general means of reducing the surface strains. This has not previously been proposed to our knowledge. The frustration observed in the crystallization of small interstitial glass volumes does not support general arguments for fully crystallized microcrystalline models.

In spite of the relative ease of detection of the microcrystalline states by SANS, the model proposed by Phillips[20] would be difficult to detect by this technique because of the small volume fraction, small dimensions, and low contrast of the zone bounding the microcrystallites. It remains to be discussed whether such thin boundary zones would be physically realistic. The strained-mixed cluster model proposed by Goodman[21] on a scale of 20 Å should be easily detectable by small-angle scattering. This is not observed in what are normally understood to be single-phase glasses. Diffraction methods are not particularly sensitive to microcrystalline structures below 20 Å unless these are present in high-volume fractions. Differential diffraction from glassy and microcrystalline solids as used in these nucleation studies is not normally applicable. High-resolution (in real space) diffraction could change this in the next decade. Spectroscopic methods, on the other hand, can clearly identify local order and geometry, but can make no observation about crystallinity.

CONCLUSIONS

Something approaching a submicrocrystalline state can be created from selected multicomponent glasses by a long-term heat treatment at or below T_g. The submicrocrystalline particles are invariably disordered, either chemically or structurally. This disorder is thought to be an essential feature stabilizing the nucleation of very small crystallites. The early stages of formation are relatively easily detected by SANS, but would normally pass undetected by diffraction. A high-density of very small crystallites subdivides the residual matrix and could frustrate further crystallization where the unrestricted matrix dimensions are less than the critical value for nucleation.

REFERENCES

1. STOOKEY, S. O. 1959. Ind. Eng. Chem. **51**(7): 805.
2. BEALL, G. H. 1985. *In* Glass: Current Issues. A. F. Wright & J. Dupuy, Eds.: 31–48. Martinus Nijhoff. Dordrecht.
3. BARRY, T. I., J. M. COX & R. MORRELL. 1978. J. Mater. Sci. **13:** 594–610.
4. TURNBULL, D. 1956. *In* Solid State Physics. F. Seitz & D. Turnbull, Eds. Vol. **3:** 225. Academic Press. New York, NY.
5. JAMES, P. F. 1985. J. Non-Cryst. Solids **73:** 517–540.
6. KALWEIT, M. 1970. *In* Physical Chemistry: Solid State. W. Jost, Ed. Vol. **10:** 719–758. Academic Press. New York, NY.
7. WRIGHT, A. F., A. N. FITCH, J. B. HAYTER & B. E. F. FENDER. 1985. Phys. Chem. Glasses **26**(4): 113–118.
8. DUMAS, T. 1984. Ph.D. dissertation. University of Paris VI. Paris.
9. DURVILLE, F., B. CAMPAGNON, E. DUVAL, G. BOULON, F. GAUME, A. F. WRIGHT & A. N. FITCH. 1984. Phys. Chem. Glasses **25**(5): 126–133.
10. WRIGHT, A. F. 1981. *In* Congrés de la Société Francaise de Physique. J. Hieblot, Ed.: 149–156. Les Editions de Physique. Paris.
11. WRIGHT, A. F. 1985. *In* Glass: Current Issues. A. F. Wright & J. Dupuy, Eds.: 21–30. Martinus Nijhoff. Dordrecht.
12. HAYTER, J. B. & J. PENFOLD. 1981. Mol. Phys. **42:** 109.
13. HAYTER, J. B. & J. PENFOLD. 1981. J. Chem. Soc. Faraday Trans. 1 **77:** 1851.
14. RAVAINE, D. & A. F. WRIGHT. 1983. Phys. Chem. Glasses **24:** 31–34.
15. SCHREYER, W. & J. F. SCHAIRER. 1961. Z. Kristallogr. **116:** 60.
16. WRIGHT, A. F. & M. S. LEHMANN. 1981. J. Solid State Chem. **36:** 371–380.
17. WRIGHT, A. F. & A. J. LEADBETTER. 1975. Philos. Mag. **31:** 1391.
18. JAL, J. F., C. FERRADOU, J. DUPUY, P. CHIEUX & A. F. WRIGHT. 1983. J. Chem. Phys. **88:** 4170–4173.
19. PORAI-KOSHITS, E. A. 1959. Glastech. Ber. **32:** 450.
20. PHILLIPS, J. C. 1982. *In* Solid State Physics. H. Ehrenreich, F. Seitz & D. Turnbull, Eds. Vol. **37:** 93–171. Academic Press. New York, NY.
21. GOODMAN, C. H. L. 1983. *In* Structure of Non-Crystalline Materials. P. H. Gaskell, J. M. Parker & E. A. Davis, Eds.: 151–163. Taylor & Francis. London.

DISCUSSION OF THE PAPER

A. NAVROTSKY (*Princeton University, Princeton, NJ*): Can you estimate the fraction of the original glass that is in the crystallites?

WRIGHT: Estimation of the size of the crystallized fraction is difficult due to the very small particle size, but the fraction is probably between 10 and 20%.

G. M. FAHY (*American Red Cross, Bethesda, MD*): I'm interested in the opposite of your topic: how disordered can a glass remain near its T_g? I assume your systems involve doubly unstable glasses ($T_H > T_g$). What is known concerning the temperature dependence of nucleation rates for systems having $T_H < T_g$?

WRIGHT: If the nucleation temperature is situated well below T_g, as is the case for many commercial glasses, then, in the absence of heterogeneous nucleation, the glass will remain stable and uncrystallized even on prolonged heating at T_g or above.

C. T. MOYNIHAN (*Rensselaer Polytechnic Institute, Troy, NY*): You might add to the list of liquids that crystallize into metastable structures the ZrF_4-based heavy metal fluoride glass compositions. For example, $62ZrF_4 \cdot 33BaF_2 \cdot 5LaF_3$ glasses on reheating almost always crystallize into the metastable β-$BaZrF_6$ and β-$BaZr_2F_{10}$, rather than into the stable α forms.

WRIGHT: It is indeed interesting to note that the heavy metal fluoride glasses based on ZrF_4 crystallize by mechanisms that involve metastable phases. This would suggest that these glass systems belong to a class in which the crystallization can be controlled at the nucleation stage.

M. WEINBERG (*Jet Propulsion Laboratory, Pasadena, CA*): 1. When the very small crystallites are grown at higher temperatures wouldn't one expect dissolution because of the increases in the critical radius with rising temperature?

2. You stated that the metastable crystallites transform (in ice I believe) to the stable phase once they exceed a certain size. Is it possible that a transformation doesn't take place but that those crystallites serve as nucleating sites for the stable size? Also, can one experimentally distinguish between these two possibilities?

WRIGHT: 1. One would expect some dissolution of very small crystallites as a result of a temperature increase if the latter were to be suddenly applied. A steady increase in temperature, on the other hand, gives time for growth to occur so that the crystal radius always exceeds the critical radius as the temperature rises.

2. In the case of ice, metastable cubic ice crystallites transform to the stable hexagonal phase once a certain size is exceeded. This is the conclusion drawn from diffraction data, which are quite unambiguous. It is possible, however, that in other cases the metastable crystallites serve as nucleating sites for the stable phase, but one would normally expect dissolution of the former as growth proceeds and the supersaturation is reduced.

C. A. ANGELL (*Purdue University, West Lafayette, IN*): I think it would be very interesting to have some calorimetric information on the extent to which the normal heat of crystallization has been released in the process of forming crystals of the extremely small size that your work implies. Have you any plans for solution calorimetry experiments on the glasses crystallized by the different procedures? Even if the precise extent of crystallization were not known for the case of your submicroscopic crystals, the calorimetric information would still be of great interest, especially if a systematic study of heat content versus mean crystal size could be made.

WRIGHT: Your suggestion that calorimetric measurements would be a useful complement to structural studies is well founded. I could prepare samples nucleated to varying degrees for subsequent measurements of heat of crystallization and measure the particle size afterwards. This might be a more direct method than solution calorimetry.

How Ordered Can a Glass Be?[a]

P. H. GASKELL

Cavendish Laboratory
University of Cambridge
Cambridge, CB3 0HE England

HOW CAN WE THINK OF "ORDER" IN DISORDERED MATERIALS?

At first sight, the question suggested by the conference organizers as the title for this paper seems to have an obvious answer. A material that is recognizably ordered, that is, crystalline, by techniques such as X-ray diffraction or transmission electron microscopy is not a glass, so that the limiting order corresponds to perhaps 1.5 nm: the diameter of a microcrystallite that might be detectable by X-rays. A solid that is not ordered over this critical distance scale can be termed an amorphous solid. This answer is of course based on what we commonly regard as the definition of a glass—a noncrystalline solid—but the description, nonetheless, has several drawbacks.

What do we mean by order? In the context of this paper, I suggest that the most useful definition is based on the notion that each atom in an ordered structure fits into a locally defined pattern and that this pattern is indefinitely repeatable. Individual atoms need not be identical, of course, but it must be possible to move from a given site and to find several equivalent sites elsewhere in the solid. As these sites are equivalent, the view from each is indistinguishable. The sites may be related translationally, in which case the material is crystalline, or orientationally, giving a multiply twinned structure or a quasi crystal.

This notion of superposition of structural domains centered on equivalent sites is more flexible than that based on the presence or absence of crystallites of a given size. We can recognize equivalence between domains that are orientationally related rather than translationally periodic. We can also incorporate further flexibility by allowing other operations such as reflection about a plane of symmetry within the domain or suppression of differences between left- and right-handed forms of the structure and differences due to minor geometrical distortions so that any topological similarity is preserved.

A simple example might involve domains consisting of four SiO_4 units connected in left- and right-handed helices, and which also may have a distribution of bond lengths (FIG. 1a). A structure containing regular arrangements of such domains could be considered to be ordered even though it may be neither translationally nor orientationally ordered.

A two-dimensional structure could be represented as shown in FIGURE 1b, where the symbols represent characteristic directions in domains of different sizes with left- and right-handed forms. It is assumed that a uniform interfacial structure can be constructed for modules of arbitrary orientation, but to give the notion sufficient

[a]Generous financial support from Pilkington PLC is gratefully acknowledged.

flexibility to deal with amorphous solids, we could also require an intergranular "matrix" phase that does not have the symmetry of any of the domains.

Using this description, a quantitative measure of order in an amorphous solid could be given in terms of the volume fraction of the ordered domains, the mean and the standard deviation of the distribution of domain sizes, the extent to which atomic sites within each domain are regular ("lattice-like," perhaps), and the internal symmetry elements of the individual domains.

These ideas may be of some use in conceptualizing the problem inherent in defining

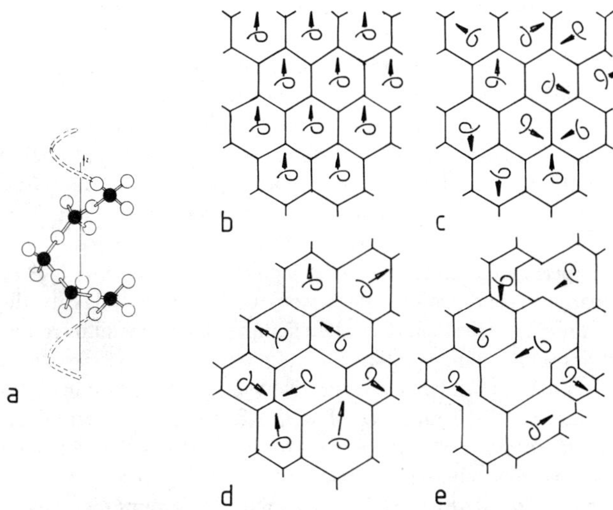

FIGURE 1. Schematic representation of order in amorphous materials. (**a**) A hypothetical four-membered SiO_2 helix, which forms the basis of the structures represented in subsequent diagrams. (**b**) An ordered crystalline arrangement. Each hexagonal cell represents an arrangement containing several helical units. The symbol within each cell is a shorthand representation of the elementary helix with the *z* axis and the handedness represented in an obvious way. (**c**) A structure containing orientationally related modules. Note the presence of left- and right-handed helical units. (**d**) A structure with topological order but (extreme) geometrical disorder. Each cell is to be thought of as containing a constant number of atoms: the individual units therefore suffer geometrical distortion. (**e**) A structure similar to **d** where the domain size is also allowed to vary. In diagrams **c** and **d** it is assumed that a coherent interface can be created. More generally, a separate interfacial region may be necessary.

order in a notionally disordered material. They may be of zero practical value unless the domains are sufficiently large, well ordered, and contain only minor geometrical distortions. Even assuming that it is possible to obtain partial pair correlation functions rather than total functions (or that the material is monoatomic) for the examples shown in FIGURE 1d, variable geometry alone could foil any attempt to decipher the scrambled code resulting from conventional diffraction measurements, and equivalent domains would not be recognized as such. Electron microscopy would be less limited because averaging over many domains may not be necessary with very thin specimens, but so far the technique has only been of limited use in detecting relatively straight lattice planes. Scanning tunneling microscopy appears to have enormous potential for studies of amorphous solids, but there are no results to date.

Moreover, it would not be reasonable to constrain bond length fluctuations to small values: such fluctuations represent one of the most obvious and important characterizing features of the glassy and liquid states. Fluctuations in geometry are explicitly built into random models for glasses and act to make the model more realistic. It is equally essential that fluctuations are included in models built from ordered clusters. These fluctuations stem from the interfacial strains inherent in coupling units together, but with few exceptions this strain has been ignored by model builders so that several possible "microcrystallite" structures have been dismissed because of inadequate evidence.

The somewhat unsatisfactory first conclusion to emerge then in answer to the question "how ordered can a glass be?" is "probably more than we shall ever know!" The directly perceived structural organization must be less than, or equal to, the actual degree of order. The best that we can do is to build realistic models based on a postulated structure and test for consistency with experimental data. (Order, even in a model of the structure, can pass unnoticed. A recent example is that described in Weaire and Wooten's[1] model for amorphous Si.)

Given the experimental difficulties what kinds of order could we look for?

In a monoatomic solid such as amorphous Ge, we would classify the structural question according to three length scales, defining local, medium-range, and long-range structural organization, involving characteristic radii up to 0.6 nm, between 0.6 and 1.5 nm, and greater than 1.5 nm, respectively. Under the heading of local structure we look for evidence of the presence of well-defined local structural units: the Ge tetrahedron or the tetrahedral arrangement of metal atoms thought to dominate the structure of amorphous monoatomic metallic glasses. Medium-range organization then implies any pattern that nature uses to connect the groups together.

In polyatomic systems these questions of positional order remain valid but are further complicated by considerations of compositional organization. For amorphous alloys, impressive progress has been made though measurements of the Bhatia–Thornton partials,[2] specifically of the concentration-concentration structure factor, $S_{CC}(Q)$, and its Fourier transform, the radial concentration correlation function. Extraction of Faber–Ziman partials and the Warren–Cowley[3,4] or Cargill–Spaepen[5] short-range order parameters can also be used to characterize the extent of chemical clustering.

For covalent solids, chemical or compositional ordering has not attracted the same degree of quantitative analysis, with the honorable exception of work on "wrong bonds" in amorphous III–V compounds[6] and on S–S and Se–Se bonds in Ge–S(Se)-containing glasses.[7] For most of the amorphous oxides the question of compositional ordering does not appear in the same guise: there is no evidence in silicates, for example, that Si atoms can ever be nearest neighbors, or that the number of oxygen neighbors is less than four. The question of compositional ordering at next-nearest neighbor level, however, is central to the description of the structure of the alkali silicate glasses, where some oxygens are bridging (forming strong Si–O–Si bonds) and others are nonbridging, carry a formal negative charge, and are ionically linked to cations such as Na^+, that is, $Si–O^-Na^+$. The various possible arrangements of compositional and spatial order are shown in FIGURE 2.

THE EXTENT OF ORDER AT LOCAL STRUCTURAL LEVEL

If the question "how ordered can a glass be?" were restricted to organization at local level the answer would be unambiguous: order may be almost perfect. For example, the SiO_4 tetrahedron in amorphous SiO_2 is thought to be almost perfectly tetrahedral. The bond length is similar to that found in α-cristobalite, the distribution of Si–O bond lengths is almost insignificant ($\sigma = 0.0035$ nm),[8] and the bond angle is tetrahedral with a standard deviation of 0.6 degrees.[9] The sharpness of the Si–O first peak shown in FIGURE 3 graphically illustrates the extent of geometrical order. As mentioned earlier, the compositional order appears to be essentially perfect.

Well-defined local coordination around network-forming atoms in oxide glasses is not difficult to observe and apparently extends to minority components such as the coordination polyhedron around Ti ions in vitreous $K_2O \cdot TiO_2 \cdot 2SiO_2$[11] where the

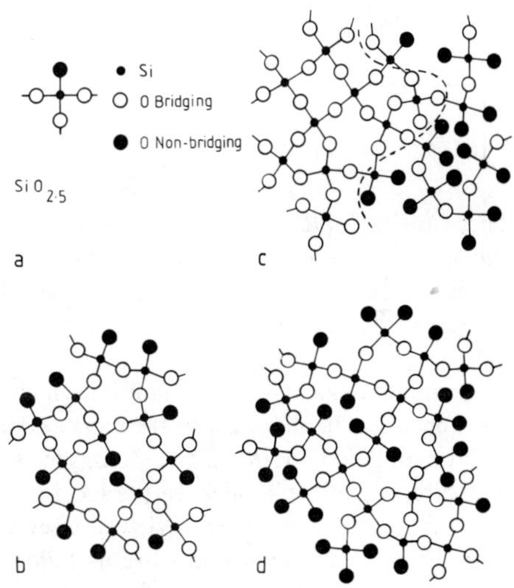

FIGURE 2. Schematic representation of chemical or compositional order in a glass of disilicate composition. (**a**) The disilicate anion with three bridging oxygens (Q_3 in NMR terminology). (**b**) A positionally disordered but compositionally ordered network. Positionally and compositionally disordered networks are shown in **c** and **d**: **c** shows segregated Q_4 and Q_2 regions whereas **d** represents a random mixture of Q^4–Q^1 anions.

coordination number is observed to be five and evidence for a square pyramidal configuration is obtained from the split first peak with components at 0.165 and 0.196 nm and with weights of 0.75 and 4.1. The standard deviation of the first Ti–O distance distribution is so small that it can be considered to be effectively zero.

Compositional order is apparently high, in the sense described in the previous section, for some vitreous alkali silicates. For example, recent magic angle ^{29}Si NMR results on glassy sodium silicates[12,13] suggest that in the stoichiometric compositions corresponding to SiO_2 or $Na_2Si_2O_5$ or Na_2SiO_3, only a single Si site is observed, corresponding to four bridging bonds (Q_4) in SiO_2, and Q_3 and Q_2 sites in disilicates and metasilicates, respectively. Intermediate compositions can be represented as the compositionally weighted sums of Q_4 and Q_3 (from SiO_2 to $Na_2Si_2O_5$) or of Q_3 and Q_2 otherwise.

These comments should not be taken to apply as a general rule in covalent glasses.

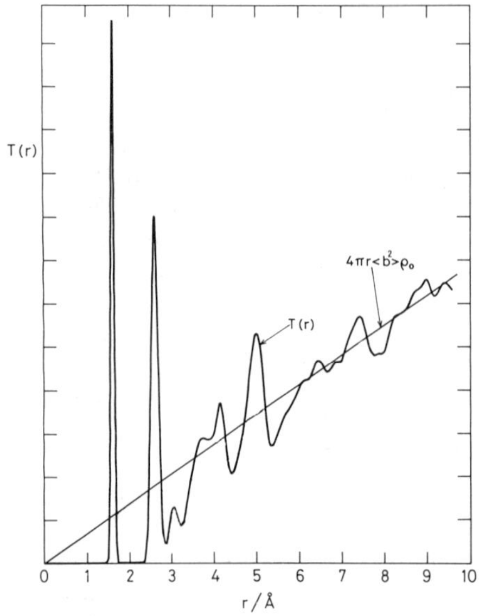

FIGURE 3. Reduced radial distribution function, $T(r)$, in units of barns $Å^{-2}$, corresponding to a model for vitreous silica that produces a neutron interference function, $Qi(Q)$, in good agreement with experimental data; $T(r)$ as represented here gives a more accurate impression of the narrowness of near-neighbor correlations in the glass because the peaks are not broadened by effects originating from the limited range of the Fourier transform of an experimental $Qi(Q)$ function. The data are from Gaskell and Tarrant.[10]

There are examples, among the Ge–S and Ge–Se alloys, for example, where the local chemical order does not appear to conform to any simple rule.[7] It is nonetheless often found that a useful starting point for any investigation of the local structure is a comparison of the environment of the glass and a compositionally equivalent crystalline phase.

In metallic glasses, the system for which we have the most structural information also appears to be the one with the most evident structural organization. The transition metal-metalloid (TM-M) glasses and, specifically, the nickel-containing alloys appear to exhibit organization at both the local and medium-range levels. The evidence for short-range order has been reviewed elsewhere,[14] and only a brief account need be given here. The salient points are as follows: 1) In those TM-M glasses for which accurate partial pair correlation functions exist, the metal coordination number around the metalloid is nine to a good approximation (as in crystals). Moreover, the coordination number does not appear to be composition dependent and the standard deviation of the TM-M bond length is very small. 2) For compositions up to 25 atomic % of the metalloid, chemical ordering is thought to be very high: in the best-testified case, that of $Ni_{81}B_{19}$,[15] the Cowley–Warren short-range order parameter is one. 3) The symmetry of the environment of the B atom in $Ni_{1-x}B_x$ alloys resembles the 9-atom-capped trigonal prismatic unit found in several crystalline compounds (such as Ni_3B and Ni_4B_3) but is distinctly different from the 10-atom-capped Archimedean antiprism found in crystalline Ni_2B.[16]

There seems every reason to assume that the structure of this glass is based on well-defined local structural units with trigonal prismatic symmetry. Exactly how these units couple together is less certain, but there is fragmentary evidence the two distinct types of edge-sharing observed in Fe_3C and Fe_3P lattices are also found in the corresponding glasses.[17]

THE EXTENT OF MEDIUM-RANGE ORDER IN TM-M GLASSES

Evidence for medium-range ordering is more speculative and insecure. The most direct evidence comes from the B–B and P–P partial correlation functions determined by Lamparter *et al.* in $Ni_{81}B_{19}$[15] and $Ni_{80}P_{20}$.[18] In $Ni_{81}B_{19}$, the B–B partial continues to oscillate to distances greater than 1.2 nm. Because boron atoms, it is argued above, lie at the centers of trigonal prisms, the B–B partial pair distribution function implies a correlated arrangement of trigonal prisms out to about 1.2 nm.

The same conclusion is reached by two other routes. First, it can be argued that medium-range organization in TM-M glasses is a necessary condition for the existence of well-defined trigonal prismatic units—at least over the composition range 0–25 atomic % of the metalloid. Briefly, the reasoning is as follows.[19] Consider the constraints operating in a glass of composition Ni_3B, and assume that the local structural unit consists of nine atoms surrounding the boron in the form of a capped trigonal prism (TP) (although the exact symmetry of the coordination polyhedron is not crucial). At this composition, each Ni atom must be shared between three boron-centered TPs. FIGURE 4 shows a representation of the structure of the cementite lattice: the corresponding crystalline form of Ni_3B. The edge- and vertex-sharing arrangement is obvious from FIGURE 4, but, in addition, atoms in the plane shown are also capping atoms for TPs in planes above and below.

The presence of one fully capped trigonal prism thus requires a correlated arrangement of about six neighboring units. These are constrained by, and in turn constrain, other neighbors, and so on. The point is that in any close-packed system, amorphous or crystalline, which is so extensively interconnected, experimental evidence for the existence of relatively well-defined local structural units implies a correlated arrangement of such units. This statement would not apply if the structure contained extensive voids that effectively relax some of the constraints. But there is ample evidence that the structure of the glassy phase has almost the same density as

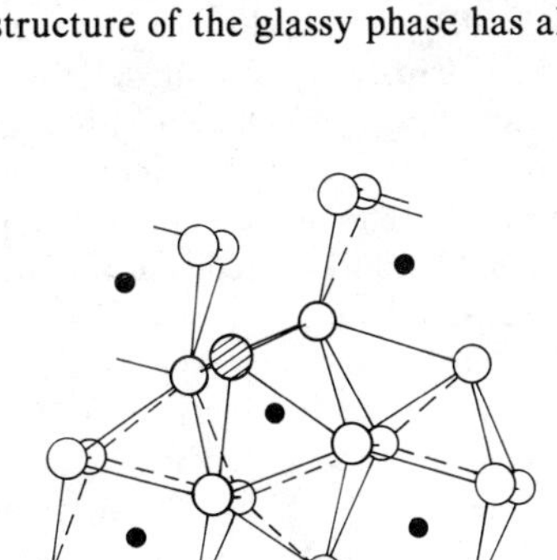

FIGURE 4. A trigonal prismatic unit consisting of nine metal atoms (large circles) surrounding a metalloid (small filled circles) must be extensively connected to other prisms by edge and vertex sharing in typical metallic glasses containing about 20 atom % of the metalloid. If the neighboring trigonal prisms are packed efficiently, only tetrahedral and half-octahedral interstices exist. The possible positions of neighboring local structural units are severely constrained, as the sheet of prisms indicates. Note that a similar sheet of prism exists above and below the plane linked by capping atoms such as the one shaded. The result is that for a well-defined trigonal prism to exist at all the medium-range structure is also defined within relatively narrow limits.

the crystal, and the packing density of crystals containing small metalloid atoms (which do not dilate the circumscribing TPs) is only a little less than cubic close packing.[20]

A more direct approach to the problem is to construct a model for amorphous Ni–B alloys in which both local and medium-range organization is included. This has been reported recently by Dubois *et al.*[21] Specifically, the models contain about 1800 atoms and are composed of positionally ordered domains that resemble, but are not identical with, any crystalline phase occurring in the Ni–B system. Layers of packed trigonal prisms exist as shown in FIGURE 4, but an additional close-packed plane of metal atoms separates the sheets. Connections with each domain are prescribed by a relatively well-defined set of rules: 1) Each Ni atom is either a vertex atom or a capping atom for a TP centered on the B sites and/or caps a neighboring prism. 2) TPs are interconnected either by vertex or edge linkages. 3) An atom not at the vertex of any prism lies at the common vertex of two half-octahedra that cap neighboring prisms.

A neighboring domain consists of an identical unit, but this unit has a characteristic vector normal to the sheets of TPs that is rotated such that atoms at the boundary between domains continue to fulfill the rules given above. Atoms in the interfacial region are thus even more constrained than atoms within domains because they must satisfy rules appropriate to two or more neighboring domains. Clearly, continuity of the structural operations across domain boundaries can be possible only for a very limited set of orientational relationships. After energy minimization, using a Lennard–Jones 6:12 potential, the elastic strain generated at domain boundaries extends over the entire structure. There are only limited changes in topology, however, indicating that the initial model is fairly close to a local potential energy minimum.

Computed partial pair correlation functions are shown in FIGURE 5 together with experimental data of Lamparter *et al.*[15] Agreement is, on the whole, good. The most crucial test of models for these glasses is not the Ni–Ni or Ni–B correlations, which can be fitted relatively well by models of several types, but the B–B partials. Poor fits to these data have eliminated two earlier types of model: the randomly packed trigonal prism model of Gaskell[22] and the soft-sphere loose-packing model described by Lamparter *et al.*[15] The present model shows much better agreement with the experimental B–B partial function. The split first peak is correctly reproduced, and the amplitude of the long-range oscillations is predicted (although the positions of peaks become inaccurate beyond about 0.7 nm).

This work demonstrates that the experimental partial pair correlation functions for amorphous $Ni_{81}B_{19}$, which are among the most accurate data in existence at the moment, are not inconsistent with domain models with extremely well-defined order extending over a distance of about 1.2–1.5 nm. The disagreement with experimental data is not insubstantial, however, suggesting that the precise arrangement of trigonal prisms is, as yet, less than adequate. Specifically, the real glass appears to be still more ordered than the models!

The conclusions to be drawn from this section are that for glasses for which the network is extensively interconnected, especially the close-packed metallic alloys with strong bonds between dissimilar elements, but also including the close-packed regions of other glasses such as the silicates, local structural order and medium-range organization are not necessarily independent. The first may constrain the second. Organization of the local coordination shell around B in Ni–B alloys or around, say, Li in silicate glasses may represent the strongest (medium-range) structure-building force to which other medium-range structural requirements must then become subservient. When other steric constraints are taken into consideration, connectivity of the local units or connectivity of other local structural units in the glass such as the SiO_4 tetrahedra in silicates, the result may be that to satisfy local requirements, possible medium-range structures must be constrained to a set of arrangements that allows such favorable local organization. Dent–Glasser[23] has shown that although silicate crystals are common in nature, there are nonetheless some surprising gaps where "nonexistent" silicates might be expected to occur. She has argued that the crystals are not stable because the constraints imposed by the local structure: packing of oxygen ions around highly charged cations, say, are inconsistent with the

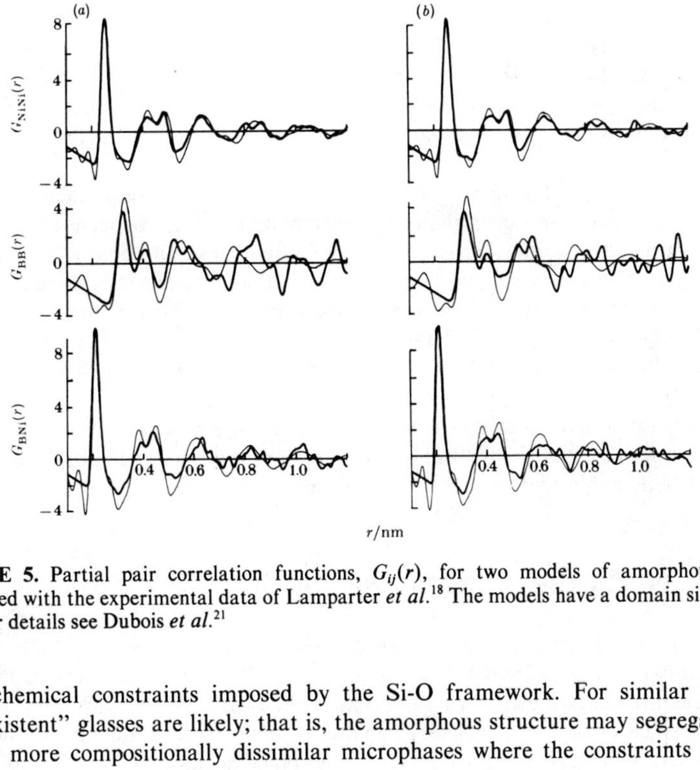

FIGURE 5. Partial pair correlation functions, $G_{ij}(r)$, for two models of amorphous Ni_4B compared with the experimental data of Lamparter *et al.*[18] The models have a domain size of 1.2 nm. For details see Dubois *et al.*[21]

stereochemical constraints imposed by the Si-O framework. For similar reasons "nonexistent" glasses are likely; that is, the amorphous structure may segregate into two or more compositionally dissimilar microphases where the constraints are not inconsistent.

MEDIUM-RANGE ORDER AND CRYSTALLINITY

This argument does not necessarily imply that suitable members of the constrained set of possible medium-range structures must necessarily be crystal-like arrangements. Indeed the frustration argument provides a cogent argument for the existence of local and medium-range structures that are noncrystallographic.[24] The energetically favorable packing of metallic atoms based on polytetrahedral arrangements leads to the notion that in a liquid, where long-range influences are less important than short-range forces, a polytetrahedral cluster of atoms may form, which by virtue of its fivefold symmetry will be inconsistent with the requirements of space filling in three-dimensional space. Crystals therefore are excluded from the set of acceptable medium-range structures.

For the examples considered here, all we can say, at present, is that domain models represent one of a possibly large constrained set of acceptable medium-range structures. But we can affirm the notion that medium-range structures of some more or less defined form must exist as a consequence of the well-defined local structural organization.

This existence of alternative medium-range structures for amorphous $Ni_{81}B_{19}$ is demonstrated by the fit to experimental data obtained by the molecular dynamics simulations of Beyer and Hoheisel.[25] A 2048-atom model was computer generated and relaxed using a Lennard–Jones 6:12 potential. The model provides a good fit to the split B–B first-neighbor peak and a reasonable agreement to the longer range oscillations up to about 0.7 nm; that is, the fit to experimental data is almost exactly equally good (or bad) as that of the domain models of Dubois *et al.*[21]

Beyer and Hoheisel have not examined the correlations present in their model, but they have supplied atomic coordinates. An examination of their model has been started by Dubois and the author. From a very preliminary examination, it is clear that ninefold coordination is the most prevalent and that recognizable capped trigonal prisms (often extensively distorted) outnumber any other kind of first-coordination polyhedron around B atoms.[22] The interprism connections observed in cementite and built into the models of Dubois *et al.*[21] and Gaskell,[22] however, are not readily identifiable.

The choice between possible candidate medium-range structures must therefore be made on other grounds: the realism of the potential functions used in each case or the extent of small angle neutron or X-ray scattering, for instance.

STRUCTURAL ORDER AND THE GLASS TRANSITION

Another view of the question "how ordered can a glass be?" is gained by a consideration of the processes of glass formation. Turnbull[26] has argued that the central requirement for glass formation is the avoidance of crystallization. A liquid or vapor must be quenched sufficiently rapidly to prevent the formation of nuclei. A defining feature of glasses, as distinct from amorphous solids in general, is that a glass transition is observable. If we restrict discussion to glasses, we must therefore enlarge the range of acceptable structures to include not just that configuration which survives as the glass below the glass transition temperature, T_g, but the other configurations, locally accessible from that of the glass and into which the system can transform above T_g to give liquid-like properties. How many such configurations must there be at T_g, and how should the number change with temperature above T_g? Introducing Turnbull's criterion, as the accessible configurational space is explored in the supercooled liquid phase, the configuration corresponding to the crystal is to be avoided. At first sight it would seem that avoidance of crystallization for temperatures just above T_g requires one or both of two preconditions. First, that the number of acceptable medium-range structures should be large so that the one definitely acceptable structure, the crystal, has a low probability of formation. Second, that the set of acceptable liquid-like states falls into a different region of configurational space from that of a crystal nucleus, so that the two are not easily interconvertible without surmounting a considerable activation energy barrier.

The question becomes compounded by (and yet unified by) considering the time scale for structural reorganization as represented by the viscosity of the melt. Adam and Gibbs[27] have argued that at high temperatures, in the vicinity of the melting temperature, T_m, conversion of one local configuration into another may involve little more than the local group of atoms. The ensemble of local configurations is effectively decoupled. As the temperature is reduced towards T_g, it becomes necessary to involve cooperative motions of neighboring groups before the transformation can occur, so that the number of atoms involved in a cooperative transformation (z^*) increases rapidly. In the region of the glass transition, Adam and Gibbs argue that z^* increases until the whole specimen volume is included. The system effectively jams, no further rearrangement is possible, and at all lower temperatures, liquid-like properties are extinguished and the viscosity corresponds to that of a solid. Alternatively expressed, near T_m, the number of accessible amorphous configurations, N_c, is very large and falls rapidly to a single configuration at T_g.

As discussed earlier, the number of configurations is a function of the connectivity of the lattice, and this number is small for extensively interconnected networks. In the TM–M system $TM_{1-x}M_x$, and assuming a structural model based on nine-atom units, the connectivity is a function of the concentration, x, of the metalloid element and is high for values of x near 0.25 (a rough limit to the range of edge-linked prisms with no B–B first neighbors) and falls to zero as x approaches about 0.1, corresponding to an isolated trigonal prism. The number of members of the constrained set of acceptable medium-range structures is thus small at high values of x and low for $x \simeq 0.1$.

Both temperature and composition control the number of configurations available to the system in the region $T_g < T < T_m$, and it is the detailed interplay of these variables that determines whether the temperature interval, between $T > T_m$ where the amorphous (liquid) state is thermodynamically stable and $T < T_g$ where the amorphous (glassy) state is kinetically stable, can be crossed without crystallization intervening. We can discuss two limiting cases in terms of the density (or probability) of accessible states as a function of the energy, E, relative to that of the crystal.

Extensively Ordered Medium-Range Structures

Where the local structural units are extensively interconnected, the number of members in the set of constrained medium-range structures must be relatively small so that the probability of an exchange costing little activation energy must also be small. The ratio of the number of aperiodic structures to the number of periodic structures is thus relatively unfavorable, at least for the idealized, defect-free structures that we postulate are important for $T \simeq T_g$. At higher temperatures, defective aperiodic structures become possible with missing atoms, incomplete coordination, geometrically distorted local structural units, limited chemical short-range ordering, etc. This may be the situation for the liquid at $T \simeq T_m$. On cooling, defect states are progressively eliminated and the configuration evolves along increasingly complex and increasingly cooperative reaction paths leading to locally ordered periodic or aperiodic structures. As the number of accessible aperiodic structures becomes small, the probability of a defective region (which might be a defective crystalline state) transforming into a periodic rather than an aperiodic nucleus progressively increases. Whether it succeeds in doing so will depend on kinetic factors, specifically, on the form of the N_c/E curve, which controls the viscosity. In FIGURE 6, we postulate that for the highly interconnected material, N_c drops rapidly to values approaching zero corresponding to a limited number of defect-free aperiodic structures. The constraints are less important for

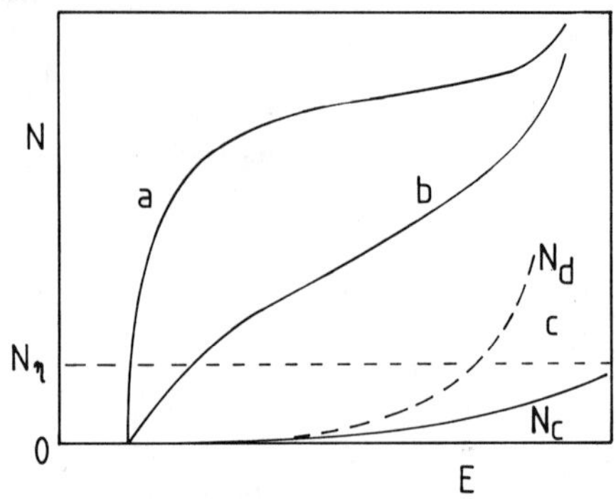

FIGURE 6. Schematic representation of the number of available configurations as a function of energy, E, relative to that of the crystal at T_g; N_c corresponds to the number of available noncrystalline configurations with the majority of atoms having perfect local structural order; N_d represents the number of defective states with atoms missing from first coordination shells. Curves a, b, and c correspond to supercooled liquids that have low, medium, and severe constraints, respectively, and in the TM-M glasses, these are represented by compositions with low, medium, and high atomic fractions of the metalloid, respectively. Curves a and b give the variation in the number of configurations ($N = N_c + N_d$), and the form of the N_d curve is taken to be unchanged in all cases so that $N \simeq N_c$ for cases a and b; T_m lies near the rapid rise in N_c due to the highly defective states; N_η represents the number of configurations necessary to give the viscosity corresponding to the glass transition.

defective structures so that the number of these constraints, N_d, decreases more slowly with E. The situation in the liquid somewhere below T_m is that the defective liquid state has a viscosity fixed by z^* through the Adam–Gibbs relationship

$$\log \eta(T) = \overline{A} + z^*\Delta\mu/kT$$

where $\Delta\mu$ corresponds to a potential energy term associated with the cooperative rearrangement of a mobile atom or local structural unit and $\overline{A}$ is a constant. The viscosity can equally be represented in terms of the total number of accessible configurations $N = N_c + N_d$ because $\log N \propto z^{*-1}$; that is, $\log \eta = \overline{A} + c\Delta\mu/T$, or $\log N = \overline{A} + c\Delta\mu/TS_c$ where S_c is the configurational entropy. The liquid at a viscosity less than 10^{12} Pa · sec thus has the choice of either nucleating to an aperiodic structure or a crystal. Because there are relatively few configurations corresponding to the former, crystallization may result unless the quenching rate is high, or the liquid is separated into very small droplets.

Limited Medium-Range Ordering

For compositions where local structural units are unconnected or the connectivity is low, the number of available medium-range structures is very large. Indeed, the concept of medium-range organization may cease to have meaning; N_c remains high to low values of E and is large in relation to the number of defect structures (which also tend to lose their meaning). The many configurational states available at relatively low energy (temperature) thus implies a low viscosity though the Adam–Gibbs relationship or, alternatively, that the value of $\eta(T) = 10^{12}$ Pa · sec corresponding to T_g is reached at low values of $T = T_g$, the process being controlled, perhaps by the $\Delta\mu$ term. The fluid will therefore be kinetically vulnerable to crystallization. Because the size of stable critical nuclei varies as $(T_m/(T_m - T))^3$, small critical nuclei and therefore a high probability of nucleation are expected for large differences $T_m - T_g$.[25] Nucleation and crystallization will thus be circumvented, again, only by extremely rapid quenching.

Between these two compositional extremes we postulate, in line with Phillips's[28] arguments, that a balance between the two countervailing tendencies may be struck. Specifically, that medium-range organizational constraints exist but are not very restrictive so that available aperiodic medium-range structures exist, their number being relatively large in relation to N_d. The viscosity corresponding to the glass transition is reached at a relatively high value of T_g/T_m and, if Turnbull's criterion ($T_g/T_m \simeq 2/3$) is approached, glass formation will be possible without extremely rapid quenching.

Converting these simple notions into a predictive theory that could allow some delineation between the regions of easy and difficult glass formation is difficult in the absence of a more quantitative theory of constraints such as those devised by Phillips[28] and Thorpe[29] for covalent amorphous solids. One measure might be the average metal atom connectivity, which is given by the number of capped trigonal prisms meeting at each vertex, this value being three for the amorphous Ni_3B discussed earlier. Alternatively we could introduce a scale of very hard, hard, and soft constraints related, say, to edge linkage, vertex linkage, and shared capping atoms. On this basis a glass of composition Ni_3B, with an overall metal connectivity of three and with two thirds of the metals involved in very hard edge linkages and one third with hard vertex linkage, should lie at the constrained end of the spectrum and therefore near the edge of glass formation. On the other hand, a glass of composition Ni_9B with an average connectivity of one and all atoms involved in only very soft metal–metal linkages must lie at the unconstrained end of the spectrum of glass formation. These two might therefore bracket the range of glass formation, but clearly the situation is not quite so straightforward. The simple bond-counting argument assumes that the bond strength is insensitive to composition, and this is unlikely to be correct. Such variations enter through the term $\Delta\mu$ in the Adam–Gibbs relationship. A liquid with a high value of $\Delta\mu$ will jam at relatively small values of z^*.

THE OBSERVATION OF MEDIUM-RANGE STRUCTURES IN GLASSES

If we wish to look for evidence of pronounced medium-range structure in glasses, where should we look? The arguments of the previous section suggest that the region of composition corresponding to highly interconnected structures provides the surest

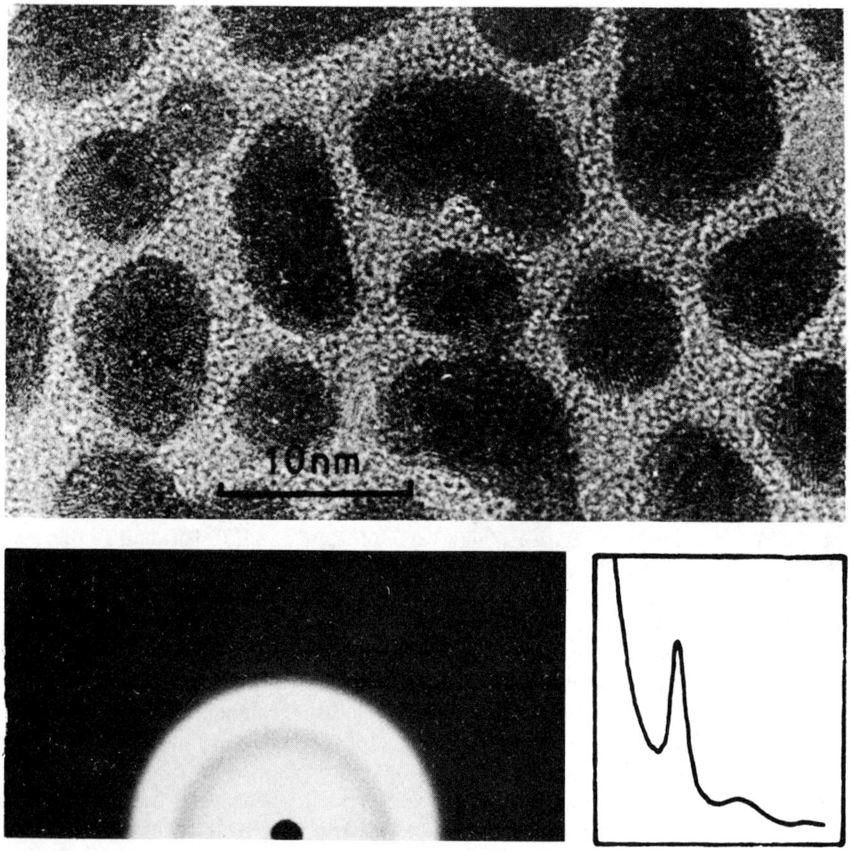

FIGURE 7. High-resolution transmission electron micrograph of an amorphous Pd–Si alloy prepared by flash evaporation. Particles of the amorphous metal are supported on an insulating film. Near the edges of several of the particles, groups of lattice fringes are seen. Lower diagrams show electron diffraction from this specimen and a trace of the relative intensity as a function of the scattering angle. For further details see Gaskell and Smith.[30]

possibility. Specifically, well-annealed glasses with strong nearest-neighbor forces (not necessarily directional) leading to well-defined local structural units effectively constrain the number of available medium-range structures. Although this statement has not been confirmed by any planned series of experiments as a function of composition, it is consistent with the evidence from the neutron scattering studies mentioned earlier.

Materials quenched from liquids at the constrained end of the spectrum are likely to crystallize, however, so that the prospects for producing a well-annealed, defect-free glass necessarily decrease as the constraints multiply.

The classical experiments of Turnbull[26] on quenching small liquid droplets suggest an answer, though. Crystalline nuclei that form within an isolated droplet cannot grow beyond the surface and can be discounted. The structure of the remaining, noncrystalline droplets should therefore represent an approximation to maximum ordering at that composition and, by the arguments given above, the combination of experimental conditions that is most likely to induce medium-range structure in that glass-forming system. Some support for this comes from the high-resolution electron micrographs of thin Pd–Si alloy films produced by flash evaporation of an Si-rich alloy.[30] Droplets of the amorphous alloy are formed on an insulating substrate (probably SiO_2, formed by oxidation of Si by residual O_2). It is possible to produce films in which some droplets have extensive crystal-like lattice fringes; in others lattice fringes are confined to relatively small regions and with a considerable number of (not obviously correlated) fringe systems constituting each particle, some being shown in FIGURE 7. The latter are thus qualitatively different from polycrystalline particles. Such direct images suggest that medium-range order is certainly more prominent in this alloy (and similar work on other metallic alloys corroborates the statement) than those seen in amorphous semiconductors and oxide glasses with lower connectivity.

The converse of this is not necessarily true; that is, under-constrained glasses, or those of the central region corresponding to good glasses, are not necessarily devoid of medium-range structure. The point is that there is no built-in scaling parameter in the arguments presented in the previous section. Although $\Delta\mu$, for example, can be thought of as the energy or chemical potential associated with an elementary motion of a single segment in a correlated sequence of such motions, and one normally thinks in terms of atomic motions as the elementary step, there is no *a priori* reason for this to be so. The argument would work equally well with an elementary motion involving a group of atoms, large or small. Consequently, although it seems less likely that medium-range structures exist in good glasses such as SiO_2, we cannot exclude the possibility.

Medium-range structure might be expected, even in good glasses, if the precursor supercooled liquids are thought to contain structural units with noncrystallographic or curved-space structures. Again, such clusters can be small aggregates of atoms such as the "vitrons" of Tilton[31] or the "amorphons" of Grigorovici and Manaila[32] or the large fivefold arrangements of trigonal prisms described by Gaskell,[19] Janot *et al.*,[33] and Sadoc and Mosseri.[34]

REFERENCES

1. WEAIRE, D. & F. WOOTEN. 1985. *In* Physics of Disordered Materials. D. Adler, H. Fritzche & S. R. Ovshinsky, Eds.: 1. Plenum. New York, NY.
2. BHATIA, A. B. & D. E. THORNTON. 1970. Phys. Rev. B **2:** 3004.
3. WARNER, B. E. X-ray Diffraction. Addison–Wesley. New York, NY.
4. COWLEY, J. M. 1950. Phys. Rev. **77:** 667.
5. CARGILL, G. S. & F. SPAEPEN. 1981. J. Non-Cryst. Solids **43:** 91.

6. THEYE, M. L., A. GHEORGHIU & H. LAUNOIS. 1981. J. Phys. C **13:** 6569.
7. BRESSER, W. J., P. BOOLCHAND, P. SURANYI & J. P. DENEUFVILLE. 1981. Phys. Rev. Lett. **46:** 1689.
8. MOZZI, R. L. & B. E. WARREN. 1969. J. Appl. Crystallogr. **2:** 164.
9. GRISCOM, D. L., E. T. FRIEBELE, G. H. SIGEL & R. J. GINTHER. 1977. *In* Structure of Non-Crystalline Materials. P. H. Gaskell, Ed.: 113. Taylor & Francis. London.
10. GASKELL, P. H. & I. D. TARRANT. 1980. Philos. Mag. **42:** 265.
11. YARKER, C. A., P. A. V. JOHNSON, A. C. WRIGHT, J. WONG, R. B. GREEGOR, P. W. LYTLE & R. N. SINCLAIR. 1986. J. Non-Cryst. Solids. **79:**117.
12. GRIMMER, A. R., M. MAGI, M. HAHNERT, H. STADE, A. SAMOSAU, W. WIEKER & E. LIPMAA. 1984. Phys. Chem. Glasses **25:**105.
13. DUPREE, R., D. HOLLAND, P. W. MCMILLAN & R. F. PETTIFER. 1984. J. Non-Cryst. Solids. **68:** 399.
14. GASKELL, P. H. 1983. *In* Glassy Metals II. H. Beck & H. J. Guntherodt, Eds.: 4. Springer–Verlag. Berlin.
15. LAMPARTER, P., W. SPERL, S. STEEB & J. BLÉTRY. 1982. Z. Naturforsch. **37A:** 1223.
16. PANNISOD, P., I. BAKONYI & R. HASEGAWA. 1983. Phys. Rev. B **28:** 2374.
17. GASKELL, P. H. 1981. Nature **284:** 474.
18. LAMPARTER, P. & S. STEEB. 1985. *In* Rapidly Quenched Metals. V. S. Steeb & H. Warlimont, Eds.: 459. Elsevier/North-Holland. Amsterdam.
19. GASKELL, P. H. 1985. J. Non-Cryst. Solids **75:** 329.
20. GASKELL, P. H. 1981. Acta Metall. **29:** 1203.
21. DUBOIS, J.-M., P. H. GASKELL & G. LE CAER. 1985. Proc. Roy. Soc. London Ser. B **402:** 323.
22. GASKELL, P. H. 1979. J. Non-Cryst. Solids **32:** 207.
23. DENT-GLASSER, L. S. 1979. Z. Kristallogr. **149:** 29.
24. TURNBULL, D. 1985. J. Non-Cryst. Solids **75:** 197.
25. BEYER, O. & C. HOHEISEL. 1983. Z. Naturforsch. **38A:** 859.
26. TURNBULL, D. 1969. Contemp. Phys. **10:** 473.
27. ADAM, G. & J. H. GIBBS. 1965. J. Chem. Phys. **43:** 139.
28. PHILLIPS, J. C. 1979. J. Non-Cryst. Solids **34:** 153.
29. THORPE, M. F. 1984. J. Non-Cryst. Solids **57:** 250.
30. GASKELL, P. H. & D. J. SMITH. 1980. J. Microsc. (Oxford) **119:** 63.
31. TILTON, L. W. 1957. J. Res. Natl. Bur. Stand. **59:** 139.
32. GRIGORIVICI, R. & M. MANAILA. 1967/68. Thin Solid Films **1:** 343.
33. JANOT, C., G. PIECUCH, D. MARCHAL, D. TEIRLINCK & M. VERGNAT. 1983. *In* The Structure of Non-Crystalline Materials. P. H. Gaskell, J. M. Parker & E. A. Davis, Eds.: 234. Taylor & Francis. London.
34. SADOC, J. F. & R. MOSSERI. 1984. J. Non-Cryst. Solids **61** & **62:** 487.

Pressure Melting below the Glass Transition

A New Way of Making Amorphous Solids[a]

EDWARD WHALLEY, OSAMU MISHIMA,[b]
Y. PAUL HANDA, AND D. D. KLUG

Division of Chemistry
National Research Council of Canada
Ottawa, Canada K1A 0R6

INTRODUCTION

Amorphous solids have been made in many ways, some of which may be classed as thermodynamic, in that equilibrium is at least approximately maintained during the amorphization, and some of which may be classed as essentially irreversible. Among the irreversible processes we may include methods in which a crystal is pumped to a disordered state by particle bombardment, grinding, or some other related method. The principal thermodynamic methods are the cooling, more or less rapidly, of a liquid, which was discovered in distant antiquity and has allowed us to let the light into our houses while keeping the cold out, and the condensation of a vapor at a low-enough temperature, which was invented by Tammann and Starinkevitsch in 1913.[1] A possible method that is missing from this list is the direct thermodynamic conversion, or "melting," of the crystal to the glass.

It is missing because all or almost all crystals melt to a fluid rather than "melt" to a glass at their zero-pressure melting point. There seems to be no fundamental reason, however, why a crystal and a glass cannot be in thermodynamic equilibrium, and perhaps some silicate or polymer "melt" will be a glass at the ordinary melting temperature. For most substances, however, the viscosity decreases with increasing temperature along the melting curve, and so solids "melting" to glasses would be very common if they could take high negative pressures. For example, glycerol was estimated to "melt" to its glass at a negative pressure of a few tens of kilobars.[2] Unfortunately, materials are usually too weak, and fracture before such pressures are reached.

For several crystalline solids, however, the melting temperature falls as the pressure increases because the volume decreases on melting. All these falling melting lines reach a triple point where another solid phase becomes stable, although there seems to be no reason in principle why this must happen. Nevertheless, as the melting line is a first-order line, it exists beyond the triple point as a metastable line, and must, at least in principle, exist until zero temperature. If such a solid can be compressed at a

[a]National Research Council of Canada No. 26147.

[b]Present address: National Institute for Research in Inorganic Materials, 1-1 Namiki, Sakura-mura, Niihara-gun, Ibaraki 305, Japan.

low-enough temperature to prevent transformation to another crystalline phase, it must eventually melt to the liquid, and, if the liquid is below its glass transition, it should "melt" to a glass. Because the "melting" would occur at high pressures, the resulting "melt" should be much denser than the ordinary amorphous forms made by quenching the liquid or condensing the vapor.

The project began when Mishima, who worked in our laboratory for a few years, asked what would happen if ice I were compressed at 77 K. It was a brilliant question! He had been a student of Kawai's, and Kawai had argued that melting lines on a phase diagram like the phase diagram of ice in FIGURE 1 are usually concave downward, and that they should always reach a maximum temperature and then fall.[3] This is probably true for solids that have a low coordination number, like ice III and some other phases of ice as long as they do not transform to a denser phase, but may not be true in general. Mishima's question was inspired by Kawai's argument, and, once he asked it, the answer was obvious: either the crystal "melts" at ~10 kbar to a glass, because 77 K is well below the glass transition of water at zero pressure and almost certainly still well below it at 10 kbar, or becomes a crystal that is metastable relative to its own liquid. It seemed more likely that the crystal would "melt," and so the experiment was done.

THE CRYSTAL-GLASS TRANSITION: "MELTING" BELOW THE GLASS TRANSITION

The crystal-glass transition was sought experimentally by squeezing ~1.2 g of ice I held in an indium cup in a steel piston-cylinder apparatus having an inside diameter of

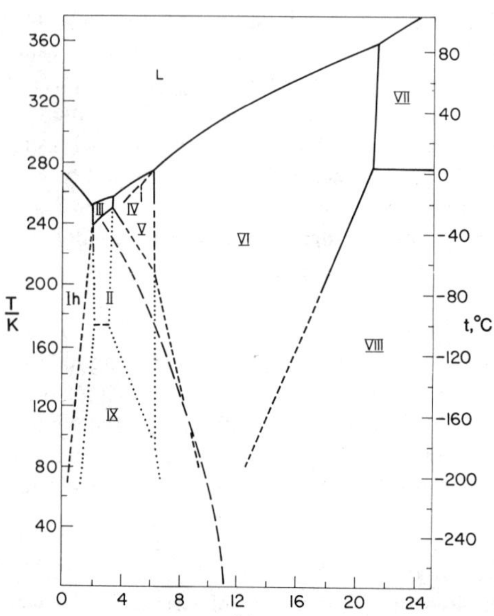

FIGURE 1. Phase diagram of ice in the region between 0 and 380 K and between 0 and 25 kbar. The liquid I line is extrapolated to zero temperature as the heavy dashed line.

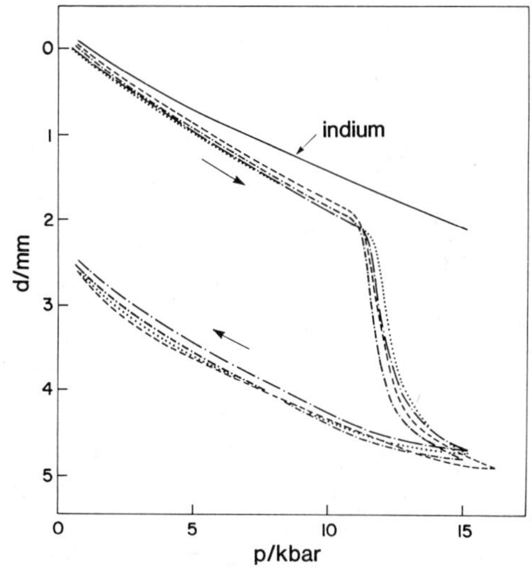

FIGURE 2. The displacement d of the pressurizing piston during four independent compressions of ice I at 77 K plotted against the nominal pressure p, without allowing for friction. The compression of indium having the same volume as the sample plus cup is also plotted and was used to convert the displacement to the change of volume of the sample using the known compression of indium in this range.

12.5 mm.[4] The displacement d of the piston in four independent runs is plotted in FIGURE 2, along with the compression of indium as a calibration.[5] Below ~10 kbar the ice compresses elastically, as was verified by measuring decompressions starting at various pressures. At a real pressure of 10 kbar a sudden and large change of volume occurs, which is about two-thirds complete within 0.7 kbar and is quite complete by 15 kbar, as FIGURE 2 shows. During the transformation the volume decreases by about 24%. On removing the pressure, the transformation does not reverse, and the transformed sample can be recovered at 77 K, having expanded only elastically from its state at 10 kbar. Both the permanent compression that is left at zero pressure and the buoyancy of the sample in liquid nitrogen show that the density is 1.17 g · cm^{-3}, which is 25% greater than the density of the original ice I.

The nature of the new phase was determined by X-ray diffraction at 95 K after heating to several temperatures, and microphotometer traces[4] are plotted in FIGURE 3. The diffraction pattern labeled 95 K in FIGURE 3 was taken at that temperature without heating the sample to higher temperatures. It is very broad, and the sample is undoubtedly amorphous. After heating to 112 K and to 135 K and cooling to 95 K, the pattern sharpens, and that obtained after heating to 135 K resembles the pattern of the usual form of low-density amorphous ice, or low-density amorph (lda in the figures), that is made by condensing the vapor onto a cold surface.[6]

The relation of the specific volume of the new phase at 10 kbar to the specific volume of the liquid at the same pressure, as extrapolated from measurements in the temperature range 275–350 K, is shown by line c in FIGURE 4. The specific volume of the liquid on the liquid I line is extrapolated in the same way. The specific volume of the high-density amorphous ice, or high-density amorph (hda in the figures), is certainly consistent, within the uncertainties of the extrapolations in FIGURE 4, with it being the specific volume of the "liquid" at 77 K and 10 kbar.

Ice I transforms to the high-density amorph at 77 K and 10 kbar so readily compared with other transitions in ice at these conditions that the amorph is probably not formed by nucleation within the sample. Rather the transformation resembles a

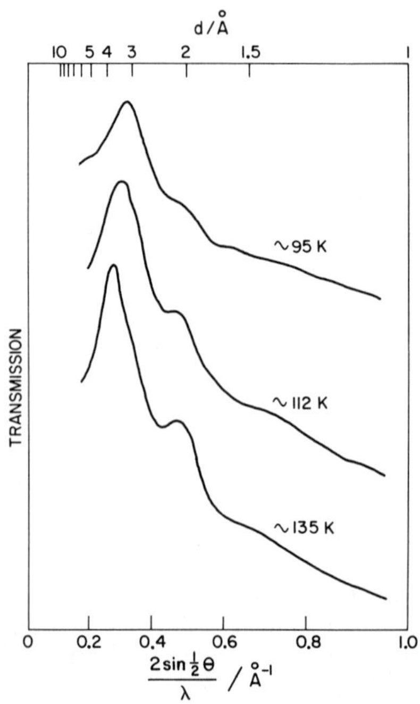

FIGURE 3. Representative microphotometer tracings of the diffraction patterns of the phase made by compressing ice I to 10 kbar at 77 K. The samples were mounted in a flat-plate powder camera and were cooled to ~95 K with flowing nitrogen gas. The sample had been heated to the temperatures attached to the curves.

"melting" transition, and probably occurs when the surface of the crystal becomes unstable and collapses to the "melt." Solids cannot usually be superheated, or superpressed, into the region of their melt when the melt is a liquid. It seems now that they cannot be taken into the region of their melt even when the "melt" is a glass. The reverse transition, from the "melt" to the crystal, almost certainly occurs by nucleation. But nucleation at these temperatures is very slow, and so the transition is not reversible on normal laboratory time scales.

The thermal effect of the transformation has been investigated[7] in more detail in an

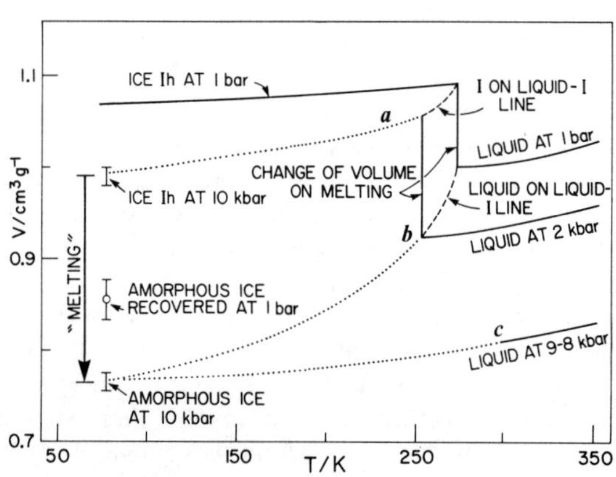

FIGURE 4. Graph of the specific volume of ice I and liquid water on the liquid-to-I line, labeled a and b, respectively, and of the liquid at 9.8 kbar, labeled c. The specific volumes at 10 kbar and 77 K of ice Ih and the high-density amorph are also plotted.

automated Tian–Calvet heat-flow calorimeter.[8] The result of a scan through the range from high-density amorph to hexagonal ice I, or ice Ih, at 10 K · hr^{-1} is drawn in FIGURE 5. For our interests today, the most important part is the first peak, which is caused by the transformation from the high-density to the low-density amorph. There is a clear exothermic effect starting at ~85 K, at which 224 and 231 J · mol^{-1} was slowly evolved in two independent runs until the temperature reached ~105 K. The temperatures are, of course, not precisely defined points and varied somewhat from run to run.[7] The heat evolution is attributed to a slow annealing of the high-density amorph to a less-dense form. A sharp exothermic peak starts at temperature T_i, which is taken at ~105 K, and finishes at temperature T_f, which is about 123 K, and 515, 535, and 544

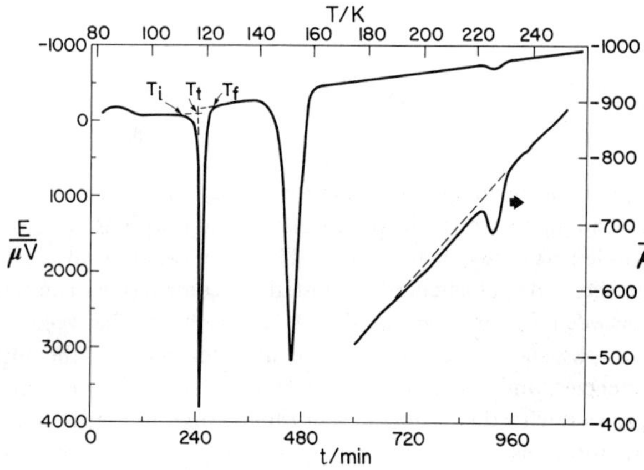

FIGURE 5. The voltage generated by the differential thermopile during the heating of a sample of the initially high-density amorph at the rate of 10 K · h^{-1}. The upper abscissa represents the temperature of the sample corresponding to the time, which is linear on the lower abscissa. The left-hand ordinate belongs to the principal curve, and the right-hand ordinate belongs to the right-hand part of the curve that is plotted on an expanded scale. The first peak represents the exothermic effect of the transition from the high- to the low-density amorph. The second and third peaks represent, respectively, the exothermic effect of the transition from the low-density amorph to ice Ic and from ice Ic to Ih.

J · mol^{-1} was evolved in these runs. It is undoubtedly caused by the remaining part of the transition from the high-density to the low-density amorph. The two exotherms at higher temperature are caused by the transition from the low-density amorph to cubic ice I, or ice Ic, and from ice Ic to ice Ih, respectively. The heat evolved at the transition from the low-density amorph to ice Ic in three independent runs was 1363, 1376, and 1425 J · mol^{-1}. The differences from run to run are probably real and may be due to incomplete and variable conversion of ice I to the high-density amorph.[4] This explanation is consistent with the equality of the ratio of the heats of transformation from the low-density amorph to ice Ic and from the high-density to the low-density amorph, which is 0.375 ± ~0.008, and the uncertainty is about the same as the consistency of the measurements described by Handa *et al.*[8]

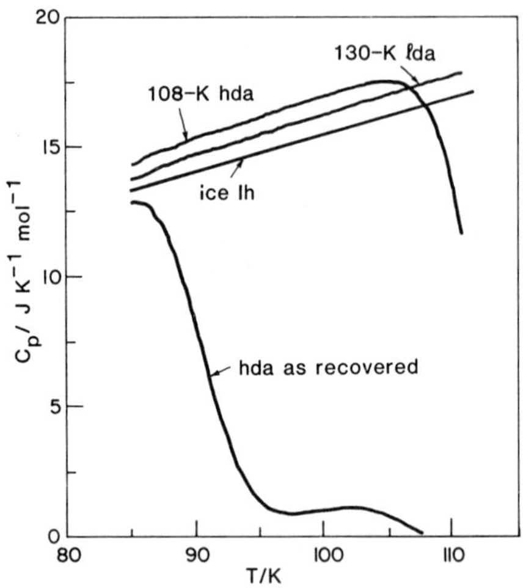

FIGURE 6. The apparent heat capacity of the high-density amorph as recovered at 77 K, showing the exothermic effect of the annealing, and the heat capacities of a sample of the high-density amorph that had been heated to 108 K, of a sample of the low-density amorph that had been heated to 130 K, and of ice Ih.

The heat capacity of the high-density amorph, after it has been heated to 108 K, is plotted in FIGURE 6 along with the heat capacity of the low-density amorph made by heating the high-density amorph to 130 K. The heat capacity of ice Ih measured independently in the same calorimeter[8] is plotted for comparison. The increase in the heat capacity between Ih and the low-density amorph and between the low- and high-density amorphs shows clearly that the disorder reduces the intermolecular vibrational frequencies, and hence the intermolecular force constants, and the Debye frequencies agree well with the frequencies of the maxima in the Raman spectrum of the translational vibrations.[7]

THE GLASS-GLASS TRANSITION: "MELTING" OF THE LOW-DENSITY AMORPH

If ice I can "melt" to a high-density amorphous solid at 77 K and 10 kbar, it probably does so because the surface becomes unstable. The low-density amorph may do the same because it has a similar structure, although a more disordered one.

The relation between the Gibbs functions of ice I and the low- and high-density amorphs is as shown qualitatively in FIGURE 7, because the specific volumes, which are the slopes of the curves, are higher for the low-density amorph and ice I than for the

FIGURE 7. Sketch of the relations between the Gibbs function of the low- and high-density amorphs of ice and of ice Ih.

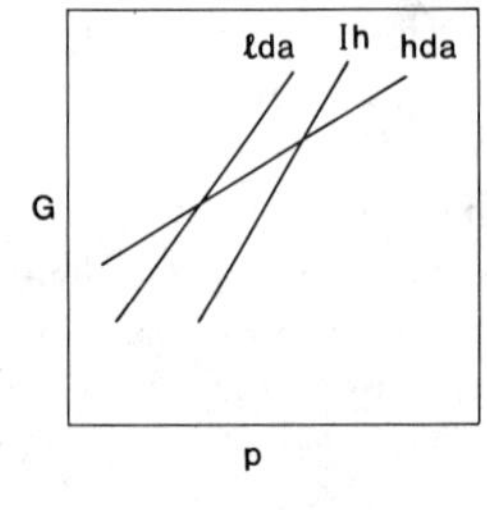

high-density amorph, and because the low-density amorph is metastable relative to ice I. The transition from the low- to the high-density amorph, if it occurs, should therefore occur at a lower pressure than the transition from ice I to the high-density amorph.

This was tested by Mishima *et al.*,[9] who first made the high-density amorph by squeezing ice I in an indium cup at 77 K and ~15 kbar. The pressure was reduced to zero, the apparatus was heated to ~132 or ~136 K, and the transformation to the low-density amorph was followed by the piston displacement. When the transformation was complete, the sample was cooled to 77 K and pressurized. The displacement of the piston is plotted against the pressure in FIGURE 8. There is a sharp transition at 6 kbar to a denser phase, and further compression to 18 kbar causes only elastic compression. The density of the new phase at 6 kbar, as determined from the piston displacement of the sample and of an equal volume of indium, is 1.26 ± 0.02 g · cm^{-3}, and the density at zero pressure is $1.19 \pm \sim 0.02$ g · cm^{-3}. The density at zero pressure is the same (within the experimental uncertainty) as that of the phase made by melting

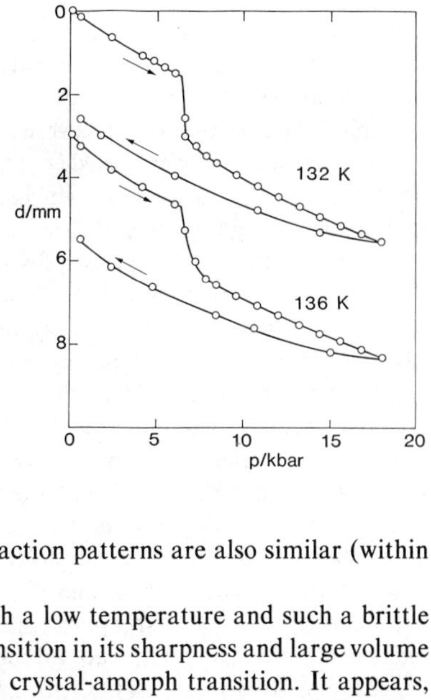

FIGURE 8. Compression of two samples of the low-density amorph at 77 K. The samples had been heated to various temperatures, as shown. The arrows show the direction of pressure changes.

ice I at 77 K and 10 kbar, and the X-ray diffraction patterns are also similar (within the experimental resolution).[4,9]

The transition is remarkably sharp for such a low temperature and such a brittle material and clearly resembles a first-order transition in its sharpness and large volume change. It is, however, not reversible, like the crystal-amorph transition. It appears, therefore, to resemble a first-order transition that goes easily in the forward direction, perhaps because the low-density amorph becomes unstable at its surface, but slowly in the reverse direction, perhaps because nucleation of the original phase is difficult at so low a temperature.

The difference at zero pressure of the Gibbs functions of two phases that are in equilibrium at pressure p_e is

$$\Delta G(p = 0) = -\Delta V(p = 0)p_e$$

if the difference of compression of the two phases is neglected. From the observed transformation pressures of ice I and the low-density amorph to the high-density amorph, the differences of Gibbs function at zero pressure and 77 K are

$$\Delta_{\text{Ih}}^{\text{hda}} G(p = 0) = 4.1 \pm \sim 0.3 \text{ kJ} \cdot \text{mol}^{-1}$$

and

$$\Delta_{\ell\text{da}}^{\text{hda}} G(p = 0) = 2.5 \pm \sim 0.3 \text{ kJ} \cdot \text{mol}^{-1}$$

so that

$$\Delta_{\text{I}}^{\ell\text{da}} G(p = 0) = 1.6 \pm \sim 0.3 \text{ kJ} \cdot \text{mol}^{-1}$$

where I means ice I.

RAMAN SPECTRUM

The Raman spectrum of the high-density amorphous phase and the phases it transforms to on heating has been obtained.[10] A 5 mol % solution of H_2O in D_2O and of D_2O in H_2O were frozen in indium cups containing a copper channel[11] and compressed to ~16 kbar, which is well beyond the 10 kbar of the transformation to the high-density amorph.[4] The channels were recovered under liquid nitrogen and cleaned of extraneous ice. The ice surfaces were polished smooth, and the channel was mounted in an Air Products Displex closed-cycle refrigerator and its temperature measured with a thermocouple of gold-iron and chromel that was attached to the base of the channel.

A few spectra of the uncoupled O-D bands of a 5 mol % solution of D_2O in H_2O amorphous ice were recorded at 12 K after heating the samples to various temperatures, as shown in FIGURE 9. The sample as recovered at 77 K has a broad band that is asymmetric of the high-frequency side. The Raman intensity of the uncoupled O-D band is probably not greatly affected by the hydrogen bonding, in contrast to the infrared intensity, which is strongly affected. The intensity of the Raman band at any frequency is, therefore, approximately proportional to the density of oscillations at that frequency. Furthermore, although the frequency is affected by many interactions, the O-O distance probably has the strongest effect by far. The next strongest effect is that of the O-H---O angle, but, as the effect of bending this angle is even in the bending, the lowest term is quadratic in the angle and can probably be neglected in a first approximation. In a first approximation, therefore, the relative intensity of the uncoupled band at any frequency measures the relative intensity of O-H oscillators and hence the relative density of O-O distances. The large band width tells us that there is a broad distribution of distances, and the greater intensity at high than at low frequency tells us that bonds that are longer than average are more common than bonds that are shorter than average. This is, of course, just what would be expected because of the negative anharmonicity of O-O bonds; it is easier to stretch bonds than to compress them. Thus, there are more stretched than compressed bonds.

The uncoupled O-H frequency of water molecules has recently been measured over the pressure range 20–189 kbar or a range of O-O distances of 2.6–2.9 Å.[12] As the O-O distance goes to infinity, the frequency of the uncoupled O-H oscillator goes to the vapor value for HDO, that is, 3707.5 cm^{-1}. The frequencies[12] can then be represented as a function of O-O distance by the equation

$$\nu_{OH}/\mathrm{cm}^{-1} = 3707.5 - 3409.4 \exp(-R/1.12\ \text{Å}) + 683490 \exp(-R/0.896\ \text{Å}) \qquad (1)$$

The Raman frequencies of the O-H oscillators in the amorphous phases can be scaled to those of O-D oscillators by dividing them by 1.353.[13–16]

The peak frequency of the upper curve in FIGURE 9 corresponds to an O-O distance of 2.82 Å; the frequencies at half the peak height correspond to 2.73 and 2.93 Å; and the extreme frequencies of the band, which are somewhat arbitrarily defined, correspond to about 2.7 and 3.1 Å. The mean O-O distance in ice I at the same temperature is 2.75 Å, and the orientational disorder causes a distribution of distances

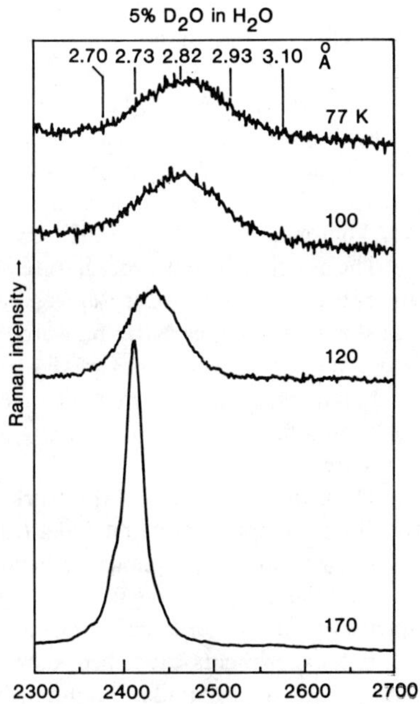

FIGURE 9. Raman spectrum at 12 K of a 5 mol % solution of D_2O in the H_2O high-density amorph after heating to various temperatures, as shown.

having a half-width of a few hundredths of an angstrom.[13] In the high-density amorph the mean distance is about 0.07 Å longer, although the density is about 25% greater, and the half-width is about 0.2 Å. The increased density is no doubt caused by an increased first-neighbor coordination number, which itself causes an increased O-O distance, and the increased disorder causes the increased half-width. A full discussion is being prepared.

When the sample is heated to 100 K at 5 K · min^{-1}, kept there for 5 min, and then cooled to 12 K, slight changes occur, and the band narrows greatly when it is heated to 120 K where the low-density amorph is formed. The density of this 120 K phase is 0.94 ± 0.02 g · cm^{-3},[4] and it is undoubtedly closely related to the low-density amorph. The spectrum of the low-density amorph confirms this.[17]

IMPLICATIONS AND PROSPECTS OF THE NEW AMORPHOUS FORMS

As was mentioned in the Introduction, both fluid phases of matter, liquid and vapor, have been converted directly to glasses by thermodynamic methods, the liquid since distant antiquity and the vapor since over 70 years ago. Now, both solid phases of matter, crystalline and amorphous, have also been directly converted to amorphous phases by transitions that resemble first-order transitions.

Until now, only a limited range of amorphous phases of ice have been made, and that only by condensing the vapor at low temperature[18] or by rapidly quenching the liquid.[19] A review of the properties of these phases has been published.[17] A small range of properties can apparently be obtained by varying the rate of deposition of the vapor and the temperature of the substrate, and by varying the rate of quenching of the liquid. But the range of properties that can be obtained is much smaller than can be obtained from our high-density amorph, which can be transformed controllably to amorphous phases having any density, at zero pressure, between 1.17 and 0.94 $g \cdot cm^{-3}$. Phases having the same density but different properties may be made by annealing for suitable times at various temperatures. In addition, amorphous phases of even higher density might also be obtained by compressing ice I or a low-density amorph at even lower temperatures than 77 K. Near zero temperature a pressure of $\sim$12 kbar may be needed for the transformation.

The apparently first-order nature of the transition from the low- to the high-density amorph suggests that we try squeezing amorphs of various densities to see if they also transform by an apparently first-order transition even when the difference in density between the phases becomes small.

Finally, several measurements need to be made on the high-density amorphs, the most important measurements being X-ray and neutron diffraction. Both projects are in progress.

The principal findings of this work may be summarized as follows: 1) A new kind of transition, an apparently first-order transition from a crystal to a high-density amorph. 2) A new form of amorphous ice having a high density. 3) A range of amorphs of ice covering the density range 0.94–1.17 $g \cdot cm^{-3}$. 4) A second new kind of transition, an apparently first-order transition from a low- to a high-density amorph.

The experiments described show that both crystalline ice and the low-density amorph "melt" to similar high-density amorphs when squeezed at 77 K and zero pressure. It seems reasonable to speculate that other crystals will behave in the same way. The congruent melting point of tetrahydrofuran clathrate hydrate also falls under pressure,[20] and so should "melt" to an amorphous form at low temperature. Several samples have been compressed to $\sim$15 kbar at 77 K. They undergo a transition at 10 kbar (O. Mishima and E. Whalley, unpublished results) similar in steepness and volume change to that of ice I, and the transformed phase can be recovered at 77 K and zero pressure. But its X-ray diffraction pattern taken in a cold stream of gas at $\sim$85 K was that of the original clathrate. The high-pressure phase is, therefore, quite unstable relative to the original clathrate and readily transforms to it. Solid-solid reactions are probably very slow at these temperatures, so a phase separation of the ice and the tetrahydrofuran is unlikely to have occurred at the transformation. Therefore, a phase that is either crystalline or amorphous, but has the same composition as the original clathrate, was formed under pressure. The transformation of the clathrate to the high-density form is sharp enough to be a melting transition, and more experiments are being done in a low-temperature anvil apparatus.

Other materials whose melting temperature falls when the pressure increases should behave in the same way if other crystalline phases do not intervene. Obvious examples are ammonium fluoride I,[21] indium antimonide,[22] and germanium,[23] which may transform at 77 K to an amorphous phase when squeezed to ~20, ~50, and ~170 kbar, respectively. Other possible candidates for amorphization in this way are graphite, diamond, silicon, cubic and hexagonal boron nitride and phosphide, and graphite-like boron nitride. A transformation to a dense amorphous phase may limit the maximum force that can be applied to diamond, and so limit the pressures obtainable in diamond-anvil apparatuses.

The transformations of ice I and the low-density amorph may be important in the planetary system. Some of the outer planets were probably made by the agglomeration of small particles of ice I or, more likely, its low-density amorph.[24] If this ice agglomerates to form a planet, the high-density amorph will form at the center when the planet has grown to a radius of 2 Mm.

SUMMARY

The melting temperature of some solids, such as ice, falls as the pressure rises, and the melting line usually reaches a triple point, at which a new solid phase appears. The melting line is first order, however, and must in principle exist beyond the triple point as a metastable line, and can stop only at zero temperature. The viscosity of the fluid in equilibrium with the solid usually increases as the equilibrium temperature falls, and it will usually become very high as the so-called glass transition of the liquid is passed. Below this temperature the solid may exist, in principle, in equilibrium with the glass. This suggests a new way of making amorphous solids, by "melting" the crystal below the glass transition of the melt. A new kind of high-density amorphous ice, of density 1.17 $g \cdot cm^{-3}$, has been made in this way. When heated, the high-density amorph changes gradually to lower density, and a continuous range of amorphs having densities between 1.17 and 0.94 $g \cdot cm^{-3}$ can be made. The new amorphs have been studied by X-ray diffraction, calorimetry, and Raman spectroscopy.

REFERENCES

1. TAMMANN, G. & J. STARINKEVITSCH. 1913. Z. Phys. Chem. **85:** 573–579.
2. JOHARI, G. P. & E. WHALLEY. 1972. Faraday Symp. Chem. Soc. **6:** 23–41.
3. KAWAI, N. & Y. INOKUTI. 1970. Jpn. J. Appl. Phys. **9:** 31–48.
4. MISHIMA, O., L. D. CALVERT & E. WHALLEY. 1984. Nature **310:** 393–395.
5. P. W. BRIDGMAN. 1945. Proc. Am. Acad. Arts Sci. **76:** 9.
6. NARTEN, A. H., C. G. VENKATESH & S. A. RICE. 1976. J. Chem. Phys. **64:** 1106–1121.
7. HANDA, Y. P., O. MISHIMA & E. WHALLEY. 1986. J. Chem. Phys. **84:** 2766–2770.
8. HANDA, Y. P., R. E. HAWKINS & J. J. MURRAY. 1984. J. Chem. Thermodyn. **16:** 623–632.
9. MISHIMA, O., L. D. CALVERT & E. WHALLEY. 1985. Nature **314:** 76–78.
10. KLUG, D. D., O. MISHIMA & E. WHALLEY. 1986. Physica **139** & **140B:** 475–478.
11. WONG, P. T. T. & E. WHALLEY. 1972. Rev. Sci. Instrum. **43:** 935–937.
12. KLUG, D. D. & E. WHALLEY. 1984. J. Chem. Phys. **81:** 1220–1228.
13. BERTIE, J. E. & E. WHALLEY. 1964. J. Chem. Phys. **40:** 1637–1645.

14. IKAWA, S.-I. & S. MAEDA. 1968. Spectrochim. Acta **24A:** 655–665.
15. FORD, T. A. & M. FALK. 1968. Can. J. Chem. **46:** 3579–3586.
16. IOGENSEN, A. V. & M. S. ROZENBERG. 1978. Opt. Spectrosc. (USSR) **44:** 49–53.
17. SCEATS, M. G. & S. A. RICE. 1982. Water: A comprehensive treatise. F. Franks, Ed. Vol. **7:** 83–214. Plenum. New York, NY.
18. BURTON, E. F. & W. F. OLIVER. 1935. Proc. Roy. Soc. London Ser. A **153:** 166–172.
19. BRÜGELLER, P. & E. MEYER. 1980. Nature **288:** 569–571.
20. GOUGH, S. R. & D. W. DAVIDSON. 1971. Can. J. Chem. **49:** 2691–2699.
21. KURIAKOSE, A. K. & E. WHALLEY. 1968. J. Chem. Phys. **48:** 2025–2031.
22. MERRILL, L. 1977. J. Phys. Chem. Ref. Data **6:** 1205–1252.
23. CANNON, J. F. 1974. J. Phys Chem. Ref. Data **3:** 798–824.
24. KLINGER, J. 1983. J. Phys. Chem. **87:** 4209–4214.

DISCUSSION OF THE PAPER

M. THORPE (*Michigan State University, East Lansing, MI*): I was trying to speculate, during your talk, on the structure of your high-density amorph. As the nearest O-O distance is roughly constant and as the density increases by 25%, presumably the oxygen coordination to nearest oxygens increases from 4 to 6–8. Is there any evidence for this?

WHALLEY: According to the Raman spectrum of the uncoupled O-D stretching vibrations at 12 K in FIGURE 9 of our paper, the most probable hydrogen-bonded O-O distance is 2.82 Å. As the most probable distance in ice I at the same temperature is 2.75 Å, if the O-O coordination number were to remain constant at 4, the high-density amorph would be about 7% less dense than ice I. As it is about 25% denser, the effective O-O coordination number must increase from 4 to about 5.4.

A similar increase in coordination number occurs in liquid water when it is compressed. The first evidence for this was the Raman spectrum of the uncoupled O-H stretch (Walrafen, G. E., *J. Solution Chem.* **2:** 159, 1973), which shows that the rate of decrease of the O-H---O distance with pressure is much less than one-third of the compressibility (Whalley, E., *J. Chem. Phys.* **63:** 5205, 1975—the value $-$ 0.0030 Å $kbar^{-1}$ quoted in this paper should be 0.002 Å $kbar^{-1}$). This, in turn, shows that the compression is largely accomplished by an increase in the coordination number, and the increase has since been confirmed by both neutron (Wu, A. Y., E. Whalley & G. Dolling, *Mol. Phys.* **47:** 603, 1982) and X-ray (Gabella, G. A. & G. W. Neilson, *Mol. Phys.* **50:** 97, 1983) diffraction. The high-density amorph at 77 K and zero pressure has about the same density as liquid water at room temperature and 6 kbar. Although the two states are in some ways quite different, their coordination numbers seem to be about 5.4 and about 5.1 (Wu *et al., ibid* and Gabella & Neilson, *ibid*), respectively. There is, therefore, a rough consistency between the coordination numbers of the high-density amorph and of liquid water at the same density.

A NAVROTSKY (*Princeton University, Princeton, NJ*): A test of whether the amorphous to dense amorphous ice change is really a first-order transition would be to prove its reversibility. Has this been done, or could one envision a feasible thermodynamic cycle to do so?

WHALLEY: If a first-order transformation is reversible in practice, there must be mechanisms that allow the transformation to go in both directions in laboratory times. Ice I cannot be heated above its melting point by heating its surface, presumably because the superheated surface is unstable and immediately melts. The interior can be superheated by irradiating with infrared light, but then the ice melts as soon as heat is conducted to imperfections. This melting, which is apparently caused by a surface instability, produces artificial Tyndall flowers.

On the other hand, the transformation of pure liquid water to ice I, if no ice I is present, is quite difficult because nucleation must occur. Nucleation is difficult, and, because of this, vast amounts of water as micrometer-sized droplets occur in the upper atmosphere at temperatures as low as –40° C. A transformation is experimentally reversible, therefore, only if there are mechanisms that allow the transformation in both directions to proceed at a reasonable rate. Although the transformation of ice to water goes readily, the reverse transformation is hard to start if no nuclei are present.

The transformation of ice I to the high-density amorph at 77 K goes rapidly at constant pressure. According to experiments done by my colleague M. A. Floriano, using a piston-cylinder apparatus having a diameter of 50 mm, ice I transforms to the high-density amorph at a pressure that is constant to ±5 bar. Very likely, it goes by a surface instability. On the other hand, in order for the high-density amorph to transform back to ice I, a major molecular reconstruction must occur to form nuclei of ice I, and these nuclei must grow by adding molecules. The rate of both of them seems to be exceedingly slow at the low temperatures of the experiment. This is, of course, closely related to the metastability of the high-pressure phases of ice at 77 K and zero pressure even when far from their ranges of stability. The metastability does not prove that the transformations are not first order, but only that the reverse transformations are very slow.

The apparent continuity of the pressure and temperature of the transformation with the pressure and temperature along the melting curve supports the proposal that the transformation is a first order "melting," but does not, of course, prove it.

S. R. NAGEL (*University of Chicago, Chicago, IL*): You have found there to be two different glass phases, one at high pressure (and high density) and one at low pressure (and low density). A phase transition between two amorphous phases has been seen before. It is the transition between a liquid and a vapor that also has a difference (although larger) between the densities in the two phases. In what ways is your transition similar to and different from this other well-known type of transition?

WHALLEY: You are quite right, or course, to point out that first-order transition between amorphous fluid phases, that is, liquid and vapor, have long been known. The principal novelty of our experiments is an apparently first-order transformation, with a large volume change, between two amorphous solids, that is, the low-density and high-density amorphs.

G. M. FAHY (*American Red Cross, Bethesda, MD*): Is there a clear upper temperature limit below −22° C above which pressure-induced melting or vitrification of ice Ih should no longer be possible?

WHALLEY: The transformation from ice I to the high-density amorph at 77 K and 10 kbar is a highly metastable transformation, in the sense that ice II, IX, IV, V, and VI are all more stable than the high-density amorph. As the temperature is increased, the transformation to one or other of these crystalline phases becomes more likely, and eventually the "melting" of ice I no longer occurs. We have not yet tested how high in temperature we could still produce the amorphous solid. It might be more interesting to try the transformation at lower temperatures first, because the melting pressure rises as the temperature falls, and even denser amorphous phases should be produced.

C. T. MOYNIHAN (*Rensselaer Polytechnic Institute, Troy, NY*): You have suggested a first-order phase transition between two amorphous phases, namely, the low-density amorphous (LDA) H_2O and the high-density amorphous (HDA) H_2O. Is it possible that you are really seeing a structural relaxation process, that is, that HDA H_2O is produced in a high free energy state and on depressurization and heating relaxes to LDA H_2O and vice versa? There are some experimental tests that can and should be performed to resolve this question, as well as some additional questions about the nature of the transition that need answering. If it really is a true phase transition, is it accomplished by a nucleation and growth process? You should be able to stop the transition partway, either during heating of the HDA H_2O at 1 atm or during pressurization of the LDA H_2O at 77 K. Can you distinguish one phase from the other in the partly transformed specimen? Likewise, if this is a true phase transition, it should be reversible during pressurization of the LDA H_2O at 77 K if you stop short of complete transformation, so that some of the LDA H_2O phase is left to nucleate the reverse transition on depressurization.

WHALLEY: Your suggestion that if the transformation from the low-density to the high-density amorph is first order there should be a sharp boundary between the two phases in a partly transformed sample, whereas if it were a relaxation there should be a gradual transformation because the pressure is approximately homogeneous, is a good one. We have not done the experiment, but we will.

The transformations of ice Ih and the low-density amorph appear to be of similar kinds for two reasons. One is that ice Ih and the low-density amorph are similar in structure, the low-density amorph being, of course, more disordered, and the other is that the transformations themselves are very similar, the principal difference being the low pressure of the amorph-to-amorph transformation. The low pressure is, of course, to be expected if the transformations are first order, as described in the paper.

Further evidence for the first-order nature of the amorph-to-amorph transformation is the sharpness of the transformation. If it were a relaxation, the volume of activation would have to be $\gtrsim 100\ cm^3 \cdot mol^{-1}$ from the present measurements, which seems rather large for such a low temperature. Experiments are being done to explore the sharpness of the transformation in more detail, and should answer this question.

S. A. RICE (*University of Chicago, Chicago, IL*): Amorphous solid water made by vapor deposition has been observed in two forms: a very common low-density (0.93 $g \cdot cm^{-3}$) form and a rare high-density ($\sim$1.1–1.2 $g \cdot cm^{-3}$) form. The X-ray diffraction data show that the nearest neighbor O-O separation is the same as in ice Ih, namely, 2.76 Å. The changes in density arise from bending of O-O-O bonds. On the other hand, you interpret your data to give a longer near-neighbor O-O separation than ice Ih in both your low- and high-density polymorphs. It seems, if all data are correct, that there are at least four amorphous solid forms in this system. It seems that the free energy surface is very complex and that many more polymorphs will be found.

WHALLEY: There is, in fact, in principle a continuous distribution of amorphs of ice that have densities at zero pressure in the range $1.17 - 0.94\ g \cdot cm^{-3}$. These amorphs can be made by controlled annealing of the high-density amorph. The variables that control the annealing are the temperature and the time, and annealing to the same density using different temperatures and times might produce different amorphs. I expect that amorphs of even higher density can be made by pressure "melting" at lower temperatures. We are preparing to do this.

The most probable hydrogen-bonded O-H---O distance in the high-density amorph at zero pressure and 12 K is 2.82 Å, which is 0.07 Å longer than in ice I. In a naive model, in which the density varies inversely with the cube of the O-H---O distance and linearly with the coordination number, the coordination number of the high-density amorph at zero pressure is ~5.4.

C. A. ANGELL (*Purdue University, West Lafayette, IN*): Is it, by any chance, possible to do the inverse experiment in which the metastability of ice VII at ambient pressure is relieved by the formation of amorphous rather than hexagonal ice at very low temperatures? This should be the analogue of the transformation of stishovite into amorphous silica on careful annealing.

WHALLEY: Our results (Handa, Y. P., D. D. Klug & E. Whalley, *J. Chem. Phys.* transforms rapidly to VIII when cooled below about 0° C. An approximate extrapolation of the melting line of ice VII and an estimate of the melting line of ice VIII suggests that ice VIII would melt at ~200 K at zero pressure if it could be taken there. It would, no doubt, transform to other phases long before then. Ice VIII has often been recovered as a metastable phase at 77 K and zero pressure, and when heated it always transforms to ice Ic at about 125 K (Bertie, J. E., L. D. Calvert & E. Whalley, *Can. J. Chem.* **42:** 1373, 1964. Ice VII was later shown to transform to ice VIII when cooled below 273 K, so the measurements described in this paper were made on ice VIII.)

H. SUGA (*Osaka University, Osaka, Japan*): Could you tell me the value of enthalpy change associated with the irreversible cubic-hexagonal transition? The literature values are diverse in magnitude, owing to their significant contribution from the surface energy effect. Because your specimen is a bulk specimen, as judged from the preparation method, it would be useful to know the value.

WHALLEY: Our results (Handa, Y.P., D. D. Klug & E. Whalley, *J. Chem. Phys.* **84:** 7009, 1986) are not yet completely analyzed, but, tentatively, about 50 J · mol^{-1} is evolved at the transition of ice Ic to Ih at ~200 K when the ice Ic has almost no free surface.

Molecular Dynamics Computer Simulations of the Supercooled Lennard–Jones Liquid[a]

MARCIA H. GRABOW[b,c] AND HANS C. ANDERSEN[d]

[c]*Department of Materials Science and Engineering*
and
[d]*Department of Chemistry*
Stanford University
Stanford, California 94305

Real glass-forming liquids of various types (atomic, molecular, ionic, and polymeric) exhibit some universal or nearly universal properties near and below their glass transition, T_g. These include a drop in the heat capacity and coefficient of thermal expansion at the transition; nonexponential relaxation; nonlinear relaxation; hysteresis and history dependence of macroscopic properties; and temperature-dependent activation enthalpies for transport and structural relaxation. A central theoretical problem in glass science is that of understanding the basis for these phenomena at the atomic and molecular levels.

The molecular dynamics computer simulation methods and the closely related Monte Carlo method have become important theoretical tools for studying glass-forming liquids cooled to low temperatures. These methods have been used to investigate the thermodynamic, structural, and relaxation properties of various types of supercooled liquids,[1–11] and many similarities have been found between the computer simulation results and the properties of real materials in the laboratory.

The time scales of laboratory experiments and those of computer simulations, however, are significantly different. In the laboratory, an experiment on a real glass may last hours, days, or even years. On a computer, even relatively long computer runs correspond to calculating the time evolution of a material for real times on the order of 10^{-10} sec. In the laboratory, the fastest rate at which a liquid can be cooled below its T_g is on the order of 10^7 K/sec whereas, on a computer, the effective quench rate is typically at least five orders of magnitude faster. Angell and Torell[12] have discussed the relationship between the glass transition seen in computer simulations and that seen in the laboratory. A major conclusion of this analysis is that the much shorter time scale of the simulation causes the glass transition to occur at a higher temperature and to be much broader than the corresponding transition in the laboratory.

We have investigated the relationship between the glass transition observed in the laboratory and that observed in a simulation by performing systematic molecular dynamics calculations for a two-component Lennard–Jones material to study its

[a]This work was supported by Grants CHE81-07165 and CHE84-10701 from the National Science Foundation and by the Stanford Center for Materials Research.

[b]Present address: AT&T Bell Laboratories, 600 Mountain Avenue, Murray Hill, New Jersey.

thermodynamic and relaxation properties at and below the glass transition. These calculations are longer than those typically reported and have been repeated to decrease the statistical error of the calculation. Because of the length of the runs, the simulated material can equilibrate at temperatures so low that we can observe the onset of non-Arrhenius temperature dependence to transport. Moreover, by studying a wide range of cooling rates, we can test whether a computer-simulated glass has the universal relaxation properties characteristic of real glasses.

We have performed molecular dynamics computer simulations at constant pressure of a two-component Lennard–Jones alloy, $A_{80}B_{20}$. On cooling from an equilibrium liquid, we find a broad transition from a high-temperature liquid to a low-temperature solid state. We observe many features of glass-forming systems, including quench-rate dependence of certain properties (for example, density and enthalpy); non-Arrhenius

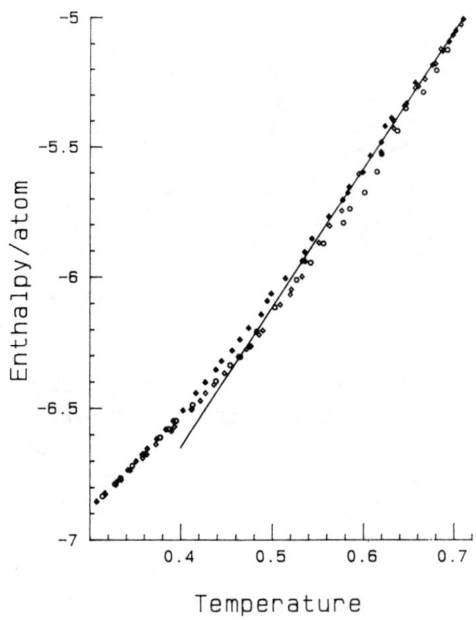

FIGURE 1. A plot of enthalpy/atom versus temperature for cooling followed by reheating (no equilibration at low temperature). Cooling and heating rates, q, in reduced units, were maintained by stochastic collisions. Filled diamonds: cooling at $q = -0.001$ (average of five runs); open diamonds: reheating at $q = +0.001$ (average of two runs); open circles: reheating at $q = +0.004$ (average of two runs); solid line: equilibrium liquid line.

temperature dependence of the self-diffusion coefficient as T approaches the glass transition region on cooling; hysteresis of the enthalpy and density on reheating the glass (FIG. 1). We saw no evidence of crystallization in any of these simulations. We have shown that two thermodynamic parameters (for example, temperature and density) are sufficient to characterize the equilibrium state and determine the other properties of the system (for example, enthalpy) in the high-temperature-liquid region, but three order parameters (temperature, density, and n_{BB}) are still not sufficient to characterize the system in and below the glass transition region.

In addition, we are able to use the Moynihan phenomenological description of low-temperature relaxation[13] (applicable to many laboratory glass transitions) to fit much, but not all, of our cooling and heating data.

Although we were unable to determine the degree(s) of freedom that fall out of equilibrium in the glass transition region, we conclude that the transition that we have observed on the computer at $q \approx 10^{10}$ K/sec has the same characteristics as the glass transition observed in the laboratory for a wide range of materials.

REFERENCES

1. RAHMAN, A., M. J. MANDELL & J. P. MCTAGUE. 1976. J. Chem. Phys. **64:** 1564.
2. MCTAGUE, J. P., M. J. MANDELL & A. RAHMAN. 1978. J. Chem. Phys. **68:** 1876.
3. WENDT, H. R. & F. F. ABRAHAM. 1978. Phys. Rev. Lett. **41:** 1244.
4. CLARKE, J. H. R. 1979. J. Chem. Soc. Faraday Trans. 2 **75:** 1371.
5. ABRAHAM, F. F. 1980. J. Chem. Phys. **72:** 359.
6. WOODCOCK, L. V. 1981. Ann. N. Y. Acad. Sci. **371:** 274.
7. STILLINGER, F. H. & T. A. WEBER. 1979. J. Chem. Phys. **70:** 4879.
8. SOULES, T. F. 1979. J. Chem. Phys. **71:** 4570.
9. BRAWER, S. 1980. J. Chem. Phys. **72:** 4264.
10. FOX, J. R. & H. C. ANDERSEN. 1981. Ann. N.Y. Acad. Sci. **371:** 123.
11. FOX, J. R. & H. C. ANDERSEN. 1984. J. Phys. Chem. **88:** 4019.
12. ANGELL, C. A. & L. M. TORELL. 1983. J. Chem. Phys. **78:** 937.
13. MOYNIHAN, T. C., A. J. EASTEAL, J. WILDER & J. TUCKER. 1974. J. Phys. Chem. **78:** 2673.

DISCUSSION OF THE PAPER

F. H. STILLINGER (*AT&T Bell Laboratories, Murray Hill, NJ*): Since your calculations on the binary alloy model were intended to represent real metal-metalloid mixtures, it might be appropriate to ask about the phase diagram corresponding to your choice of atom-atom potentials. Is this choice consistent with existence of a congruently melting crystal with composition A_3B as commonly observed in real materials? Would you expect to have a deep eutectic near the 80–20% composition used in your simulation?

ANDERSEN: We have not calculated the phase diagram for this potential, and so we do not know if the crystal with composition A_3B is thermodynamically stable for this potential. I should add that in the present calculation the potentials used were motivated by the properties of real metal-metalloid mixtures, but it was not our intention to create an accurate model for these materials.

S. R. NAGEL (*University of Chicago, Chicago, IL*): In your computer simulation of Lennard–Jones glasses you kept the pressure constant and saw evidence of the glass transition in the slope of the enthalpy versus temperature, that is, in $C_p(T)$. Would you have seen the glass transition if you kept the volume constant instead and measured $C_v(T)$?

ANDERSEN: I think that a glass transition would also be seen in constant volume quenches, but we have not studied this.

J. R. STEVENS (*University of Guelph, Guelph, Ontario, Canada*): In discussing the glass transition I would like to separate the effects of the frequency ω of the probe-experimental technique (that is, the ω of the ultrasonic or dielectric excitation, even down to very low ω) from the rate at which the temperature or pressure is changed. For example, in an ultrasonic or Brillouin experiment one obtains a break in the sound velocity-versus-temperature plot that is independent of the probe frequency ω (but dependent on the rate at which the temperature is changed) *and* a higher temperature "transition" characterized by the temperature at which $\omega\tau \sim 1$ where τ is one of a range of relaxation times in the material studied. This later transition depends upon both the probe frequency ω and the rate at which the temperature is changed. In your simulation experiment, can (or are) these two effects separable or are your transitions only comparable to the temperature at which $\omega\tau \sim 1$, where the two effects are combined?

ANDERSEN: In the results presented here, we monitored the glass transition by observing quantities like the enthalpy and volume; such monitoring is not quite the same as probing the system with a disturbance that is characterized by a nonzero value of the frequency. In a sense, our probes are "zero-frequency probes." In the context of your question, I think this means that the two "effects" you refer to are combined in the things we observe.

There is no reason in principle why probes at nonzero frequency could not be used in computer simulation studies. In practice, I suspect that the signal-to-noise ratio for nonzero frequency probes would be larger than for zero frequency, and this might make the transition more difficult to observe.

C. T. MOYNIHAN (*Rensselaer Polytechnic Institute, Troy, NY*): This is a very nice piece of work. I have one comment and two questions. The comment is that if the relaxation time has a highly non-Arrhenius temperature dependence, one expects a higher activation energy from the T_g-versus-cooling rate experiment than from a fit to a relaxation experiment. I. M. Hodge has seen this for some polymers. The first question is whether you have observed any difference in relaxation behavior of enthalpy and volume? Irreversible thermodynamics predicts that they should relax at different rates, although the difference will generally not be too large. Second, have you been able to characterize the microscopic changes accompanying structural relaxation (for example, changes in the number of AB pairs, in coordination numbers, etc.)?

ANDERSEN: We have not observed any difference in relaxation behavior of enthalpy and volume, but we have not performed studies specifically designed to see if there are any differences.

We have not yet been able to characterize the microscopic changes accompanying structural relaxation. In principle, there should be a structural difference between materials prepared by rapid cooling and slow cooling, but the statistical fluctuation in results for a single cooling rate is larger than any systematic variation with cooling rate. It could be that we have not been examining the appropriate structural parameters. With the statistical fluctuation we typically see, it would take very many expensive runs to reduce the noise to a value low enough to see the microscopic basis of the relaxation.

C. A. ANGELL (*Purdue University, Lafayette, IN*): You speak of diffusion and structural relaxation as if they are somewhat different mechanisms based on the different activation energy obtained from a variation of T_g with cooling rate from that for diffusion (extrapolated to the same temperature). Yet you obtain, via a separate route, an agreement between the activation energy for diffusion and that for structural relaxation when you show the reproduction of the enthalpy-versus-temperature plots using the Narayanaswamy–Scherer analysis. Another way of putting your observations would be that there are discrepancies between two ways of evaluating the structural relaxation activation energy and that only one of them, the latter, is in accord with the diffusion activation energy. Don't you think that the success of the second analysis removes the basis for asserting with confidence that diffusion and structural relaxation processes are to be regarded as distinct?

ANDERSEN: In principle, the following three calculations should be describable by the same relaxation theory: 1) T_g (that is, the enthalpy fictive temperature for the low-temperature material) as a function of cooling rate; 2) the enthalpy as a function of temperature on heating and cooling; and 3) the relaxation of enthalpy in a temperature-jump experiment. We have not found a phenomenological theory that fits all three sets of data with the same parameters. The failure of the usual analyses to fit our results is probably due to some extent to the considerable broadness of our transition and the temperature dependence that must be ascribed to the parameters of the theory. The heating and cooling curve data can be described with a temperature-dependent relaxation rate that is consistent with the temperature dependence of self-diffusion, but, in the absence of a theoretical fit to all the enthalpy relaxation data, I am reluctant to draw any conclusions. We don't have a model that explains all the enthalpy relaxation data, so we can not assert with any confidence that diffusion and enthalpy relaxation have a different temperature dependence or that they have the same temperature dependence.

Specific Heat and Ultrasonics as Dynamic Probes of the Glass Transition[a]

NORMAN O. BIRGE, YOON H. JEONG,
AND SIDNEY R. NAGEL[b]

James Franck Institute
and
Department of Physics
University of Chicago
Chicago, Illinois 60637

Specific heat and sound propagation measurements are historically of particular importance to the study of the glass transition. Indeed, the glass transition can be viewed either as the change that occurs as a liquid begins to support a shear and thus becomes able to propagate transverse sound or as an anomaly in the specific heat, c_p, of the supercooled liquid. The ability to support a shear is, after all, what distinguishes a solid from a liquid. On the other hand, the specific heat anomaly, where c_p falls from a value appropriate to a liquid to a value close to that of a crystal as the temperature is lowered, is the basis of the Kauzmann paradox,[1] one of the clearest indications that some sort of transition must take place between the liquid and the glass. If there were no specific heat anomaly, then, since the specific heat of a supercooled liquid is larger than that of the crystal, the entropy of the liquid would eventually become much less than that of the crystal as the temperature is lowered. This, of course, has never been observed because the glass transition invariably intervenes before such a bizarre situation has the chance to arise.

There is another way, however, to think about the glass transition. As the temperature of the supercooled liquid is lowered the viscosity, η, rapidly becomes very large. The glass transition temperature, T_g, is often taken to be the temperature at which $\eta \sim 10^{15}$ poise. Since $\eta = \tau/G$ (where τ is a relaxation time and G is a modulus), this means that the relaxation time is on the order of 1 day—the time of a fairly long experiment. It is the dramatic increase of these relaxation times as the temperature is lowered that is also a characteristic signature of the glass transition. How can these three seemingly different views of the glass transition (specific heat anomaly, resistance to shear, and diverging time scales) be reconciled and related to one another? Is there a true underlying phase transition as some theories have concluded[2-5] or is there no true phase transition at all?[6-10] Why do the relaxation times tend to diverge near T_g? Is there more than one relaxation time seen in different experimental probes of the transition? Is there a cooperative effect between the molecules giving rise to the freezing, or is there only single-particle activation over barriers? If there is a cooperative effect, is there a divergent length scale in the problem as there is in other

[a]This work was supported by Grant DMR 82-01182 from the National Science Foundation and Grant DMR 82-16892 from the National Science Foundation Materials Research Laboratory.
[b]To whom correspondence should be addressed.

phase transitions? If so, how can one measure this correlation length? These are just some of the questions one can ask about the glass transition. We will address only a few of them here.

In this paper we will report the results of two different experiments that show the relation of sound propagation and specific heat to the divergent relaxation times at the glass transition. We also hope to show that one characteristic set of relaxation times is important for both the ultrasonics and the specific heat measurements at the glass transition in glycerol.

In order to measure the specific heat we have developed[11] a new technique that allows us to measure the frequency dependence of this quantity over almost six decades of frequency: 0.01 Hz $< \nu <$ 6.0 kHz. It is not obvious that c_p should be frequency dependent. It is usually considered to be a static thermodynamic quantity rather than a dynamic one.[12] We define the frequency-dependent specific heat to be the oscillating response in temperature of the sample to an oscillating heat flux. In the limit that the frequency approaches zero, this measurement yields the ordinary thermodynamic quantity. Indeed, for most samples there will be no visible frequency dependence to c_p in the range of accessible frequencies because the sample usually equilibrates on a microscopic scale much more quickly than a period of oscillation of the heat flux. At the glass transition, however, the situation is different because some of the relaxation times are very long and do not equilibrate in one period of oscillation. A frequency dependence will appear in the measurement if the period of oscillation is comparable to the relaxation times.

In order to do this specific heat spectroscopy we do a nonadiabatic measurement.[11] The heater and thermometer are the same circuit element: a resistor with a large temperature coefficient of resistance. When an oscillating current is passed through the resistor, it heats the sample, which in turn changes the temperature of the resistor. By measuring the voltage drop across the resistor, one can measure the change in resistance and therefore the change in temperature of the sample. If the current varies at frequency $\nu/2$, the heat and temperature will oscillate at frequency ν and the voltage will have a variation at $3\nu/2$. For the resistor in the shape of a ribbon, one can solve the heat diffusion equation and show that the amplitude of the voltage signal at $3\nu/2$ is proportional to $(\nu\, c_p\, \kappa)^{-1/2}$ where κ is the thermal conductivity. This measurement thus provides a measure of the product of c_p and κ. By changing the geometry of the resistor to that of a thin wire, one determines another combination of c_p and κ so that with both measurements one can measure the two quantities separately. The wire experiment is being carried out at present. Preliminary data suggest that there is some frequency dependence to κ that occurs in a higher temperature range although most of the dependence is in c_p.

We emphasize that our measurements are not the same as those[13] that look at the cooling rate dependence of c_p. In those measurements one is measuring the temperature at which the sample falls out of equilibrium[c] at any given cooling rate. The departure from equilibrium always happens just before one reaches the "interesting" region of temperature. Our measurements, on the other hand, are always done in equilibrium because we are investigating the linear response of the system in the

[c]Equilibrium here refers to ergodicity in a restricted region of phase space excluding the crystalline phase.

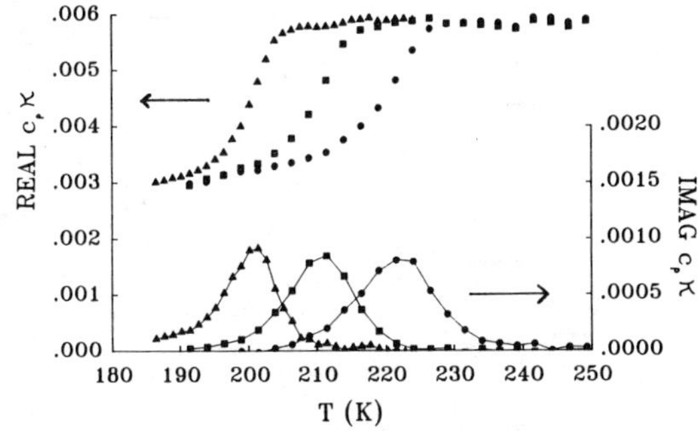

FIGURE 1. The real and imaginary parts of $c_p\kappa$, in $J^2/(cm^4 \cdot sec \cdot K^2)$, in glycerol as a function of temperature. The measurement frequencies are $\nu = 0.62$ Hz (triangles), $\nu = 34$ Hz (squares), and $\nu = 1100$ Hz (circles). The lines are guides to the eye.

supercooled liquid state. The response will, in general, be a complex number indicating whether the temperature rise is in phase (the real part) or out of phase (the imaginary part) with the heat flux.

In FIGURE 1 we show the real and imaginary parts of the product $c_p\kappa$ as a function of temperature for three different frequencies. For each frequency the data for the real part of $c_p\kappa$ drops, as expected, from a value characteristic of the liquid to one characteristic of the glass. As the frequency is lowered, the temperature at which this drop takes place also decreases. In the region of the frequency dispersion there is a peak in the imaginary part of $c_p\kappa$. This is what one expects from the Kramers–Kronig relations.

One can see the relationship between the real and imaginary part of $c_p\kappa$ more clearly if one plots $c_p\kappa$ versus frequency at constant temperature. This is shown in FIGURE 2. The low-frequency region contains contributions from all degrees of freedom of the equilibrium liquid. The high-frequency end contains contributions only from those degrees of freedom that equilibrate quickly enough and therefore is characteristic of a glass. The solid lines are the fits to the data obtained with a Williams–Watts function (that is, the Fourier transform of $d[\exp(-t/\tau)^\beta]/dt$) with $\beta = 0.65 \pm 0.05$.

In FIGURE 3 we show data from the wire geometry experiment. We see that in this case, as well, the phase of the signal varies dramatically in the region where c_p has dispersion. The analysis of this data will be presented elsewhere.[14]

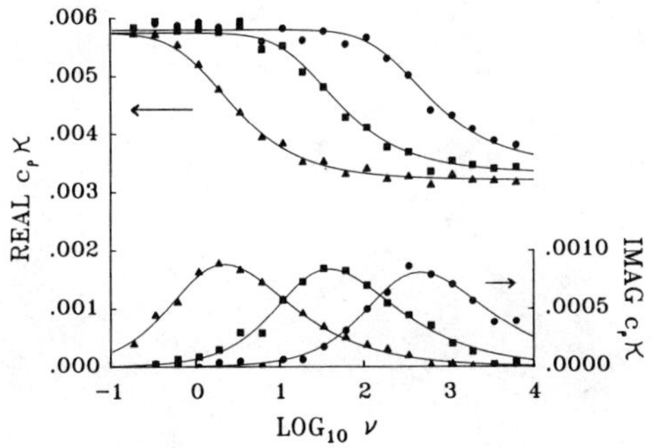

FIGURE 2. The real and imaginary parts of $c_p\kappa$, in $J^2/(cm^4 \cdot sec \cdot K^2)$, in glycerol as a function of frequency. The temperatures are $T = 203.9$ K (triangles), $T = 211.4$ K (squares), and $T = 219.0$ K (circles). The solid lines are fits to the data with a Williams–Watts function with $\beta = 0.65$.

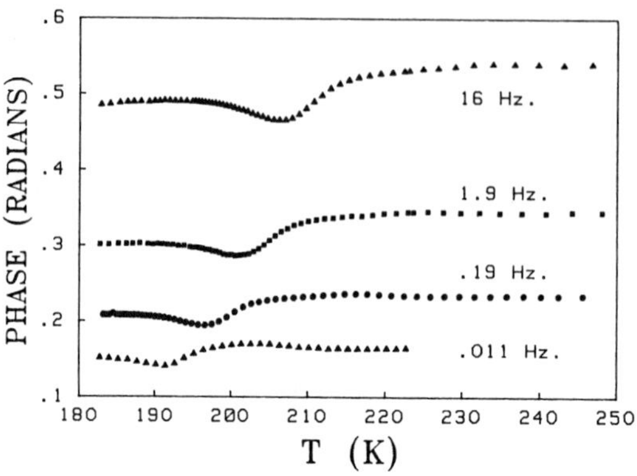

FIGURE 3. The phase of the temperature variation with respect to the heat oscillation in an experiment with a thin wire as the heater-thermometer. The phase drops in the vicinity of the temperature where structure is seen in the plane heater experiment.

We now compare our specific heat results with those obtained from longitudinal ultrasonics measurements.[15] We convert our data for velocity, v, and attenuation, α, into a longitudinal compliance, $\beta_L = \beta_L' + i\,\beta_L''$, in order to compare them with the specific heat. $\beta_L'(\omega) = [1 - (\alpha v/\omega)^2]/(\rho v^2)$ and $\beta_L''(\omega) = 2\alpha/(\rho v \omega)$ where ρ is the density. Previous data[16] only reported the modulus, which is the inverse of the compliance. The data for the compliance looks qualitatively similar to that for $c_p\kappa$. There is a peak in the imaginary part, and the real part is larger at higher temperatures (or low frequencies) and falls off as the temperature is lowered (or frequency is raised). We have normalized our data: $\tilde{\beta}_L(\omega) = [\beta_L(\omega) - \beta_L(\omega = \infty)]/[\beta_L(\omega = 0) - \beta_L(\omega = \infty)]$. In FIGURE 4 we plot all the data for $\tilde{\beta}_L(\omega)$ on a single master curve as was done by Piccirelli and Litovitz[16] by simply shifting each constant temperature curve by a rigid amount on a logarithmic scale. This allows us to line up all the peaks in $\tilde{\beta}_L''(\omega)$ but does not necessarily guarantee that the curves will all have the same width. The fact that this procedure does yield a single curve implies that the shapes of the curves are temperature independent in this range. Again, the best fit to the data accurs with a Williams–Watts form with $\beta = 0.60 \pm 0.05$. A Davidson–Cole fit ($\tilde{\beta}_L(\omega) \sim [1 - i\,\omega\tau]^{-\alpha}$) also produces good results with $\alpha = 0.42 \pm 0.05$.

One important question concerns how the temperature of the peaks, T_p, in specific

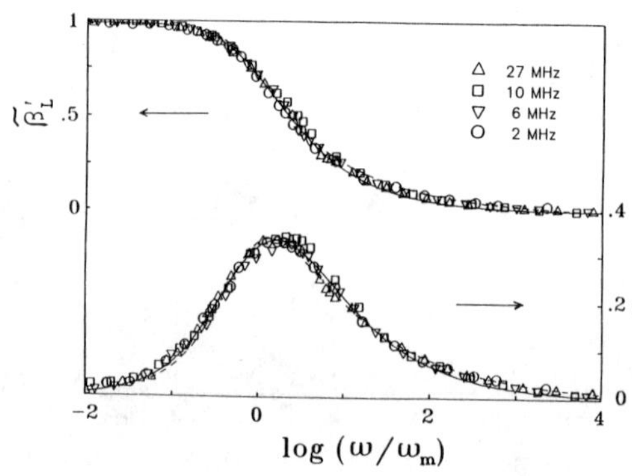

FIGURE 4. A master plot of $\tilde{\beta}'_L(\omega)$ and $\tilde{\beta}_L'(\omega)$ versus log (ω/ω_m). A scaling fit was used for ω_m. The dashed lines are Williams–Watts fits with $\beta = 0.60$. The dotted lines are Davidson–Cole fits with $\alpha = 0.42$.

heat and compliance data shift as the frequency is lowered. If there were only a single relaxation time giving rise to the peaks (that is, a Debye model), the shift of T_p with frequency would determine the temperature dependence of that relaxation time. Because the peaks we observe are broader than a Debye curve we can only determine how the average time is varying with temperature. Before graphing the specific heat and ultrasonics data on the same graph, we must remember that the ultrasonics data in this frequency range is adiabatic whereas the specific heat data is isothermal. Even in a Debye model, the difference between adiabatic and isothermal conditions in an experiment would introduce a difference between the measured values of the relaxation times in the two cases. The difference would be $\tau_s = (c_{p\infty}/c_{p0})\, \tau_T$ where τ_s and τ_T are the adiabatic and isothermal relaxation times, respectively, and $c_{p\infty}$ and c_{p0} are the infinite and zero frequency specific heat values, respectively. From our specific heat data we determine $(c_{p0}/c_{p\infty}) = 1.9$ and calculate the value of τ_T for the ultrasonics experiments. We use this value in our graph of $T_p(\omega)$. In FIGURE 5 we plot log ν (in

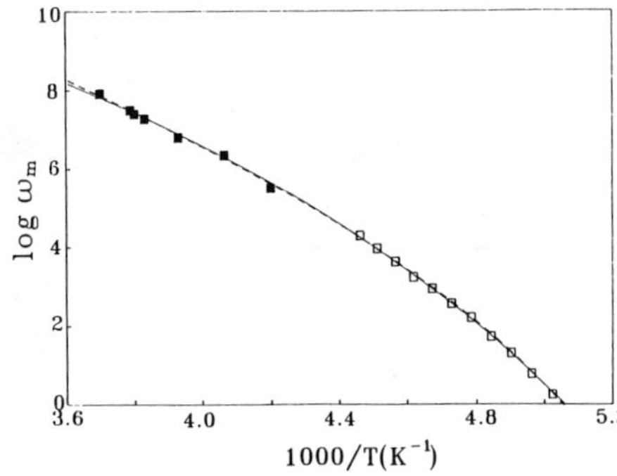

FIGURE 5. A plot of the peak of the imaginary compliance and imaginary specific heat on a log scale versus $1/T$. The solid line is a fit with the Vogel–Tammann–Fulcher law: $\omega_m = 15.7 \times 10^{14} \exp[-2310/(T-129)]$. The dashed line is a scaling law fit: $\omega_m = 15.7 \times 10^{10}[(T/175.5)-1]^{12.5}$.

sec^{-1}) versus $1/T_p$. If the relaxation time varied in an Arrhenius manner the data would lie on a straight line. As one can see the data fall on a curve. The two lines in the figure represent two possible fits to the data: a Vogel–Tammann–Fulcher law, $\omega = \omega_0 \exp\,[A/(T-T_0)]$, and a scaling law, $\omega = \omega_0[(T-T_0)/T_0]^{-\alpha}$ that has recently been predicted by various hydrodynamic theories.[6–10] The two curves fit the data equally well over nine decades of frequency. It is also apparent that the data derived from the specific heat spectroscopy lie on the same curve as those from the ultrasonics. Because the shape of the two response functions are quite similar as well (they are fit with a Williams–Watts function with $\beta = 0.65 \pm 0.05$ and $\beta = 0.60 \pm 0.05$, respectively), we can conclude that the two experiments are determined by the same distribution of relaxation times (except for the difference between adiabatic and isothermal conditions and the fact that the ultrasonics measurements probe higher frequencies than does the specific heat). The specific heat anomaly, which is so important because of the Kauzmann paradox, seems to be related to the change in sound propagation at the glass transition by the same distribution of relaxation times, which diverge as the temperature is lowered.

As mentioned above, both the fits with the Vogel–Tammann–Fulcher law and the scaling law give good agreement with the data. The exponent α, however, in the scaling law fit is 12.5, which is almost an order of magnitude higher than the initial estimates of the theory. It is not entirely clear whether or not the hydrodynamic theories are consistent with such a large exponent. On the other hand, the value of T_0 in the Vogel–Tammann–Fulcher law ($T_0 = 129$ K) is quite close to the temperature at which the Kauzmann paradox would occur ($T_K = 134$ K), so it is appealing to think that the two temperatures are indeed the same and not due to the accidental choice of picking a fitting function.

A recent theory, by Marchetti,[17] that is based on the hydrodynamic models has been used to calculate the shape of the ultrasonic modulus. Marchetti uses the experimental viscosity as input so that she does not try to fit the shape of $T_p(\omega)$. The fit of her theory to our data is quite satisfactory: it reproduces the width and the asymmetry quite well.

Finally we can put the T_p versus ω data from other experiments on glycerol onto the same graph. This is done in FIGURE 6. Again the data from all other experiments (dielectric susceptibility,[18] digital correlation spectroscopy,[19] and Brillouin scattering[20]) fit essentially the same curve as does the specific heat and ultrasonics data. The

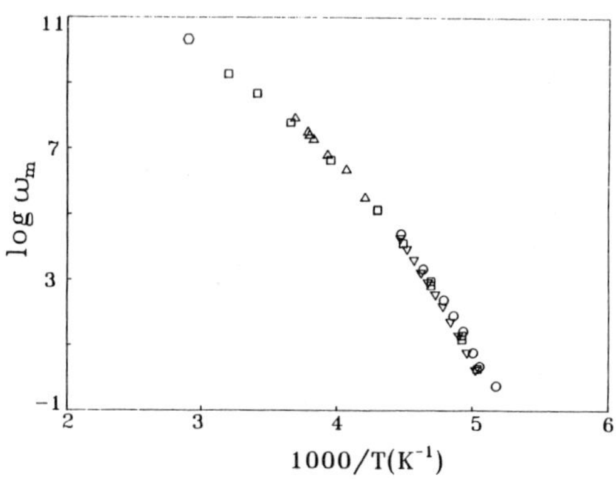

FIGURE 6. Relaxation times in glycerol. These times were obtained by different techniques, namely dielectric measurements (squares),[18] digital correlation spectroscopy (circles),[19] specific heat spectroscopy (inverted triangles), ultrasonics (triangles), and Brillouin scattering[20] (hexagon). Isothermal corrections were made for the ultrasonic and Brillouin scattering data.

most obvious difference between the results of these measurements is that the dielectric susceptibility has a much narrower distribution of relaxation times than do the other probes of the glass transition. It is appealing to hypothesize that some of the modes that account for the slow relaxation at the glass transition are not charged so that they do not couple to an electric field.

If the relaxation times do indeed diverge at some finite temperature, then, since the shape of the response function is temperature independent, the zero-frequency specific heat will be a step function as a function of temperature with the step occurring at the glass transition. Our experiments fall short of telling us whether there exists a true underlying phase transition. If there is such a transition, one might expect a divergent length scale as the critical temperature is approached as is seen in other transitions. It is possible that this diverging correlation length may be observed in a nonlinear susceptibility measurement. Such experiments were helpful in indicating the presence of a true phase transition in spin glasses.

In conclusion, we have measured the frequency dependence of the specific heat and of the ultrasonic velocity and attenuation in the vicinity of the glass transition in glycerol. Both the density and the enthalpy fluctuations have the same dependence on temperature and frequency. The temperature dependence of the relaxation times obtained from these measurements can be fit equally well by either a Vogel–Tammann–Fulcher law or a scaling law, as recent hydrodynamic theories have suggested. In the scaling law fit the exponent is 12.5, which is unexpectedly large for a true phase transition.

ACKNOWLEDGMENTS

We wish to thank C. A. Angell, J. P. Carini, L. P. Kadanoff, C. Marchetti, G. F. Mazenko, and D. Oxtoby for many useful discussions.

REFERENCES

1. KAUZMANN, W. 1948. Chem. Rev. **43:** 219.
2. GIBBS, J. H. & E. A. DIMARZIO. 1958. J. Chem. Phys. **28:** 373.
3. ADAM, G. & J. H. GIBBS. 1965. J. Chem. Phys. **43:** 139.
4. COHEN, M. H. & G. S. GREST. 1979. Phys. Rev. B **20:** 1077.
5. WOODCOCK, L. V. 1976. J. Chem. Soc. Faraday Trans. 2 **72:** 1667.
6. DAS, S. P., G. F. MAZENKO, S. RAMASWAMY & J. J. TONER. 1985. Phys. Rev. Lett. **54:** 118.
7. BENGTZELIUS, U., W. GOTZE & A. SJOLANDER. 1984. J. Phys. C **17:** 5915.
8. LEUTHEUSSER, E. 1984. Phys. Rev. A **29:** 2765.
9. FREDRICKSON, G. H. & H. C. ANDERSEN. 1984. Phys. Rev. Lett. **53:** 1244.
10. KIRKPATRICK, T. R. 1985. Phys. Rev. A **31:** 939.
11. BIRGE, N. O. & S. R. NAGEL. 1985. Phys. Rev. Lett. **54:** 2674.
12. OXTOBY, D. 1986. J. Chem. Phys. **85:** 1549.
13. MOYNIHAN, C. T., P. B. MACEDO, C. J. MONTROSE, P. K. GUPTA, M. A. DE BOLT, J. F. DILL, B. E. DOM, P. W. DRAKE, A. J. EASTEAL, R. P. MOELLER, H. SASABE & J. A. WILDER. 1976. Ann. N.Y. Acad. Sci. **279:** 15.
14. BIRGE, N. O. 1986. Phys. Rev. B **34:** 1631.
15. JEONG, Y. H., S. R. NAGEL & S. BHATTACHARYA. 1986. Phys. Rev. A **34:** 602.
16. PICCIRELLI, R. & T. A. LITOVITZ. 1957. J. Acoust. Soc. Am. **29:** 1009.
17. MARCHETTI, C. Preprint.
18. DAVIDSON, D. W. & R. C. COLE. 1951. J. Chem. Phys. **19:** 1484.
19. DEMOULIN, C., C. J. MONTROSE & N. OSTROWSKY. 1974. Phys. Rev. A **9:** 1740.
20. PINNOW, D. A., S. J. CANDAU, J. T. LAMACCHIA & T. A. LITOVITZ. 1968. J. Acoust. Soc. Am. **43:** 131.

DISCUSSION OF THE PAPER

P. W. ANDERSON (*Princeton University, Princeton, NJ*): What is the frequency dependence of thermal conductivity?

NAGEL: As far as we have been able to tell, the frequency dependence of the thermal conductivity near the glass transition has never been measured. In our studies using a plane heater we measure a signal that determines the product of the c_p and κ.

From that experiment alone we cannot tell whether the frequency dependence is in c_p or in κ. For this reason we are performing a second experiment using a wire, rather than a plane, heater sensitive to a different combination of these two parameters. By using both experiments we hope to determine the frequency dependence of c_p and κ separately. Naively one might expect most of the frequency dependence to be in c_p. After all, our data for $c_p\kappa$ shows the same general shape as is found in the cooling rate studies of c_p alone using standard calorimetric techniques. Also, one might expect that since κ is so small already at T_g there would be little change in this quantity as the glass becomes more fluid. Nevertheless, it is only by actually doing a careful measurement of $\kappa(\omega)$ that we can tell if these expectations are correct.

H. LEVINE (*General Foods Technical Center, Tarrytown, NY*): Do you plan to extend your studies on glycerol to glycerol-water mixtures, in order to obtain data that may be extrapolated to the case of pure water? Such studies might result in a useful new approach to the problem of the T_g of amorphous water, and its related value of T_0.

NAGEL: We have not considered the possibility of studying glycerol-water mixtures. I agree that the experiments you suggest may give very interesting results.

Parallels and Contrasts between Glass and Spin Glass

R. G. PALMER

Department of Physics
Duke University
Durham, North Carolina 27706

Great fleas have little fleas upon their backs to bite 'em,
And little fleas have lesser fleas, and so *ad infinitum*

—AUGUSTUS DE MORGAN

INTRODUCTION

The study of glass, and especially of glassy relaxation, is much older than the whole field of spin glass physics. Nevertheless the conceptual picture of spin glasses that has evolved in the past decade seems natural for glass too, and the methods of spin glass theory have great potential for application to glassy systems. This paper presents a nontechnical introduction to spin glasses, a discussion of some underlying concepts of broken ergodicity, an application of these spin glass ideas to glasses, and a summary of some differences between the systems. The term glass is used here to refer to structural glasses, not spin glasses.

SPIN GLASSES

Theoretical[1] and experimental[2] spin glasses deserve separate discussion. Experimental spin glass materials are normally dilute magnetic alloys (for example, *Cu*Mn) in which the magnetic ions randomly substitute for the nonmagnetic ones. FIGURE 1 illustrates a few magnetic ions (or spins), and focuses attention on three, which are labeled i, j, and k, respectively, and their interactions J_{ij}, J_{ik}, and J_{jk}. In a spin glass such a set of interactions is often frustrated, which means that the potential energies of the three pairs (for example, $J_{ij}\mathbf{S}_i \cdot \mathbf{S}_j$, and so on, in a Heisenberg model) cannot simultaneously be minimized. This occurs if an odd number of the pair interactions J_{ij} is antiferromagnetic in character. In a metallic system like *Cu*Mn ferromagnetic and antiferromagnetic interactions both arise from the oscillatory Ruderman–Kittel–Kasuya–Yosida interaction.

Spin glasses possess a phase in which the spins freeze into fixed directions without any long-range spatial order. The freezing occurs rather sharply in temperature, as shown by a cusp-like maximum in the magnetic (AC) susceptibility and the onset of irreversibility. The critical temperature T_c of the susceptibility maximum decreases slightly as the measuring frequency is decreased. Though superficially like glass in this behavior, the transition is much sharper than the glass transition, and is often considered to signal a true thermodynamic transition with critical slowing down, not just kinetic freezing. Opinions have been divided and have changed over time on this point; currently, the evidence seems to be pointing more towards a real phase transition. The frequency dependence of T_c can be fitted with a Vogel–Fulcher form

$$\tau = \tau_0 \exp\left[\frac{A}{T-T_0}\right]$$

with $\omega \sim \tau^{-1}$ and $T = T_c$, but is fitted better[3] by a dynamical scaling form appropriate to a real finite temperature transition

$$\tau = \tau_0\left[\frac{T}{T-T_0}\right]^{z\nu}$$

Below the transition, in a spin glass phase, the magnetic properties are highly irreversible and history dependent. Magnetic relaxation is slow, and may best fit a logarithm, a power law, a Kohlrausch law such as $\exp(-t^{\beta})$,[4] or an $\exp[-(\ln t)^{x}]$ law, depending on the material. When a Kohlrausch law is used the β exponents are rather small (for example, 0.1–0.4) by comparison with most glasses above the glass transition temperature T_g. A Kohlrausch law has also been used to fit spin autocorrelation functions above T_c for long-time Monte Carlo simulation data.[5]

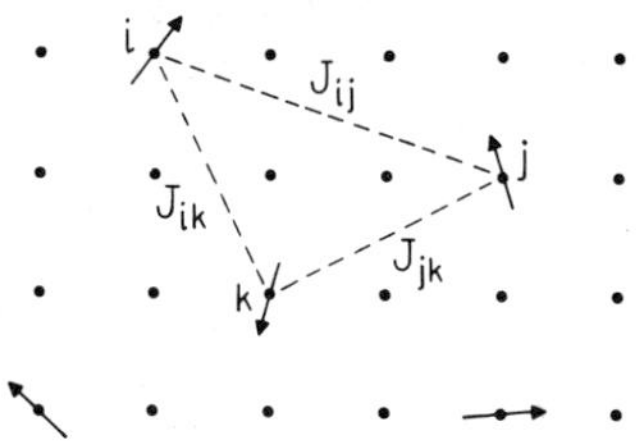

FIGURE 1. An idealized spin glass.

On the theoretical side a selective historical review is appropriate. Spin glass theory started with the 1975 paper of Edwards and Anderson.[6] They proposed treating the pair interactions J_{ij} as quenched random variables in the Hamiltonian

$$\mathcal{H} = -\Sigma\, J_{ij}\mathbf{S}_i \cdot \mathbf{S}_j$$

which describes N spins $\mathbf{S}_i$. They suggested assigning the same probability distribution $P(J)$ independently to each J_{ij}, ignoring the correlations present in a real system. Because the J_{ij} are quenched, the properties of the system (such as the free energy F) must be calculated for a particular set of J_{ij}. If, as expected, almost every set gives the same answer (in order N), the result may then be averaged over possible J_{ij} sets. Thus, for example, the free energy (ignoring broken ergodicity) may be calculated as $F = -kT\ \langle \ln Z\rangle_J$. Edwards and Anderson also proposed the order parameter $q_{\rm EA} = \langle\langle \mathbf{S}_i\rangle_T^2\rangle_J$, which is nonzero whenever the spins are frozen (in time), whether or not there is any spatial order. This order parameter was useful in early mean field work, but does not distinguish possible ordered states sufficiently to allow a correct solution.

Sherrington and Kirkpatrick[7] soon proposed an Ising version of the Edwards–Anderson model with explicitly infinite range interactions; each J_{ij} has the same probability distribution, however far apart sites i and j may be. This model has been studied intensively since 1975, and has become known as the infinite range or mean field model. Sherrington and Kirkpatrick called it the solvable model because they thought that they had found an exact mean field solution, using the replica method first introduced by Edwards and Anderson. The replica method approaches the problem of computing $\langle \ln Z \rangle_J$ by replacing $\ln Z$ by $\lim(n \rightarrow 0)\ [d(Z^n)/dn]$. $\langle Z^n \rangle_J$ can be evaluated only for integer n, for which Z^n is just the partition function of n replicas (copies) of the original system, and a continuation to $n = 0$ has to be made in an *ad hoc* manner. Unfortunately, the most natural continuation is not the correct one in the low-temperature phase, and Sherrington and Kirkpatrick's result is unphysical.[8,9]

The concept of frustration[10] introduced in 1977 drew attention to the idea that spin glasses may well have many (perhaps order $\exp[aN]$) ground states. There are often two or more degenerate local spin arrangements minimizing the energy when frustration prevents all bond interactions from being satisfied simultaneously. Such a system would be expected to require a many-component order parameter (or many distinct order parameters) in its description because the direction (or phase) of an order parameter distinguishes the frozen states of a system.[11] The successful theories for the Sherrington–Kirkpatrick model that have appeared since 1976 have all fulfilled this expectation:

1. The Thouless–Anderson–Palmer theory employed a mean field theory with N order parameters, one for each spin site.[12] It resulted, unfortunately, in a coupled set of nonlinear algebraic equations, which are almost as hard to solve as the original problem.

2. The Parisi theory suggested a new way of extending the integer n replica results to $n = 0$, via an iterative matrix scheme that introduced new order parameters in each iteration K and became accurate as $K \rightarrow \infty$.[13] The infinite set of order parameters could be replaced by an order function $q(x)$, which was mysterious in its physical interpretation. The multiple states appeared as choices in breaking symmetry in abstract replica space, and the Parisi scheme has become known as replica symmetry breaking.

3. The Sompolinsky–Zippelius dynamical theory was a time-dependent Ginzburg–Landau theory of the dynamics of an infinite number of order parameters.[14] Further, the individual relaxation times were assumed to fall on a continuum that widened indefinitely as $N \rightarrow \infty$. These times all diverged with N, but this may be regarded as an artifact of the Sherrington–Kirkpatrick model.

These three theories are essentially equivalent in their predictions (where these overlap), and are believed to be correct for the Sherrington–Kirkpatrick model. Developments since 1981 have chiefly concerned the interpretation of the theories, a better conceptual understanding of spin glass systems, and extensions to other models. Vector and Potts spins can be dealt with satisfactorily in an infinite range limit, but success in short-range (for example, nearest-neighbor) models has been lacking. Most of our knowledge of such systems stems from computer simulation.

BROKEN ERGODICITY

Most progress in spin glass theory since 1981 has concerned the elucidation of the structure of the many alternative states in configuration space. The many-ground-state/many-order-parameter picture suggested considering the structure of a configuration space of many valleys separated by free energy barriers. Each of these valleys (known also as components, phases, solutions, local minima, and bowls) represents a possible macrostate of the system, but, unless the barriers are effectively infinite, only a temporary one. Dynamically, at least on a long-enough timescale, the system flips from valley to valley, as in the dynamical theory.[14] The system is not even effectively ergodic on a physical timescale. By analogy with broken symmetry in which the system becomes stuck in one out of several symmetry-equivalent valleys, this may be called broken ergodicity. It is of particular interest when the number of valleys is large (order $\exp[aN]$), or when the observational timescale is buried in a continuum of internal system timescales.

One can think loosely of a free energy surface that has roughness on many scales (FIG. 2). The horizontal axis represents in one dimension the whole free energy space spanned by a set $\mathbf{\Phi}$ of appropriate configurational coordinates Φ_i (not necessarily bare spins). $F(\mathbf{\Phi})$ is in reality a very high dimensional surface because many Φ_i's are expected to be relevant. Given a temperature and an observational timescale, there will be certain possible valleys in FIGURE 2 that can, for all intents and purposes, confine the system. At a higher temperature, or if one is prepared to wait longer, the conceptual picture may be changed in one of two ways: either one may regard the lower free energy barriers as passable, so that the system is only confined to a larger valley composed of subvalleys (A $\rightarrow$ B $\rightarrow$ C), or (better) one may redefine the free energy space using fewer coordinates Φ_i.

An important feature of spin glasses is the slow relaxation, seen also in glasses. Relaxation functions and time-time correlation functions have significant variation for many decades of time. This suggests a continuous distribution of barrier heights between valleys, and makes difficult any quasi-static approach that regards, independent of observation time, some barriers as passable and others as impassable. A better picture must at least make passability a function of observation time and temperature, leading to a time- and temperature-dependent number of valleys and number of order parameters.

The situation is very different from that in most familiar systems, as illustrated in FIGURE 3. Here time lines (on a logarithmic scale) for various systems are shown. Relevant internal timescales are indicated by shading, and an arbitrary observational timescale is labeled τ_{obs}. Time lines a and b are for a zero-field Ising ferromagnet above and below T_c, respectively. Below T_c the time τ_{flip} for flipping from an "up" to a "down" state diverges with system size N, and so the physically relevant regime is the one

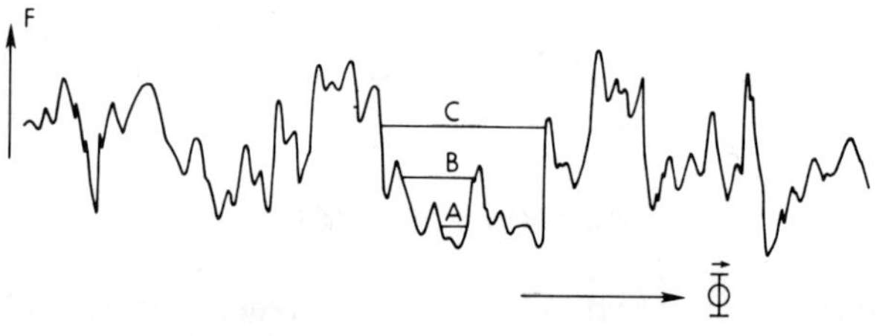

FIGURE 2. A schematic free energy surface.

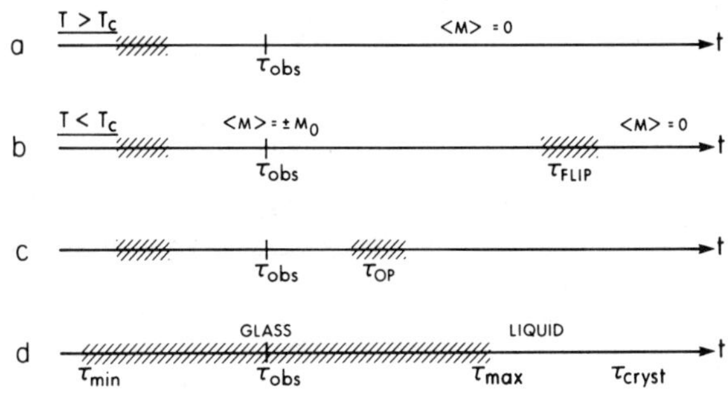

FIGURE 3. Time lines for several systems.

shown for τ_{obs}, not the ergodic one beyond τ_{flip}. One normally approaches such a system theoretically by breaking the symmetry by using an infinitesimal field, or carefully choosing boundary conditions, or selecting a particular mean field solution. These techniques effectively remove the possibility of flipping, so that methods such as phase-space averaging where $t \rightarrow \infty$ work. Time line c shows a less clear-cut case, that of molecular hydrogen with *ortho* and *para* species. The *ortho/para* conversion time τ_{OP} is on the order of years in the absence of a catalyst. For normal purposes we must regard the *ortho/para* degrees of freedom to be quenched, and treat by statistical mechanics only the fast processes (translation, rotation, and, perhaps, vibration). This is not hard because there is a timescale gap of many orders of magnitude between the "fast" and "slow" degrees of freedom. Changing τ_{obs} within the gap makes no fundamental difference. In contrast, the time line d shows what is believed to happen in spin glasses. There is no timescale gap and the meaning of fast and slow must itself be continuously τ_{obs} dependent. Quasi-static theories can only work if τ_{obs} appears as a parameter.

To make further progress one must face up to describing the set of frozen states (valleys) as a function of observational timescale τ_{obs} and external parameters (temperature, magnetic field, etc.). There is considerable evidence[11] that the number of relevant valleys increases with increased N, decreased T, or decreased τ_{obs}. The simplest model in which the number of valleys can change is the Ising ferromagnet, for which FIGURE 4 shows a state diagram. In the infinite system there are two states for $T < T_c$, corresponding to up and down spins. In a finite system at finite τ_{obs} the situation is more complicated and involves a fuzzy central region, but the same limiting states are present.

For the spin glass, with the number of states changing quasi-continuously with N, T, and τ_{obs}, it is natural to consider the extension to a tree of bifurcations[11,15] as shown in FIGURE 5. As $N \rightarrow \infty$ the bifurcation points become dense on the horizontal axis. Although labeled in terms of temperature, a very similar picture would be expected for the τ_{obs} dependence. At a given temperature and τ_{obs} there are several possible valleys (for example, A, B, and C at T in FIG. 5), which divide further as T decreases. Of course the real system cannot be expected to possess sharp bifurcations or fully defined valleys, and FIGURE 5 must be regarded as an idealization.

There is considerable evidence in support of this bifurcation tree, or hierarchical,

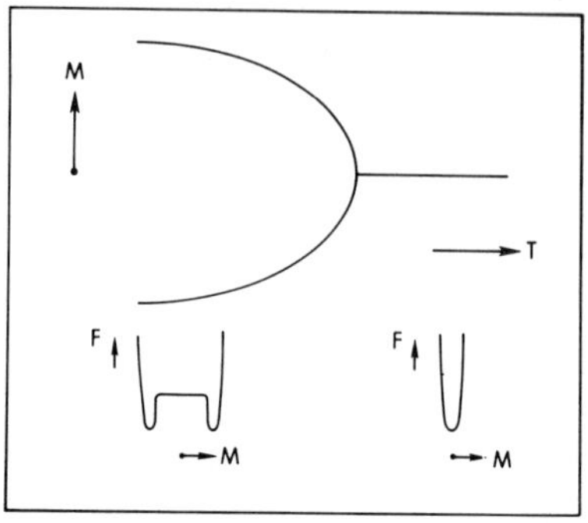

FIGURE 4. Component structure and schematic free energy surfaces for an Ising ferromagnet.

picture of the state space of a spin glass.[16] Evidence includes the behavior of the number of valleys (with N, T, and τ_{obs}) as discussed above, the existence of locally frozen clusters in simulation, the slow relaxation itself, the irreversibility in simulations, and, especially, the irreversibility signature of experimental spin glasses. Irreversibility occurs whenever the magnetic field is changed within the spin glass phase, and conditionally when the temperature is changed. FIGURE 6 shows for example the effect of temperature changes on magnetization after turning on a field h at low temperature (A → B → C). Raising T leads to an irreversible increase of M (C → D → E) up to T_c. Lowering T again at any point below T_c, however, puts the system on a reversible curve (for example, DF) that can be retraced back to CDE. This behavior can be readily understood in terms of a hierarchical state space (FIG. 5 and its finite field extension); cooling followed by heating is reversible (D → A → D in FIG. 5), but heating followed by cooling is not (A → D → B). Much more detailed correspondence can be derived.[11,16]

If we associate with each bifurcation point a free energy barrier (or corresponding transition time), the barriers and transition times between alternative states are ultrametric. This means that for any three alternative states at least two pairs will have

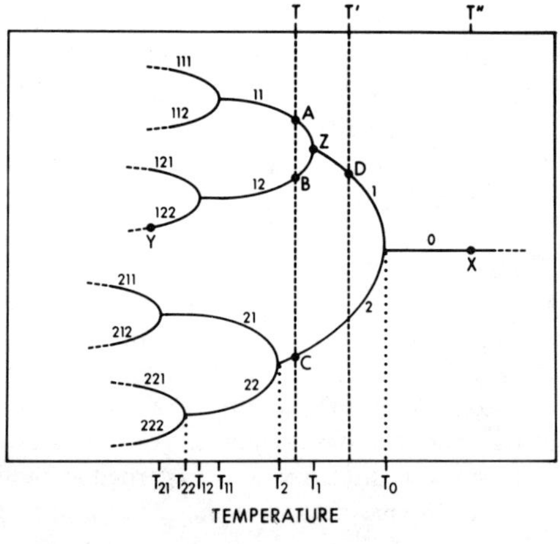

FIGURE 5. Component structure suggested for spin glasses and glasses.

the same barrier height and transition time, with the remaining pair either the same or closer (smaller barrier ΔF). If we think of barrier height or transition time as a distance between states, then the triangles formed by any three states are either acute-angled isosceles or equilateral. In FIGURE 5 for example, among the three states A, B, and C we have $\Delta F_{AC} = \Delta F_{BC} > \Delta F_{AB}$.

Ultrametricity was first introduced into spin glasses theory by Toulouse, and is now a central concept. It is always possible (though not always useful) to define an ultrametric distance function (here ΔF or τ) on a hierarchical structure, and to define a hierarchical structure from an ultrametric distance function. Recent work has proved for the Sherrington–Kirkpatrick model that the distance between valleys in configuration space (defined in terms of overlap) is ultrametric, lending further support to the hierarchical picture.[17] The same work has also succeeded in giving a physical meaning to the once mysterious $q(x)$ order function, in terms of the proportion of pairs of states a given distance apart.

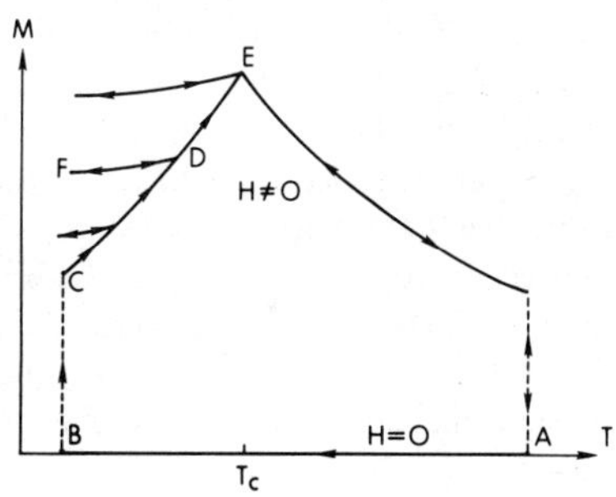

FIGURE 6. Idealized magnetization curves for typical spin glasses: the irreversibility signature.

GLASSY PARALLELS

Much of the above discussion of spin glasses can be applied to glasses. Viewed at the level of structures in configuration space the two systems show strong similarities.

First, it is obvious that the free energy surface of a glassy system is many-valleyed. In each local region there are typically several alternative atomic arrangements of comparable energy separated by energy barriers. The theory and observation of two-level systems points directly to such a structure. The residual $T = 0$ entropy suggests that there are on the order of $\exp(aN)$ such relevant valleys at low temperature.[18]

Second, there is a quasi continuum of relevant relaxation times, at least around and below T_g. This is supported by the slow relaxation, which stretches over many decades. Although it may be fitted by a single function (for example, the Kohlrausch law) containing only one timescale, when regarded in terms of microscopic transition probabilities a wide range of such probabilities is essential. Thus time line d of FIGURE 3 is the most appropriate one; certainly there is no major timescale gap as in b and c. The longest internal time τ_{max} grows rapidly larger as the temperature is reduced. It becomes larger than our maximum observation time τ_{obs} at T_g, beyond which we can no longer stay in equilibrium and are inevitably stuck in a nonergodic valley. T_g is of course determined by τ_{obs}, but is nevertheless moderately sharp because τ_{max} grows so rapidly with decreasing temperature. It is interesting to note that a similar rapid growth is seen for τ_{flip} in the Ising ferromagnet near T_c (FIGS. 3a & 3b); the flip time detaches itself from the continuum and moves to long time at a rate $d\tau_{flip}/dT$ that diverges with N. We are free to speculate[18] that τ_{max} in glass diverges to infinity at some finite temperature, perhaps according to a Vogel–Fulcher law, but we cannot gain access to this region.

Finally, but with least certainty, there are reasons to expect the state space for a glass to be hierarchical, as idealized in FIGURE 5. We know that the number of valleys increases steadily with decreasing temperature, from one (ergodic) at high temperature ($\tau_{max}(T) < \tau_{obs}$) to order exp ($aN$) at low temperature. This cannot happen all at once or we would see a first-order transition with a latent heat. It is difficult to imagine more valleys arising without a reasonably well-defined ancestry, as in FIGURE 5.

Another reason to expect a hierarchy of states comes from considering local structure and dynamics. We expect the fastest motions to be simple vibrations; then, as we look to slower and slower motions, we expect to find more and more complex atomic rearrangements, generally involving more and more atoms. Within a given "major" arrangement there will be several "minor" arrangements possible, within a given minor one there will be several $(\text{minor})^2$ ones, and so on.

We are thus led to tentatively describe a glass in configuration space in much the same way that we describe glass, represented schematically by FIGURES 2, 3d & 5. This idea is still young, and needs considerable refinement such as has been possible in spin glass. In particular we need to use the known irreversibility of glass (including pressure dependence) to constrain the picture. It should be possible to build in cooling rate dependence and nonlinearity. We may eventually be able to find experimental measures of ultrametricity.

There are several thermodynamic consequences of the present picture, stemming mainly from the need to consider a division into fast and slow degrees of freedom that itself depends continuously on τ_{obs}. One can, for example, put bounds on the Prigogine–DeFay ratio,[19] and can explicate the meaning of the residual entropy.[18,19] The Kauzmann paradox may well be resolved by this route. Here, though, we will focus on relaxation dynamics.

We cannot proceed far with analyzing the dynamics of a hierarchical state space without characterizing it in more detail. We need to know at least some of the following: 1) How is the structure of the tree to be characterized? Such a characterization includes the density of bifurcations (in T and τ_{obs}) and the typical branching ratio (2 for a bifurcation). Distributions may be needed. We may also need to get away from a pure tree (perfectly ultrametric) and allow cross-relations between branches. 2) How do the barrier heights ΔF vary from level to level? All that is essential for the hierarchical picture is that they increase as one goes towards the ergodic root of the tree. 3) How are experimental quantities to be calculated? We can certainly measure the progress of relaxation by the distance traveled in the tree (perhaps with a particular bias field), but even this notion of distance is not well defined without a proper metric on the tree.

In the face of these difficulties research has proceeded in two directions in trying to apply spin glass ideas to glass. Some work has proceeded by answering the above questions arbitrarily or generically to see what general conclusions can be drawn.[20,21]

Abstract models of configuration space dynamics have been constructed, as discussed further below. One finds that Kohlrausch-like relaxation is surprisingly natural. The other approach has been to look for systems intermediate between glasses and spin glasses, so that spin glass methods (for example, replica theory) can be applied in detail. This has led to the study of glassy crystals such as $(KCN)_x(KBr)_{1-x}$ reported elsewhere in this volume.

In the category of abstract configuration space models, the work of Ogielski and Stein[21] is the most general. They consider a tree that bifurcates regularly and synchronously (so that branches increase in number as follows: 1, 2, 4, 8, . . .), and assign an arbitrary sequence of barrier heights to the bifurcation points. They solve the resulting Markov process exactly, and can get Kohlrausch law relaxation for a particular choice of barrier heights. The most interesting result is that the process is only well defined if the barriers increase rapidly enough, and that the marginal case itself gives Kohlrausch relaxation. Huberman and Kerszberg[22] point to the many potential applications of this ultrametric diffusion.

Palmer *et al.*[20] were more specific in their assumptions and went beyond the general picture drawn from spin glasses. They emphasized the importance of constraints in the dynamics of glassy relaxation, suggesting that slow degrees of

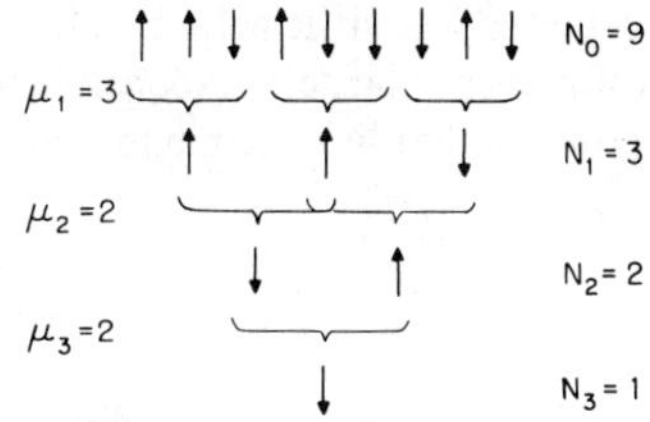

FIGURE 7. Model of hierarchically constrained dynamics.

freedom are slow precisely because they must wait for some faster ones to be in just the right state before they are able to change themselves. There can be many levels of constraints, with a wide range of timescales coming simply from the multiplied "local" constraints between levels. Constraints may also be viewed as entropy barriers on the free energy surface; there are few roads out of a valley, and many places to go within it. Of course the levels should not be regarded as discrete; a continuum limit is needed. A hierarchy of constraints eliminates the need to find separate physical processes (for example, particular bond rearrangements) to explain each relaxation time regime.

Palmer *et al.* examined a very simple toy model that embodied hierarchical constraints, as illustrated in FIGURE 7. Degrees of freedom are represented by Ising spins, which are grouped into discrete levels 0, 1, 2, . . . , and constrain each other so that a particular spin in level n is only free to flip while some μ_n spins in level $n-1$ are in one particular state (out of 2^{μ_n} possibilities). Computation of the time-time correlation function also involves a weight or number of spins N_n for each level. The two sequences μ_n and N_n become continuous functions $\mu(n)$ and $N(n)$ in the continuum limit. These are constrained by convergence requirements and the need, taken from observation, that the longest timescale τ_{max} be very much longer than the shortest, τ_{min} (FIG. 3d). Within these constraints, reasonable assumptions (almost anything not perverse) for the free functions are found to lead to Kohlrausch-like behavior. Vogel–Fulcher behavior is found for τ_{max}. If a self-similarity assumption is made for the free energy landscape, one free function remains, but the resulting relaxation is always exactly $\exp[-t^{\beta}]$. Future research will include the examination of other constraint models (such as that of Fredrickson and Andersen[23]), with the possibility of generating hierarchical behavior from local constraint conditions.

GLASSY CONTRASTS

The emphasis thus far has been that the configuration space picture of spin glass has much to lend to glass. In detail, of course, spin glasses and glasses are quite different, but we focus here on abstract theoretical understanding and analytical methods. Some important differences remain and deserve discussion.

The Spin Glass Hamiltonian

Theoretical spin glasses have well-defined Hamiltonians. For glass we are as yet far from being able to write down a Hamiltonian suitable for analysis. A Lennard–Jones potential is of little help, because we need to freeze the slowest degrees of freedom. This difference is largely responsible for the greater attraction of spin glass and the greater progress that has been made there.

Underlying Crystalline State

Most glasses have a unique, well-defined crystalline phase, lying thermodynamically well below the glass phase. This is surely irrelevant, however, to glassy relaxation; there is a timescale gap above τ_{max} before we reach the timescale for crystallization.[18] In fact the same may be true in spin glasses for which there is a stoichiometric alloy; if one could wait for atomic rearrangement an atomically ordered state would form.

Quenched Disorder

It is often stated that in the spin glass there is quenched disorder, and that in glass there is not. This is only partially true. In an actual magnetic alloy there are processes by which the spin sites can move, albeit slowly. There may be a timescale gap between the longest dynamic time τ_{max} for the spin variables $\mathbf{S}_i$ and the timescale for change of the J_{ij}. In the glass, on the other hand, the degrees of freedom that do not change appreciably within the observation time might just as well be quenched. The real difference is not so much whether truly quenched variables actually exist, but whether we are always imbedded in a timescale continuum ($\tau_{obs} < \tau_{max}$ in FIG. 3d). In at least some spin glass models the maximum time τ_{max} diverges with N and is thus effectively infinite; we are always in the continuum. In the glass, however, τ_{max} is strongly temperature dependent, and indeed passes through the observable regime right at T_g.

So a glass below T_g, or above T_g for $\tau_{obs} < \tau_{max}$, is like a spin glass below its T_c. Well above T_g and T_c glasses and spin glasses are again similar simply by virtue of being effectively ergodic.

Phase Transition

It is in the region around T_g or T_c that the systems differ most. In many spin glasses there is a sharp freezing transition, much sharper than the kinetic freezing of glass. There may in fact be a thermodynamic phase transition, as seen in some models and simulations. In glass too there could be an underlying thermodynamic transition, but not at T_g. It would presumably occur where $\tau_{max}(T)$ diverged to infinity, if this actually happens. The old objection that a perfect glass state is a contradiction in terms is laid to rest by the spin glass example, where there are many possible thermodynamically frozen states without long-range order. The transition region has not yet been satisfactorily understood in any disordered system displaying broken ergodicity, and it is as yet impossible to say how similar or dissimilar spin glasses and glasses will eventually appear to be. A configuration space picture is badly needed. The study of glassy crystals is also likely to illuminate the question. The Kauzmann paradox and the Vogel–Fulcher law are both up for explanation.

REFERENCES

1. FISCHER, K. H. 1983. Phys. Status Solidi B **116:** 357–414.
2. FISCHER, K. H. 1985. Phys. Status Solidi B **130:** 13–71.
3. SOULETIE, J. & J. L. THOLENCE. 1985. Phys. Rev. B **32:** 516–519.
4. CHAMBERLIN, R. V., G. MOZURKEWICH & R. ORBACH. 1984. Phys. Rev. Lett. **52:** 867–870.
5. OGIELSKI, A. T. 1985. Phys. Rev. B **32:** 7384–7398.
6. EDWARDS, S. F. & P. W. ANDERSON. 1975. J. Phys. F **5:** 965–974.
7. SHERRINGTON, D. & S. KIRKPATRICK. 1975. Phys. Rev. Lett. **35:** 1792–1796.
8. DE ALMEIDA, J. R. L. & D. J. THOULESS. 1978. J. Phys. A **11:** 983–990.
9. VAN HEMMEN, J. L. & R. G. PALMER. 1979. J. Phys. A **12:** 563–580.
10. TOULOUSE, G. 1977. Commun. Phys. **2:** 115–119.
11. PALMER, R. G. 1982. Adv. Phys. **31:** 669–735.
12. THOULESS, D. J., P. W. ANDERSON & R. G. PALMER. 1977. Philos. Mag. **35:** 593–601.
13. PARISI, G. 1980. J. Phys. A **13:** L115–L721.
14. SOMPOLINSKY, H. & A. ZIPPELIUS. 1982. Phys. Rev. B **25:** 6860–6875.
15. KREY, U. 1977. J. Magn. Magn. Mater. **6:** 27–37.
16. PALMER, R. G. 1983. *In* Heidelberg Colloquium on Spin Glasses. J. L. van Hemmen & I. Morgenstern, Eds.: 234–251. Springer–Verlag. Berlin.
17. MÉZARD, M., G. PARISI, N. SOURLAS, G. TOULOUSE & M. VIRASORO. 1984. Phys. Rev. Lett. **52:** 1156–1159.
18. PALMER, R. G. & D. L. STEIN. 1984. *In* Relaxations in Complex Systems. K. L. Ngai & G. B. Wright, Eds.: 253–259. U. S. Government Printing Office. Washington, DC.
19. JÄCKLE, J. 1981. Philos. Mag. **44:** 533–545.
20. PALMER, R. G., D. L. STEIN, E. ABRAHAMS & P. W. ANDERSON. 1984. Phys. Rev. Lett. **53:** 958–961.
21. OGIELSKI, A. T. & D. L. STEIN. 1985. Phys. Rev. Lett. **55:** 1634–1637.
22. HUBERMAN, B. A. & M. KERSZBERG. 1985. J. Phys. A **18:** L331.
23. FREDRICKSON, G. H. & H. C. ANDERSEN. 1984. Phys. Rev. Lett. **53:** 1244–1247.

DISCUSSION OF THE PAPER

J. T. BENDLER (*General Electric Research and Development Center, Schenectady, NY*): 1. Can one say what the simplest entropy barrier distribution is that will lead to the Kohlraush–Williams–Watts (KWW) decay function?

2. Do you extend the analysis to the glass state itself (below T_g), and if so, how does the KWW exponent behave and does the Vogel–Tammann–Fulcher effect disappear?

PALMER: 1. If we demand the Kohlrausch functional form exactly there is of course a unique corresponding distribution of free energies ΔF. The asymptotic form is $P(\Delta F) \sim \exp(-a \cdot \exp(b\Delta F))$, with $bT = \beta/(1-\beta)$. This is generated, for example, by the simple rules for $\mu(n)$ and $N(n)$ given in reference 20 of this paper. There the free energy barriers have their source in constraints and are thus purely entropic, $\Delta F = -T\Delta S$, with negative (labyrinthine) entropy barriers ΔS. But relaxation of precisely the Kohlrausch form should not be overemphasized; other functional forms can fit the data as well.

2. The model given in reference 20 of this paper does not incorporate temperature effectively, except to suggest Vogel–Tammann–Fulcher behavior for τ_{max} (not τ itself). D. Stein and I are at present investigating a model with temperature-dependent constraints.

Sequential Freezing of Quadrupolar and Dipolar Order in $(KBr)_{1-x}(KCN)_x$ Glasses[a]

A. LOIDL AND K. KNORR

Institut für Physik
Universität Mainz
D-6500 Mainz, Federal Republic of Germany

The mixed molecular crystals $(KBr)_{1-x}(KCN)_x$ show a fascinating x-versus-temperature phase diagram (FIG. 1).[1] The room temperature structure is of the NaCl type with the CN^- molecules being rotationally disordered. In this plastic high-temperature phase each CN^- ion undergoes fast reorientations. For $x \geq 0.6$ four crystallographic low-temperature phases have been reported[1,2] where the orientational degrees of freedom are reduced. Only in the orthorhombic phases, for $0.9 \geq x \geq 1$ and $T = 80$ K, is the orientational order complete.[2] All the other phases exhibit dipolar disorder characterized by a residual degeneracy with respect to head and tail down to the lowest temperatures. Crystals with $x \leq 0.6$ remain in a quadrupolar disordered state of quasi-cubic structure with frozen-in random CN^- orientations and frozen-in lattice strains.[3,4] As indicated in FIGURE 1, the low-temperature frozen-in state is regarded as a glass state. This can be justified for the following reasons:

1) The freezing is signaled by anomalies in the temperature dependence of the quadrupolar susceptibilities that show striking similarities with the spin glass transition.[5] In the spin glasses and in the alkali halide-alkali cyanide mixtures frustrated interactions are thought to play the decisive role in the freezing process. In $(KBr)_{1-x}(KCN)_x$ the frustration results from the interplay of site disorder of the CN^- molecules combined with the anisotropic quadrupolar interactions.

2) It has been found that the low-temperature thermodynamic,[6] ultrasonic,[7] and dielectric[8] properties are analogous to those found in canonical glasses and in amorphous solids.[9-11]

3) The diffraction data indicate the breakdown of true translational long-range order at the freezing temperature T_F.[4] This effect is demonstrated in FIGURE 2 by plotting the temperature dependence of the Bragg reflection (440) as measured in a single $(KBr)_{0.5}(KCN)_{0.5}$ crystal.[4] The data show a well-defined Bragg reflection at 90 K but a strong broadening of the line width as the temperature is lowered. The asymmetry that becomes clearly visible at 10 K can be explained by a bilinear coupling of rotations and translations.[3]

In studying the relaxation kinetics the reorientational motion of the CN^- ions has been described in terms of dielectric dispersion at frequencies $\nu < 1$ MHz and in terms of elastic dispersion for $\nu > 1$ MHz. Until now the question remained open as to whether both dispersion regimes belonged to the same relaxation mechanism. In this article we will present data on the elastic and dielectric dispersion that cover a common frequency range 10^2 Hz $\leq \nu \leq 10^9$ Hz and allow a direct comparison of the quadrupolar

[a]This work was supported in part by the Bundesministerium für Forschung und Technologie.

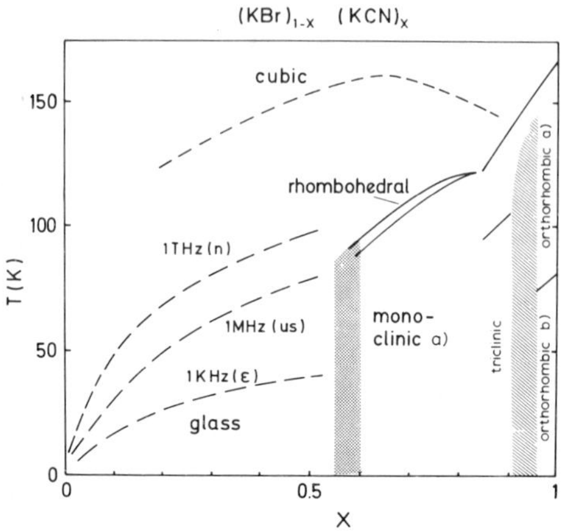

FIGURE 1. Phase diagram of $(KBr)_{1-x}$ $(KCN)_x$.[1] Phase boundaries are shown as solid lines; freezing temperatures, as dashed lines. The hatched area is a coexistence region; the dotted area, the transition region from elastic order to the glass state.

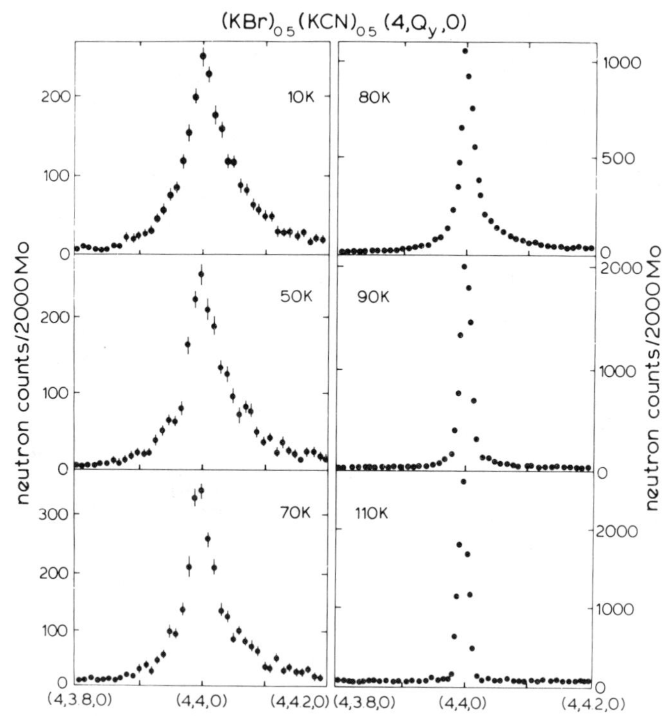

FIGURE 2. Q_y scans at zero energy transfer in $(KBr)_{0.5}$ $(KCN)_{0.5}$ through the reciprocal lattice point (440) at various temperatures.[4]

and the dipolar anomalies. In particular, it will be demonstrated that the freezing-in of dipoles and quadrupoles occurs at different temperatures.

The quadrupolar susceptibility of the orientational subsystem $\chi^Q(T, \nu)$ can be studied via the elastic constant $c_{44}(T, \nu)$ where $c_{44}(T, \nu) = c_{44}^0 \times (1 - g^2 \chi^Q(T, \nu))$. Here c_{44}^0 is the background elastic constant of the center of mass lattice and g^2 is the rotation translation coupling constant. In FIGURE 3 we show the elastic constant $c_{44}(T)$ for $(KBr)_{0.8}(KCN)_{0.2}$ as deduced from the transverse acoustic (TA) phonon dispersion measured at (2, 0.1, 0)TA in an inelastic neutron-scattering experiment.[12,13] The lower part of FIGURE 3 shows the temperature dependence of the intensity of the central peak as determined at zero energy transfers: c_{44} softens, passes through a minimum, and recovers again towards 0 K. At the minimum sound velocity the central line starts to grow. The central line intensity can be regarded as the order parameter of the glass state.[3] The appearance of a central peak immediately demonstrates that the freezing-in of the quadrupolar degrees of freedom is not a single-ion phenomenon but a

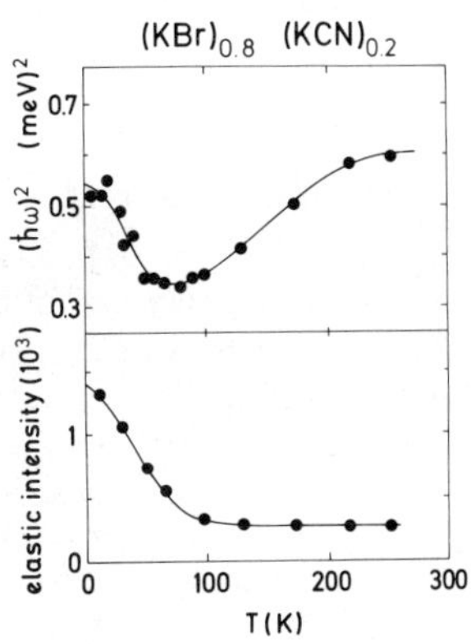

FIGURE 3. Squared phonon frequency and elastic intensity at $\mathbf{Q}$ = (2, 0.1, 0) (T_{2g} symmetry) versus temperature, displaying the freezing process in the $x = 0.2$ mixed crystal.

cooperative effect where the strain-mediated interactions between the molecules play an important role.

FIGURE 4 shows the elastic constants and the damping in $(KBr)_{0.965}(KCN)_{0.035}$ as determined from the sound wave propagation along the [100] and [110] directions.[14] The elastic constants, except c_{44}, exhibit a minimum at about 16 K. The attenuation peaks at somewhat lower temperatures. The signal of the c_{44} mode was lost at 22 K because of a large damping, but from the analysis of all the data one can deduce that c_{44} has the same quantitative temperature dependence.[14]

The temperature dependence of the elastic constant c_{44} has also been determined by a torsion pendulum with eigenfrequencies ranging from 50 Hz to 1 kHz using capacitive drive and pick-up.[15] The torque is parallel to [100]. A representative result for a crystal with the concentration of $x = 0.2$ is shown in FIGURE 5 and compared with the analogous results obtained in the ultrasonic[17] and inelastic neutron-scattering[12,13] results. In addition Brillouin results for $x = 0.19$ are indicated in FIGURE 5. These results demonstrate the well-known frequency-dependent cusps of the quadrupolar susceptibility.[18]

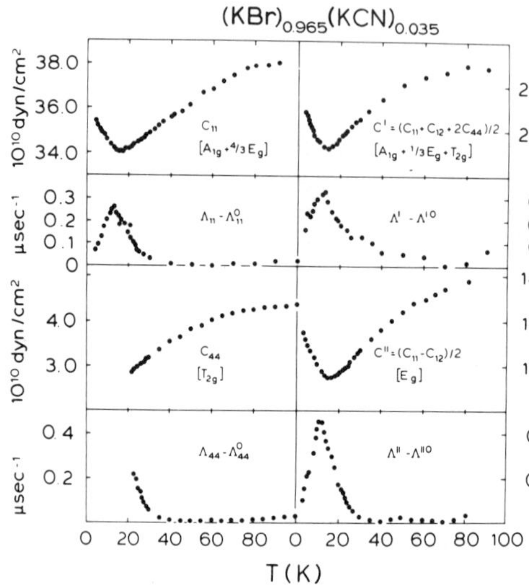

FIGURE 4. The temperature dependence of the elastic constants and the sound-wave attenuation for $(KBr)_{0.965}(KCN)_{0.035}$.[14]

Measurements of the dielectric constant $\epsilon(T, \nu)$ probe the dipolar susceptibility $\chi^D(T, \nu)$: $\epsilon(T, \nu) = \epsilon_\infty + 4\pi \times \chi^D(T, \nu)$. The results of the real part of $\epsilon(T, \nu)$ in $(KBr)_{0.8}(KCN)_{0.2}$ are shown in FIGURE 6. The low-frequency data are taken from reference 19. The high-frequency data have been obtained by determining the reflection coefficient and measuring the phase shift utilizing an automatic radio frequency impedance analyzer.[15]

Both dipolar and quadrupolar susceptibilities χ^D and χ^Q are given by $\chi^{D,Q}(T, \nu) \alpha 1/k_BT/(1 + i\omega\tau^{D,Q}(T))$. Here $\omega = 2\pi\nu$, and τ^D and τ^Q are the dipolar and quadrupolar relaxation times, respectively. As the temperature is lowered the molecular relaxation slows down, yielding dispersion in the real part and a peak loss in the imaginary part of the orientational susceptibility. The maximum in the loss appears at $\omega\tau = 1$, defining the freezing temperatures T_F^Q and T_F^D where the relaxation rates equal a given measuring frequency. A direct comparison between these freezing temperatures can only be performed in the kHz range using dielectric and torsion pendulum data. T_F^Q always appears well above T_F^D. For $x = 0.2$ $T_F^Q = 22$ K and $T_F^D = 17$ K; for $x = 0.5$ $T_F^Q = 63$ K and $T_F^D = 32$ K. Unfortunately, no reliable loss data can be extracted from the ultrasonic and the inelastic neutron-scattering data. Therefore, we took the

FIGURE 5. Normalized elastic constant $c_{44}(T, \nu)$ versus temperature as determined in $(KBr)_{0.8}(KCN)_{0.2}$ for different measuring frequencies: ●: inelastic neutron scattering studies,[12,13] THz; □: Brillouin data ($x = 0.19$),[16] GHz; +: ultrasonic results,[17] 10 MHz; ——: torsion pendulum measurements.[15]

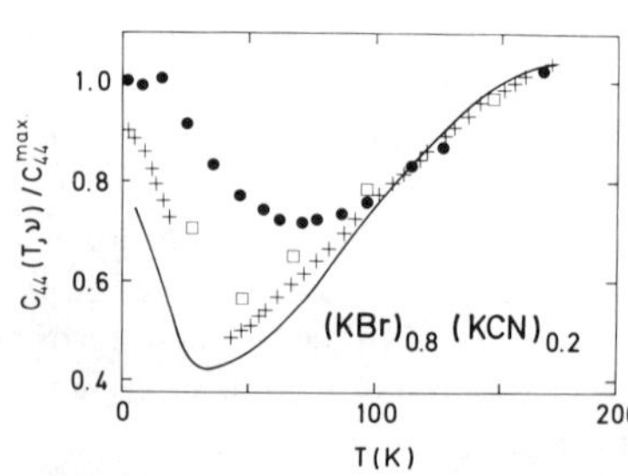

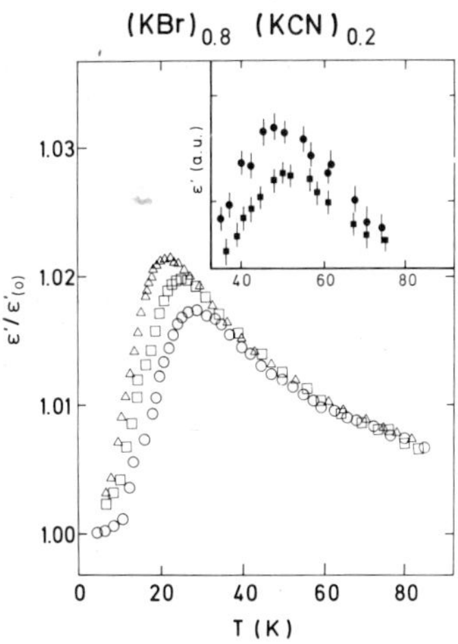

FIGURE 6. Real part of the dielectric constant in $(KBr)_{1-x}(KCN)_x$ versus temperature for different measuring frequencies: ○: 100 kHz;[19] □: 10 kHz;[19] △: 87.5 Hz.[19] The high-frequency data are shown in the insert: ●: 520 MHz;[15] ■: 940 MHz.[15]

maxima of the real part of the susceptibilities defined by min ($c_{44}(T, \nu)$) and by max ($\epsilon(T, \nu)$) as freezing temperatures T_F^Q and T_F^D, respectively. From the dielectric measurements in $(KBr)_{1-x}(KCN)_x$ we know that the maxima of the real part appear 5–10 K above the maxima of the dielectric loss.

FIGURES 5 & 6 reveal that the dispersion regimes for strain and polarization are at different temperatures. For a quantitative comparison we plotted ln (ν) versus $1/T_F^Q$ and $1/T_F^D$ for $x = 0.2$ and $x = 0.5$ (FIG. 7). $T_F^D(\nu)$ is well described by an Arrhenius law (solid line) but there exist systematic deviations from Arrhenius-type behavior in the quadrupolar relaxation. The experimentally observed frequency dependence of $T_F^Q(\nu)$ can be described using a phenomenological Vogel–Fulcher ansatz or a power law deduced from dynamic scaling theories.[21] These descriptions signal a "true" transition into a glass phase. The very limited number of data points, however, hampers a more quantitative analysis.

For both concentrations the freezing temperatures coincide only at GHz frequencies. Significant differences become apparent for lower measuring frequencies. At present, we have no explanation for the dipolar and the quadrupolar freezing

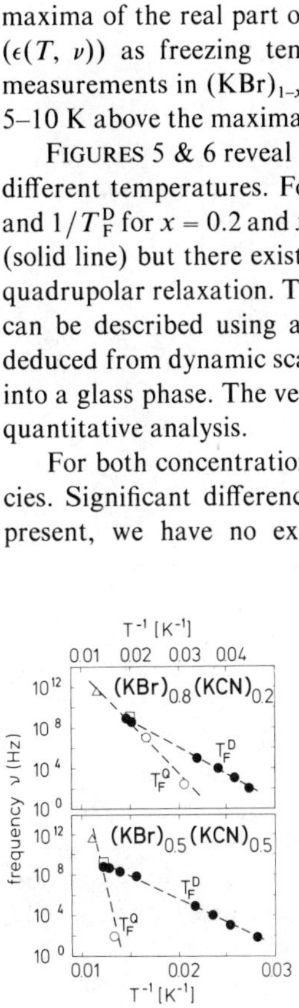

FIGURE 7. Arrhenius plots (log ν versus $1/T$) of the dipolar and quadrupolar freezing temperatures in $(KBr)_{1-x}(KCN)_x$ for $x = 0.2$ and $x = 0.5$. Full symbols represent dipolar data.[15,19] Open symbols represent quadrupolar data: △: inelastic neutron-scattering results for $x = 0.2$ (references 12 and 13) and $x = 0.5$ (references 4 and 20); □: Brillouin results;[16] ○: ultrasonic results[17] and torsion pendulum measurements.[15]

temperatures becoming identical in the low-GHz range. It is worthwhile to note that GHz frequencies are close to the critical relaxation rate where all the alkali cyanides exhibit a transition for the plastic high-temperature phase into an orientationally ordered state.[22] Hence this crossover could define the glass transition temperature T_g. Following arguments by J. M. Rowe (personal communication), the dynamics of the dipoles and the quadrupoles are essentially the same above T_g. At the T_g the onset of spontaneous random strains creates local distortions, thereby increasing the barrier height and introducing an asymmetry into the reorientational potential.[23] Now the quadrupoles spend a longer fraction of time in a preferred orientation yielding a nonzero expectation value of the order parameter of the glass state.[3] If this is the case, T_F^Q as determined at THz frequencies is relevant to the paraelastic state, whereas measuring frequencies below 1 GHz probe the glass state.

We have parameterized the quadrupolar and the dipolar susceptibilities in terms of the single-particle-potential Debye–Arrhenius model. The results, namely the hindering barrier E_0 and the attempt frequency $\tilde{\nu}_0$, are shown in FIGURE 8.

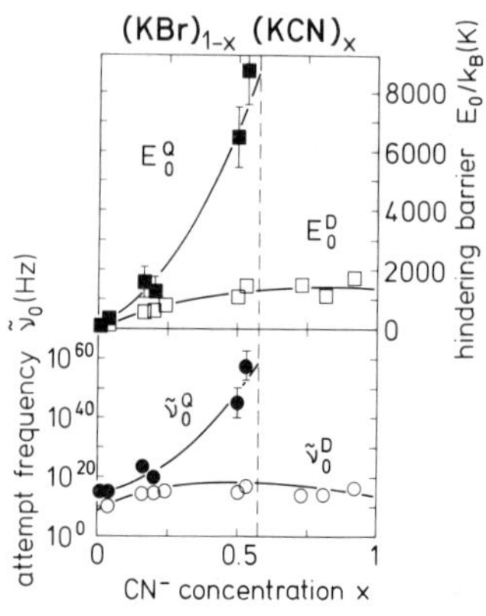

FIGURE 8. Attempt frequencies $\tilde{\nu}_0$ and hindering barriers E_0 versus x as derived from Arrhenius fits to the dipolar and quadrupolar freezing temperatures. Dipolar parameters: ○: ν_0^D; □: E_0^D. Quadrupolar parameters: ●: ν_0^Q; ■: E_0^Q.

We are aware that an Arrhenius type of analysis of the cusps of the real part of the susceptibilities (and not of the peak maxima of the loss) introduces systematic errors into the parameters E_0 and ν_0. For the dielectric measurements we performed a rigorous analysis fitting the real and imaginary parts of the dielectric constant within the framework of an Arrhenius model including a symmetric distribution of relaxation times. From this analysis we know that the dielectric parameters as shown in FIGURE 8 are overestimated: the hindering barriers E_0^D are overestimated by a factor of 1.3 to 2; the attempt frequencies ν_0^D, by two to five decades. As stated before, however, no quadrupolar loss data are available at higher frequencies: in the ultrasonic experiments the signals are lost due to heavy damping;[14,17] in the inelastic neutron-scattering experiments the loss is hidden in the temperature-dependent broadening of the phonon linewidth. An analysis of the neutron data is model dependent and limited by the experimental resolution. Nevertheless, we tried an analysis of the quadrupolar loss data based on neutron and torsion pendulum results in $(KBr)_{0.8}(KCN)_{0.2}$. Again we found that the hindering barriers and the attempt frequencies as shown in FIGURE 8 are overestimated (E_0^Q approximately by a factor of 1.5, ν_0^Q by five decades). We take this as a proof—although the absolute values of E_0 and ν_0 as shown in FIGURE 8 are systematically too high—that the trends in the quadrupolar and dipolar parameters are well established.

Only at very low concentrations may reasonable values for the quadrupolar parameters be found. With increasing CN^- concentrations E_0^Q and ν_0^Q increase and become unphysically large. It must be clearly stated that a description of relaxational phenomena in terms of Arrhenius-type behavior describes the single-ion case only. Collective effects that we think drive the quadrupolar freezing enter artifically into the free parameters yielding unrealistically high values of $\tilde{\nu}_0$ and E_0. If one accepts these two parameters as a translation of collective effects into a single-ion picture, one should be aware that $\tilde{\nu}_0$ and E_0 can be rewritten in terms of alternative parameters E and ν_0 where $\nu_0 = \tilde{\nu}_0 \times \exp(-\alpha/k_B)$ and $E = E_0 - \alpha T$. Using this formulation the collective effects are cast more naturally into a linear temperature dependence of the hindering barrier. The parameter ν_0 is a real attempt frequency that could be visualized as the rotation frequency of the free CN^- rotator or as a typical frequency of acoustic shear waves.

The high values and the strong concentration dependence of the quadrupolar single-ion Arrhenius parameters prove that the freezing in $(KBr)_{1-x}(KCN)_x$ is a cooperative phenomenon and that collective effects become more and more important as the concentration is increased. The high values of the hindering barriers demonstrate that the reorientations of a single quadrupole are accompanied by changes in the surrounding strain fields, which trigger the reorientation of neighboring CN^- molecules. The contribution of the single-ion crystalline anisotropy to E_0 is on the order of 50 K and can be neglected when $x \geq 0.1$. A possible interpretation can be given in terms of a cluster formation process: at the glass transition the CN^- ions that are exposed to the strongest random strain fields are blocked in clusters leaving only cooperative relaxation channels for the molecular reorientation. The clusters grow with decreasing temperature and thereby increase the barrier height against molecular relaxation. A distribution of clusters gives the broad distribution of relaxation times that has been observed in $(KBr)_{1-x}(KCN)_x$.[24]

Conclusions concerning the origin of the dipolar hindering barriers are more difficult to draw. For all concentrations the dipolar freezing is measured in an orientationally ordered or in a quadrupolar glass state. Hence the Arrhenius parameters are relevant to a cubic crystalline potential with local deformations only. FIGURE 8 suggests that the local environment is very similar in the quadrupolar glass and the quadrupolar long-range ordered state. There are no discontinuities at the critical concentration, that is, when $x = 0.6$. Therefore, we agree with Sethna *et al.*[25] that the energy barriers to dipolar reorientations depend predominantly on the local deformation of the single-ion potential through quadrupolar interaction forces. At the moment, however, there exist no experimental data that allow us to determine the importance of the dipolar interactions in the dipolar freezing process. These experimental results support the x-versus-temperature phase diagram as proposed by Ihm[26] on theoretical grounds.

In conclusion, we have shown that the quadrupolar and the dipolar freezing-in in $(KBr)_{1-x}(KCN)_x$ is decoupled. The quadrupolar freezing is a cooperative phenomenon whereas the dipolar relaxation is a thermally activated process over the hindering barriers created by the quadrupolar interaction energies.

SUMMARY

Dipolar and quadrupolar susceptibility measurements were reported for the molecular glass system $(KBr)_{1-x}(KCN)_x$ and covered a wide range of frequencies. The results allowed a direct comparison of the dipolar and quadrupolar anomalies and demonstrated unambiguously that the freezing-in of the dipolar and quadrupolar degrees of freedom occurs at different temperatures.

ACKNOWLEDGMENTS

We would like to thank J. M. Rowe for the helpful discussions.

REFERENCES

1. Knorr, K. & A. Loidl. 1985. Phys. Rev. B **31:** 5387.
2. Rowe, J. M., J. J. Rush & S. Susman. 1983. Phys. Rev. B **28:** 3506.
3. Michel, K. H. & J. M. Rowe. 1980. Phys. Rev. B **22:** 1417.
4. Loidl, A., M. Müllner, G. F. McIntyre, K. Knorr & H. Jex. 1985. Solid State Commun. **54:** 367.
5. Mydosh, J. A. & G. J. Nieuwenhuys. 1980. *In* Ferromagnetic Materials. E. P. Wohlfahrt, Ed.: 71–182. Elsevier/North-Holland. Amsterdam.
6. De Yoreo, J. J., M. Meissner, R. O. Pohl, J. M. Rowe, J. J. Rush & S. Susman. 1983. Phys. Rev. Lett. **51:** 1050.
7. Berret, J. F., P. Doussineau, A. Levelut, M. Meissner & W. Schön. 1985. Phys. Rev. Lett. **55:** 2013.
8. Moy, D., J. N. Dobbs & A. C. Anderson. 1984. Phys. Rev. B **29:** 2160.
9. Anderson, P. W., B. I. Halperin & C. Varma. 1972. Philos. Mag. **25:** 1.
10. Phillips, W. A. 1972. J. Low Temp. Phys. **7:** 351.
11. Hunklinger, S. 1977. *In* Festkörperprobleme XVII. J. Treusch, Ed.: 1–11. Vieweg. Dortmund.
12. Loidl, A., K. Knorr, R. Feile & J. K. Kjems. 1983. Phys. Rev. Lett. **51:** 1054.
13. Loidl, A., R. Feile, K. Knorr & J. K. Kjems. 1984. Phys. Rev. B **29:** 6052.
14. Loidl, A., R. Feile & K. Knorr. 1981. Z. Phys. B **42:** 143.
15. Volkmann, U. G., R. Böhmer, A. Loidl, K. Knorr, U. T. Höchli & S. Haussühl. 1986. Phys. Rev. Lett. **56:** 1716.
16. Satija, S. K. & C. H. Wang. 1978. Solid State Commun. **28:** 617.
17. Feile, R., A. Loidl & K. Knorr. 1982. Phys. Rev. B **26:** 6875.
18. Loidl, A., R. Feile & K. Knorr. 1982. Phys. Rev. Lett. **48:** 1263.
19. Knorr, K. & A. Loidl. 1982. Z. Phys. B **46:** 219.
20. Rowe, J. M., J. J. Rush, D. J. Hinks & S. Susman. 1980. Phys. Rev. Lett. **43:** 1158.
21. Souletie, J. & J. L. Tholence. 1985. Phys. Rev. B **32:** 516.
22. Lüty, F. & J. Ortiz-Lopez. 1983. Phys. Rev. Lett. **50:** 1289.
23. Stokes, H. T. & R. D. Swinney. 1985. Phys. Rev. B **31:** 7133.
24. Birge, N. O., Y. H. Jeong, S. R. Nagel, S. Bhattacharya & S. Susman. 1984. Phys. Rev. B **30:** 2306.
25. Sethna, J. P., S. R. Nagel & T. V. Ramakrishnan. 1984. Phys. Rev. Lett. **53:** 2489.
26. Ihm, J. 1985. Phys. Rev. B **31:** 1674.

DISCUSSION OF THE PAPER

A. NAVROTSKY (*Princeton University, Princeton, NJ*): The model for a range of activation energies rests on the binomial distribution of next-nearest-neighbor occupancies, that is, a random CN-Br distribution. Do experiments detect either clustering (positive ΔH_{mix}) or ordering (negative ΔH_{mix})? If there is local order, how seriously are the theoretical predictions affected? What would be the stable low-temperature structure, as opposed to that in glass? [NOTE: This same question is answered in the discussion of the paper by J. P. Sethna in this volume—*Ed.*]

LOIDL: The momentum transfer dependence Q of diffusive neutron-scattering data were analyzed but no indications of chemical clustering could be detected. Of course, there is a dramatic concentration dependence of the low-temperature structure in KBr:CN. Four stable elastically ordered phases have been detected in a CN^- concentration range where $0.6 \leq x \leq 1$. We believe that this polymorphism is essential to understanding the glassy properties: if the crystalline structure of KBr:CN is indifferent to shear distortions, the glass phase can be built by mixing clusters that have different structures.

Glassy Crystals

Low-frequency and Low-temperature Properties[a]

JAMES P. SETHNA

Laboratory of Atomic and Solid State Physics
Cornell University
Ithaca, New York 14853

GLASSES

Glasses are amorphous like liquids, but are rigid like crystals. To a chemist, this may seem natural: silicon dioxide, when covalently bonded into a well-relaxed random network, should resist deformations as an elastic solid just as it would in a crystalline network. To a theoretical physicist, this is more mysterious. What broken symmetry[1] causes the rigidity?

Of course, the chemist may in the end be right—the observed rigidity may be due purely to short-range order. If so, glasses must flow like a (very viscous) liquid on sufficiently long time scales. Even though it is not favorable for any individual bond to break, rearrangements of sufficiently large clusters of atoms will allow the energy to decrease or remain the same. (If the order is short range with a range R, a cluster of size larger than R will typically allow such an internal rearrangement.) At any finite temperature, these rearrangements will proceed by thermal activation at a slow but nonzero rate. Any external stress applied to the system will produce a bias in these rearrangements, producing in the end the finite shear rate characteristic of a liquid. If the glass is truly solid, it must have some sort of long-range order to produce infinite barriers to rearrangements. The study of the glass transition is the study of the phase boundary between the fluid liquid and the rigid glass. This subject is controversial, and has recently seen several interesting developments.

In this paper we will not discuss the glass transition. We will deal partly with the universal low-temperature[2] properties of glasses, which are not as controversial; indeed, in many ways they are quite well understood. The tunneling center theory[3,4] explained why glasses have specific heats proportional to temperature and thermal conductivities proportional to T^2, in contrast to insulating crystals where both properties are proportional to T^3 for low temperatures T. It predicted correctly that glasses would have a saturable ultrasonic attenuation and a "time-dependent" specific heat; the experimental specific heat varies logarithmically with the measuring time.

We will also deal with the low-frequency response of glassy materials. In the experimental phenomenology, there are two kinds of low-frequency response in glasses, with quite different signatures, which often (or always) coexist. There has been much study of the α-relaxation processes because of their relationship to the glass transition. The α-relaxations typically are observed in the supercooled liquid, and have a

[a]This work was supported in part by Grant DMR-8217227A01 from the National Science Foundation.

characteristic time scale that appears to diverge at a temperature somewhat below the temperature at which the liquid falls out of equilibrium (freezes into a glass); their time dependence often is well fit by the Williams–Watts or Kohlrausch form $\chi(t) = e^{-(t/\tau)^\gamma}$. The relaxations we will discuss almost certainly correspond to the β-relaxations in glasses, which have not been studied much. The β-relaxations are observed in the glass at lower temperatures; they often span many decades in frequency, and their characteristic time scale diverges only at zero temperature, with an Arrhenius law suggesting thermal activation.

A glass will have a completely different configuration of atoms each time it is formed. Thus very generally, whether or not the glass is truly rigid, there will always be some local rearrangements of the atoms with very small energy costs; since the atoms have no definite positions, they have flexibility in their local configurations. The tunneling center theories use these rearrangements to explain the universality of the low-temperature properties of glasses. All glasses have local rearrangements of atoms with nearly degenerate energies; quantum tunneling of these atoms between their two configurations produces many low-energy excitations with long time scales; these "two-level systems" produce the observed behavior, which is thus naturally common to all glasses. These rearrangements can also proceed by thermal activation, and thus provide a natural explanation for the low-frequency (β) relaxation.

The experimental methods used to understand the structure of crystals are largely useless in studying glasses. Although some average information about the microscopic structure of glasses such as the radial distribution function can be measured, relatively atypical regions like tunneling centers are hard to study. (Tunneling centers that contribute to the specific heat below 1 K involve maybe 1 molecular group in 10^5.) The development of the tunneling center theory has been hindered both by the lack of a microscopic picture of the centers and by the related lack of experimental or theoretical information about the distribution of tunneling center parameters. The relaxations responsible for the low-frequency properties in glasses are also largely not understood; there is little upon which to base a theory.

Thus the discovery that certain crystalline systems have many properties characteristic of glasses is an exciting opportunity. The mixed crystal $(KBr)_{1-x}(KCN)_x$ shares many properties with glasses; however, unlike glasses, its structure is well understood. In the next section, we will review some of what is known experimentally about the glassy behavior found for intermediate values of x, and give some important parameters describing the dilute ($x \approx 0$) and pure ($x = 1$) limits. In the third section we will discuss our mean-field theory[5–7] and describe its predictions for the dielectric loss peak and the time-dependent specific heat at $x = 0.5$. In the final section we will discuss the relationship with true glasses.

EXPERIMENTS ON $(KBr)_{1-x}(KCN)_x$

$(KBr)_{1-x}(KCN)_x$ forms cubic single crystals with a sodium chloride structure. The cyanide ions, CN^-, are shaped roughly like (North American) footballs, and randomly displace bromines in the lattice. Since the molecular size of the cyanide ion is close to that of bromine, mixed crystals of high quality with no apparent chemical clustering can be made over the whole range from the dilute limit $x \approx 0$ to the pure potassium cyanide limit $x = 1$ (FIG. 1). For very low cyanide concentrations ($x \lesssim 0.01$) the

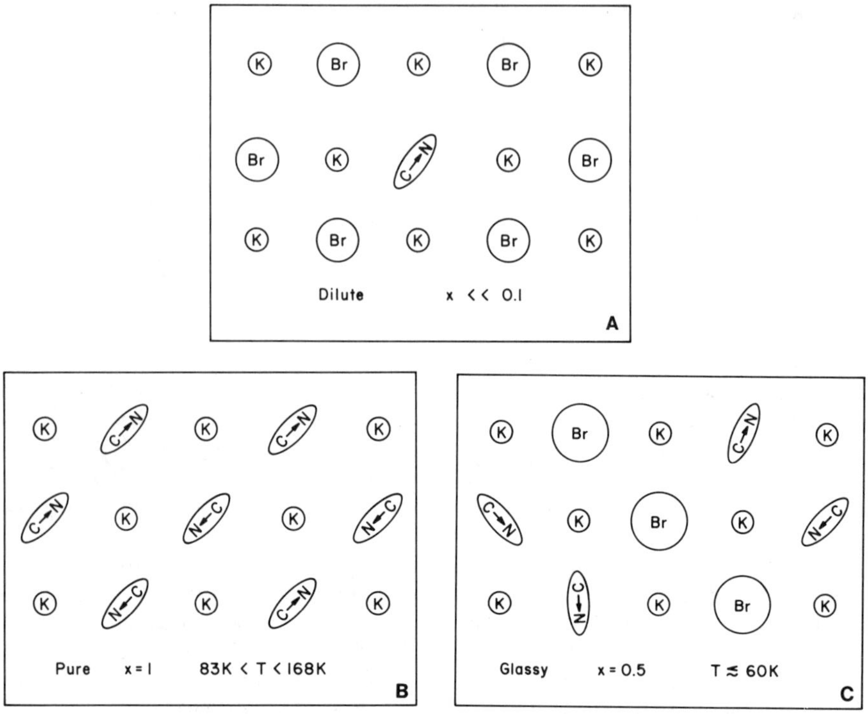

FIGURE 1. A theorist's conception of a typical (1,0,0) plane in (**A**) dilute, (**B**) pure, and (**C**) 50% $(KBr)_{1-x}(KCN)_x$. The arrows on the cyanide molecules indicate the direction of **n**. The cyanide molecule in **A** tilts out of the plane of the paper.

cyanides act as independent tunneling centers; for pure KCN the cyanides align into ordered phases at low temperatures (FIG. 2). For reasonably high cyanide concentrations ($0.6 < x < 1$) the cyanides still develop long-range orientational order, although the phase diagram becomes rather complicated. In the range $0.1 \lesssim x < 0.6$, the cyanides freeze into a disordered "orientational glass" state. In this section, we will discuss each of these ranges in turn, briefly summarizing the relevant experimental information.

In the dilute limit (FIG. 1A) each cyanide is isolated and has a cubic crystal environment. The cyanide ion has eight $\langle 111 \rangle$ equilibrium orientations,[9] with a barrier height of 36 K separating them. At low temperatures the cyanide tunnels between these equilibrium orientations, producing, for example, a Schottky anomaly peak in the specific heat and a dip in the thermal conductivity. This behavior is well understood.[10] In particular, we will use the effective moment of inertia of the cyanide ion measured in the dilute limit (2.65×10^{-39} g · cm^2);[11] this is larger than the corresponding moment of cyanide in vacuum, because of an effective mass contribution from the motion of surrounding ions during the tunneling process.[12,13] Note the other important fact that we learn from the dilute limit: the crystal field contribution to the local potential is small. In our model, we can safely ignore the 36 K contribution of the crystal field to the barrier height; the cyanides start as free rotors, and cyanide-cyanide interactions will determine the barrier heights.

Pure potassium cyanide (FIG. 1B) goes through two orientational phase transitions. The higher temperature ferroelastic transition at 168 K aligns the axes of the cyanide ions without introducing dipolar order; the lower temperature antiferroelectric transition at 83 K orders the cyanide dipoles.[14] We can conclude from this (and from measurements in the dilute limit) that the forces that distinguish between the carbon and nitrogen ends of the ion are relatively weak; we shall ignore the dipole-dipole interactions in our model, and concentrate on the stronger elastic forces, which in pure potassium cyanide act to align the quadrupolar axes of the ions. In the ferroelastic phase, each cyanide ion has two equilibrium orientations separated by 180°; thermally activated flips between these orientations produce a dielectric loss peak. From the temperature dependence of this peak estimated from combined dielectric loss, nuclear magnetic resonance, and ionic thermal conductivity measurements, we discover that the barrier height for flipping a cyanide at low temperatures is 1570 K; this gives us a measure of the cyanide-cyanide interaction strength. There is also a softening of one of the phonon modes associated with this transition. (This transition, which is first order, has been studied at length: we will not discuss it here.) The ferroelastic phase transition in a variety of alkali cyanides is well described by quadrupolar mean-field theories;[7,15–22] our model for the diluted system is a straightforward generalization of these theories. Finally, the asymmetry energy for a cyanide dipole in the low-temperature antiferroelectric phase (the energy difference between the preferred orientation and the metastable flipped orientation) has been measured to be roughly 340 K (L. Wu, Y. H. Jeong, and S. R. Nagel, personal communication); we shall use this as a crude estimate of the spread of asymmetry energies in the glassy cyanides.

For cyanide concentrations above $x = 0.6$, the low-temperature phase has long-range orientational order. The phase diagram in this region is quite complex;[23,24] the orthorhombic ferroelastic phase of the pure cyanide gives way to triclinic, monoclinic, and rhombohedral regions, which sometimes coexist. (There appears not to be any long-range dipolar order (J. M. Rowe, personal communication) in any phase

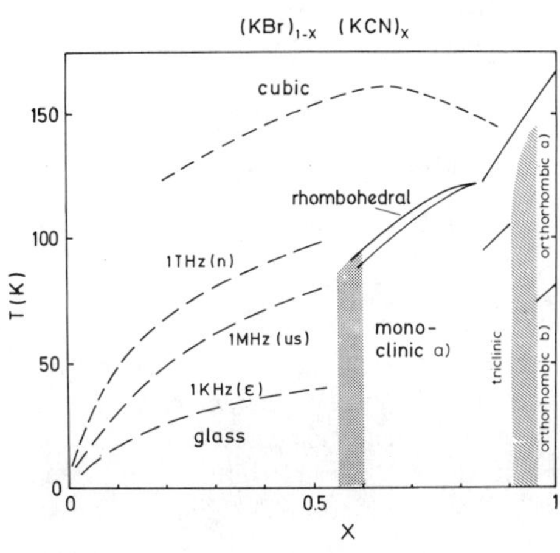

FIGURE 2. A phase diagram for $(KBr)_{1-x}(KCN)_x$. Note that the universal glassy properties extend into the region of long-range orientational disorder ($x = 0.7$). Reproduced here with permission, from Loidl and Knorr.[8]

other than the orthorhombic phase). Such complex phase diagrams are common in frustrated systems; the elastic interactions between cyanides depend on their relative positions, and can prefer either parallel or perpendicular relative orientations. We argue below, however, that this complexity is not central to the properties (low-temperature properties and dielectric loss) that we wish to understand; our model will ignore this frustration. The mean-field theory of Kanter and Sompolinsky,[25] which we will briefly discuss in the next section, is very frustrated, but ignores the dilution; their results would seem to indicate that the frustration is only partial. Finally, we will note but not discuss a recent announcement[26] that the orientational freezing transition

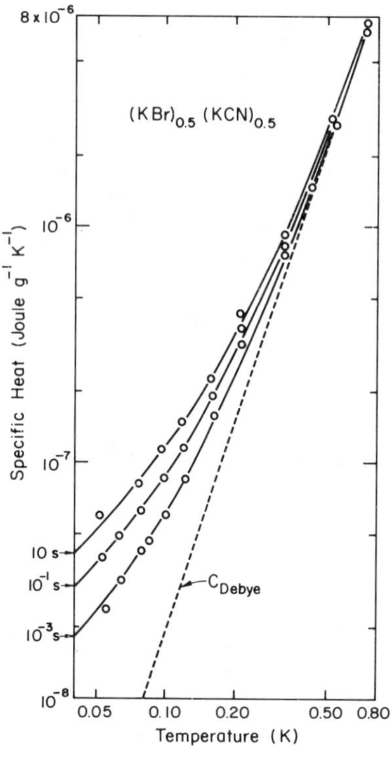

FIGURE 3. Specific heat, determined from the temperature rise of the quasi-adiabatic sample after the times t indicated. The curves are of the form $C = A(t)T + C_D T^3$, where C_D is the Debye value determined from elastic measurements. Adapted from Meissner *et al.*[6]

becomes second order at $x = 0.73$, in the middle of the range 0.6–1 of complex crystalline phases.

At low temperatures and intermediate concentrations, $(KBr)_{1-x}(KCN)_x$ shows all of the universal properties of glasses. The specific heat is proportional to T, and is time dependent (FIG. 3); the thermal conductivity is proportional to T^2, and crosses over at ~2 K into the plateau characteristic of glasses.[27] The dielectric constant is glassy.[28] Ultrasonic attenuation and phase velocities are characteristic of glasses.[29] It has recently been demonstrated that the low-temperature glassy properties persist into the crystalline region[6] to $x = 0.7$. Thus the glassy behavior is not closely related to the lack of long-range order in the cyanides; it may well be that the low-temperature properties of glasses are quite independent of the glass freezing transition.

Let us discuss the specific heat at length. Consider the specific heat measurements of Meissner and Knaak[6] for $(KBr)_{0.5}(KCN)_{0.5}$ shown in FIGURE 3. Note first that the specific heat at a fixed temperature depends on the time scale of measurement (in this experiment, on the time passed since the heat pulse was added to the sample). The specific heat near 0.1 K varies by nearly a factor of two as the measuring time changes from 10^{-3} to 10 sec. This reflects the long internal relaxation times for the cyanide tunneling centers. The tunneling centers in glasses always produce this time dependence: they have a wide range of tunneling rates, and absorb or emit energy as they come to equilibrium. As the time scale becomes longer, the measured specific heat includes more of these slow degrees of freedom, and therefore becomes larger. Note second that the specific heat crosses over from linear in T to cubic in T at a temperature around 0.5 K. The curves fit to the data are of the form $A(t)T + C_D T^3$. The linear term $A(t)$ is universally found in glasses; its magnitude and time dependence are within the normal range. The cubic term is not fit to the data; it is the Debye specific heat due to phonons calculated from the elastic constants. It gives a good fit to the data around 1 K (although beyond 1 K there are deviations). Usually in glasses the Debye contribution does not exhaust the cubic term, and there is an anomalous T^3 contribution to the specific heat. For $(KBr)_{0.5}(KCN)_{0.5}$ this anomalous term is absent within experimental accuracy.

The low-frequency behavior of $(KBr)_{1-x}(KCN)_x$ has caused some confusion in the past. There are several experimental probes of the dynamics that have been used. Ultrasonic, Brillouin, and inelastic neutron-scattering experiments were used to measure the temperature at which the sound velocity softened at high frequencies; dielectric loss measurements were used to measure characteristic relaxation rates at low frequencies. Two groups interpreted the results in terms of a single Arrhenius law—suggesting that the relaxation slows down continuously with a single activation energy and that there was no glass transition.[30,31] Not everyone accepted this view; Rowe, Rush, and Michel (personal communication) claimed that continuous slowing down was incompatible with their neutron-scattering measurements, that dielectric loss was probing a different susceptibility (in particular, was observing 180° cyanide reorientations), and that the sound velocity measurements when taken alone showed a much steeper temperature dependence (compatible with a freezing of the cyanide quadrupolar order). These ideas have now been confirmed by torsional oscillator experiments that overlap the frequency range of the dielectric loss measurements.[8,32] These measurements show that the frequency of the sound velocity anomaly does decrease very quickly with temperature, and is decoupled from the frequency of the dielectric loss peak. (It is tempting to identify the former with the α-relaxations in real glasses, but the analogy will probably not hold; the disorder here is fixed whereas in glasses the disorder is freezing in while the glass is forming.) Thus the elastic interactions between the cyanide quadrupoles freeze their long axes into a disordered configuration, but the cyanide ions remain able to flip over, just as in pure potassium cyanide.

The dielectric loss measurements gave more information than just a characteristic frequency. FIGURE 4 shows the frequency-dependent dielectric loss for $(KBr)_{0.5}(KCN)_{0.5}$, at three fixed temperatures, measured by Birge *et al.*[33] Notice three

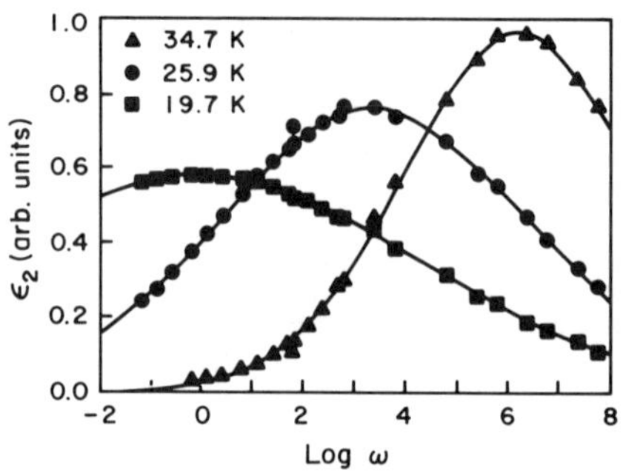

FIGURE 4. Dielectric loss as a function of frequency for $(KBr)_{0.5}(KCN)_{0.5}$. The horizontal scale is measured in decades. Note that ϵ_2 versus $\log_{10}\omega$ is shown for several temperatures with least-squares Gaussian fits (solid lines). Adapted from Birge *et al.*[33]

things about these data. First, the peaks span up to eight decades in frequency, indicating the very broad range of relaxation times characteristic also of glasses. (In contrast, the dielectric loss peak in KCN forms a good Debye peak with a width of 1.1 decades.) Second, the peak shifts to lower frequencies and broadens as the temperature is lowered. Indeed, the peak frequency follows an Arrhenius law over their experimental range; also, if the relaxation is due to a distribution of barrier heights, the peak should broaden like $1/T$ (differences in activation energies producing larger differences in rates at lower temperatures). Third, the loss peak is described beautifully by a log-normal distribution (the solid Gaussians in FIGURE 4).

Birge *et al.*[33] proposed that their log-normal loss peak was due to a Gaussian distribution of barrier heights hindering cyanide reorientations

$$P(V) = (\sqrt{\pi}\sigma_\nu)^{-1} \exp\,[(V-V_0)^2/\sigma_\nu^2] \tag{1}$$

Theories involving distributions of relaxation times have become unpopular in the study of glasses. Independent centers relaxing with a distribution of barrier heights suffer from the same criticism; almost any relaxation, whether due to independent or collective effects, can be described by a suitable distribution. It is crucial for credibility that the distribution be temperature independent. If the barrier crossing rate is Arrhenius

$$\omega(V) = \omega_0 e^{-V/k_B T} \tag{2}$$

the dielectric loss is well described by the distribution shown in equation 1 with a weak temperature dependence

$$\begin{aligned} V_0 &= 659\ \text{K} \\ \sigma_\nu &= 300\ \text{K} - 1.5\ T \end{aligned} \tag{3}$$

That is, the peak position of the loss and most of the width variation are characteristic of a distribution of barrier heights. The experimental data at different temperatures test different regions of the barrier height distribution; FIGURE 5 shows the range over which the dielectric loss measurements have checked the Gaussian form of the distribution.

Our model for $(KBr)_{1-x}(KCN)_x$ is based on the mean-field theory of ferroelastic freezing in pure potassium cyanide.[7] One could argue that it is overambitious to extrapolate from a theory of long-range crystalline order to a glassy state that is separated from it by several other crystalline phases. I claim that this is not so; the properties we are interested in are local, and do not depend strongly on frustration or nonexistence of long-range order. Despite the complexity of the phase diagram, the dielectric loss peak corresponding to the 180° cyanide reorientations in the pure cyanide broadens smoothly and continuously into the loss peak observed in the glassy material.[11] The glassy low-temperature properties, on the other hand, extend into the region with long-range crystalline order, as noted above. Thus there are good experimental reasons to believe that the low-frequency and low-temperature behavior can be described by a diluted ferroelastic mean-field theory.

MEAN-FIELD THEORY

In this section we will develop a mean-field theory for $(KBr)_{0.5}(KCN)_{0.5}$ in terms of a diluted lattice of interacting quadrupoles. There have been other theories applicable to this material;[34–37] indeed, Fischer and Klein[34] stimulated interest in it by predicting that it would have glassy low-temperature properties. The approach of Michel *et al.*[35] is probably closest to ours, although they concentrate on the glass transition and mostly do not explore lower temperature behavior.

First, we will discuss the dielectric loss. The Gaussian form for the peak will follow from the central limit theorem; its width will be explained as fluctuations in the number of nearest neighbors. The (weak) temperature dependence of the barrier height distribution will be explained from the temperature dependence of the order parameter. Briefly, the barrier height to reorienting a cyanide will be due to cyanide–cyanide interactions, and will become smaller as the surrounding cyanides become thermally disordered. Second, we will discuss the time-dependent specific heat. Extrapolating the Gaussian distribution to very low barrier heights, we can (largely independently of the mean-field theory) predict the number of active

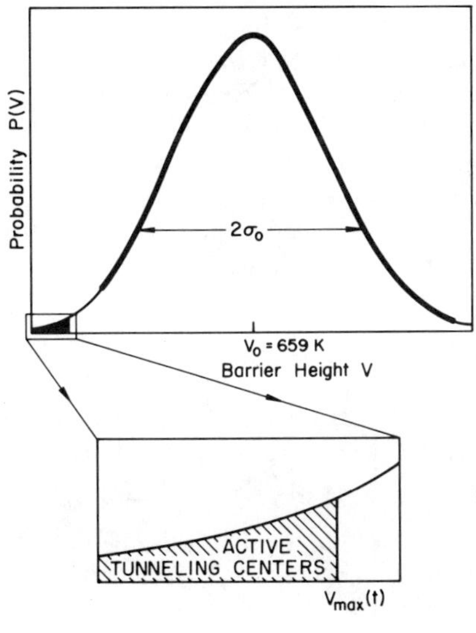

FIGURE 5. Main graph: Proposed Gaussian distribution of barrier heights for $(KBr)_{0.5}(KCN)_{0.5}$ at zero temperature. The darkened regions of the curve represent the range over which the dielectric loss measurements have tested the distribution. Expanded section: Low-barrier tail showing those tunneling centers that contribute to the specific heat below 1 K. Note that $V_{max}(t)$ is the largest barrier to reorientation that a cyanide molecule can tunnel through in a time t; V_{max} = 96 K for t = 1 sec.

tunneling centers as a function of the time scale. From this extrapolation, we are able to get a good description of the time dependence of the specific heat, and its absolute magnitude.

We could describe the orientation of a cyanide molecule by a vector **n** pointing from the carbon to the nitrogen. Since the cyanides remain free to flip by 180°, however, we should ignore the difference between the carbon and nitrogen ends of the molecule by writing Hamiltonians that are even in **n**. Thus, we shall use the quadrupole $Q^i_{\alpha\beta} = (n^i_\alpha n^i_\beta - \frac{1}{3}\delta_{\alpha\beta})$ to describe the orientation of ion i. Our Hamiltonian will give the elastic energy of the crystal in terms of the $Q_{\alpha\beta}$. It will describe the elastic forces on a cyanide due to its cyanide neighbors; these forces are mediated by lattice deformations.

There are two model Hamiltonians that have been used to study this problem. The one we will discuss at length incorporates only the effects of dilution. If we let $\chi_i = 1$ if a CN^- occupies site i, and $\chi_i = 0$ if a Br^- occupies site i, our Hamiltonian is given by

$$\mathcal{H}_D = -\frac{1}{2} J \sum_{\langle ij \rangle} Q^i_{\alpha\beta} Q^j_{\alpha\beta} \chi_i \chi_j \tag{4}$$

This Hamiltonian will generally have a ferroelastic ground state; some cyanides have more neighbors than others, but all cyanides prefer to be aligned. The mean-field theory for this Hamiltonian can easily be solved explicitly,[5] as we shall see.

The other model, studied by Kanter and Sompolinsky,[25] ignores dilution but incorporates frustration in a fashion paralleling spin glasses. Their Hamiltonian is

$$\mathcal{H}_F = \sum_{\langle ij \rangle} J^{\alpha\beta\gamma\delta}_{ij} Q^i_{\alpha\beta} Q^j_{\gamma\delta} \tag{5}$$

where the coefficients $J^{\alpha\beta\gamma\delta}_{ij}$ are, apart from obvious symmetries, independent random variables. They solve this approximately in mean field, using replica symmetry breaking. It is clearly true that their model contains more of the glassy physics; they have a multiplicity of disordered ground states. The trivial, unfrustrated way we have included the disorder produces many ground states only insofar as each cyanide has two orientations. In the microscopic energetics of the tunneling centers, however, surely the dilution is important (cyanides with few neighbors will have lower barriers), and surely the frustration is only partial (the cyanides are not dilute; all the near-neighbor cyanides will presumably have the same J_{ij}). Indeed, the width of the barrier height distribution predicted from Kanter and Sompolinsky's model is far too large, and the predicted distribution is not Gaussian.

For the diluted Hamiltonian of equation 4, the force experienced by a quadrupole at site i is given by the derivative

$$\frac{\partial \mathcal{H}_D}{\partial Q^i_{\alpha\beta}} = -J \sum_{j \text{ nbr of } i} \chi_j Q^j_{\alpha\beta} \tag{6}$$

We now make a mean-field approximation that all neighbors are equivalent to an average neighbor with field strength $\mathbf{Q}_{\alpha\beta}$.[b] Thus, if the number of neighbors is

[b] This is exact on a Bethe lattice with a random number of neighbors, if the average coordination number Z_0 is taken to infinity keeping the probability distribution $P(Z/Z_0)$ of having Z neighbors fixed.

$\Sigma_{j \text{ nbr of } i} \chi_j = N_i$

$$\left[\frac{\partial \mathcal{H}_D}{\partial Q^i_{\alpha\beta}}\right]^{MF}_T = -JN_i\mathbf{Q}_{\alpha\beta} \tag{7}$$

With the assignment $J_i = JN_i$, this gives the effective mean-field Hamiltonian

$$\mathcal{H}_{\text{eff}} = -JN_i\mathbf{Q}_{\alpha\beta}Q^i_{\alpha\beta} = -J_i\mathbf{Q}_{\alpha\beta}Q^i_{\alpha\beta} \tag{8}$$

Now the average neighbor is not an average cyanide. Those cyanides that have more cyanide neighbors affect proportionately more cyanides, and thus contribute more to the mean field. Let there be M cyanides in the sample. Then the probability of a cyanide having N neighbors is

$$P(N) = \left(\sum_{i=1}^{M} \delta_{NN_i}\right) \Big/ M \tag{9}$$

On the other hand, the probability that a neighbor j of a cyanide has N neighbors is

$$\begin{aligned} P'(N) &= \left(\sum_{i=1}^{M}\sum_{j=1}^{N_i} \delta_{NN_j}\right) \Big/ \left(\sum_{i=1}^{M} N_i\right) \\ &= \left(\sum_{j=1}^{M} N_j\delta_{NN_j}\right) \Big/ (MN_0) \\ &= \frac{N}{N_0}\left(\frac{\sum_{j=1}^{M} \delta_{NN_j}}{M}\right) \\ &= \frac{N}{N_0} P(N) \end{aligned} \tag{10}$$

where N_0 is the average number of neighbors. That is, each cyanide is counted once in $P(N)$, but counted N_i times as a neighbor in $P'(N)$. The average quadrupole moment of a cyanide depends on the number of its neighbors.

$$Q(N) = \frac{Tr(Qe^{-\beta NJ\mathbf{Q}Q})}{Tr(e^{-\beta NJ\mathbf{Q}Q})} \tag{11}$$

Thus the average neighbor moment $\mathbf{Q}$ will then be an average over N, weighted by

$$\begin{aligned} \mathbf{Q}_{\alpha\beta} &= \sum_N P'(N)Q_{\alpha\beta}(N) \\ &= \sum_N \frac{N}{N_0} P(N)Q_{\alpha\beta}(N) \\ &= \int \mathrm{d}J \frac{J}{J_0} P(J)[Q^J_{\alpha\beta}]_T \end{aligned} \tag{12}$$

where J_0 is the mean of the distribution $P(J)$.

By a change of coordinates, $\mathbf{Q}_{ij}$ can be diagonalized. We assume $\mathbf{Q}$ is uniaxial, with major axis in the z direction. We can then write the traceless symmetric tensor $\mathbf{Q}$ in terms of a single function $\alpha(T)$

$$\mathbf{Q}_{ij} = \begin{pmatrix} -\alpha(T) & 0 & 0 \\ 0 & -\alpha(T) & 0 \\ 0 & 0 & 2\alpha(T) \end{pmatrix} \tag{13}$$

In terms of α, $\mathcal{H}_{\text{eff}}$ is then

$$\mathcal{H}_{\text{eff}} = J\alpha(T)(1 - 3n_z^2) \tag{14}$$

and the barrier height for a 180° flip of a cyanide orientation is

$$V = 3J\alpha(T) \tag{15}$$

Thus within our model, the distribution of coupling constants J is directly scaled into a distribution of barrier heights V by a temperature-dependent order parameter $\alpha(T)$. We must now discuss the form of $P(J)$. The coupling of an ion to the mean field, and therefore the barrier hindering the dipolar reorientation, will be a sum of contributions from the (randomly distributed) neighboring cyanides. Our model assumes that these contributions are uncorrelated; since each cyanide has 12 near neighbors, the distribution is given by the binomial distribution

$$P(N) = 2^{-12}\binom{N}{12} \tag{16}$$

This distribution is very nearly Gaussian; if we assume[c] that next-nearest-neighbor interactions are large enough to smear out the discreteness in the tails, the central limit theorem implies a Gaussian form for $P(J)$

$$P(J) = (\sqrt{\pi}\sigma_J)^{-1} \exp\left[-(J - J_0)^2/\sigma_J^2\right] \tag{17}$$

where the mean coupling J_0 will scale linearly with the cyanide concentration x. (This also implies that the transition temperature scales roughly with x,[7,35] which is pretty accurate (FIG. 2).) Using equation 9, this leads to a barrier height distribution of precisely the form seen in equation 1 with V_0 for $(KBr)_{0.5}(KCN)_{0.5}$ half that of the 1570 K Arrhenius barrier height for the dielectric loss peak in pure KCN. Thus the mean-field prediction for the mean coupling strength in $(KBr)_{0.5}(KCN)_{0.5}$ is 20% higher than the experimental value $V_0 = 659$ K. The ratio of the width to the peak of

[c]This is a crucial assumption, as the entire contribution to the specific heat comes from cyanides with "between zero and one neighbors." On the one hand, we can see in FIGURE 5 that the dielectric loss peak remains Gaussian and smooth to quite high barriers; if it is symmetric and represents the barrier height distribution our extrapolation is justified. On the other hand, $P(J)$ is convolved with a Debye peak to produce the dielectric loss, which may smooth out effects of discreteness. There is a recent report of evidence for some kind of discreteness in Brillouin scattering off $(KBr)_{0.5}(KCN)_{0.5}$.[38]

the (nearly Gaussian) binomial distribution is $1/\sqrt{6}$; this prediction is 12% smaller than the experimental ratio $\sigma/V_0 = 300\text{ K}/659\text{ K}$.

In the rest of the paper, we will use the experimental values for the zero temperature barrier height distribution (equation 3) to determine the distribution of coupling constants (rather than using our extrapolations from the pure KCN). Thus since $\alpha(0) = 1/3$

$$J_0 = V_0 = 659\text{ K}$$

$$\sigma_J = \sigma_V(T = 0) = 300\text{ K} \tag{18}$$

(See FIGS. 4 & 5.)

We can find the temperature-dependent order parameter $\alpha(T)$ by solving the mean-field theory using the standard self-consistent calculation. The mean alignment

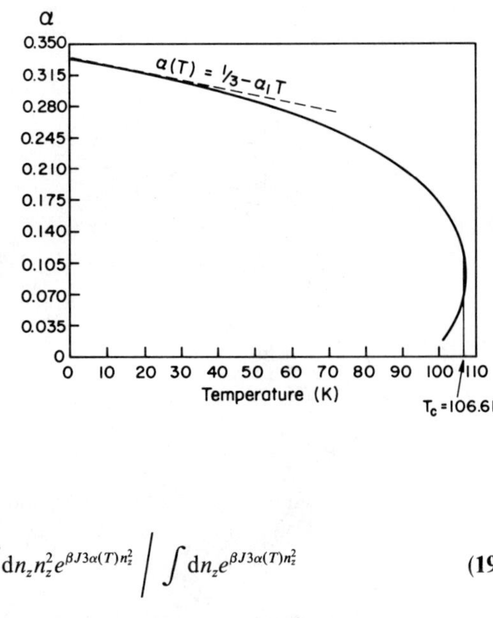

FIGURE 6. Mean-field order parameter α as a function of temperature, for the Gaussian distribution of coupling strengths given by the experimental parameters of equation 18. The ferroelastic transition temperature T_c obtained from minimizing the free energy is 106.6 K. (This temperature need not correspond to the glass transition temperature; our model does not describe the glass transition.)

of a given cyanide

$$\langle (n_z^J)^2 \rangle = \int \mathrm{d}n_z n_z^2 e^{\beta J 3\alpha(T) n_z^2} \bigg/ \int \mathrm{d}n_z e^{\beta J 3\alpha(T) n_z^2} \tag{19}$$

depends on the strength $\alpha(T)$ of the mean field. The mean-field strength

$$\alpha(T) = \frac{1}{2} \int \mathrm{d}J P(J) \frac{J}{J_0} \left\langle (n_z^J)^2 - \frac{1}{3} \right\rangle_T \tag{20}$$

depends in turn on the alignment of the cyanides. In FIGURE 6, we may see the results of our self-consistent calculation for $\alpha(T)$. (Details can be found in Sethna and Chow.[5])

At low temperatures, $\alpha(T)$ is approximately linear in T. As for pure KCN,[7,33] we can see that the peak frequency of the dielectric loss will continue to have an Arrhenius form. The peak J_0 of the coupling constant distribution corresponds to a temperature-dependent peak $V_0(T) = 3J_0\alpha(T)$ of the barrier height distribution. If $\alpha(T) = (1/3) -$

$\alpha_1 T$, then barrier crossing over the temperature-dependent barrier $V_0(T)$ is indistinguishable from crossing the zero temperature barrier $V_0(0) = J_0$ with a different prefactor

$$\omega_0 e^{-V_0(T)/k_BT} = \omega_0 e^{-3J_0\alpha(T)/k_BT} = (\omega_0 e^{3J_0\alpha_1/k_B})e^{-J_0/k_BT} \tag{21}$$

Thus the measured Arrhenius slope should be temperature independent, as is observed experimentally. The width of the barrier height distribution (and thus of the dielectric loss peak), however, scales directly with $\alpha(T)$

$$\sigma_V = 3\alpha(T)\sigma_J \tag{22}$$

The value of α_1 leads to a prediction $\sigma_V(T) = \sigma_V(0) - 0.708T$; our mean-field theory predicts roughly half of the experimental temperature dependence (equation 3).

Thus fluctuations in the number of nearest neighbors plausibly explains the Gaussian form, width, and peak position of the dielectric loss in $(KBr)_{0.5}(KCN)_{0.5}$. The temperature dependence of the width is explained within a factor of two by describing the thermal disorder in the neighboring cyanides using a mean-field theory with a

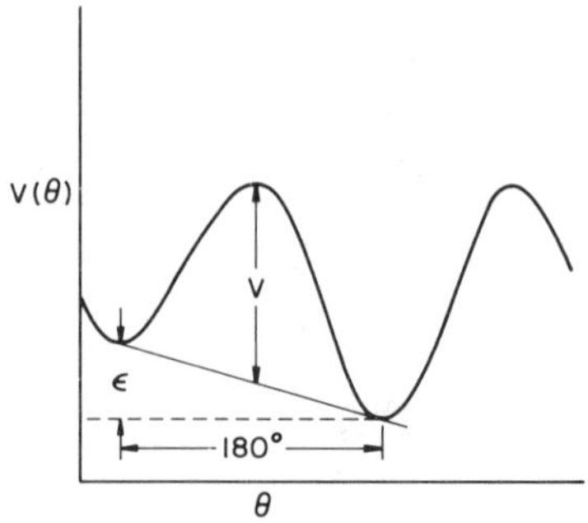

FIGURE 7. A typical double-well potential plotted against θ, the orientation angle of the cyanide molecule. The barrier height V and asymmetry ϵ are functions of the local environment of the cyanide.

distribution of coupling constants. The low-frequency dielectric loss is due to cyanides independently crossing over a distribution of barriers; because the barriers are due to cyanide-cyanide interactions, they become smaller as the neighboring cyanides become thermally disordered.

We have explained dielectric loss in terms of 180° cyanide reorientations that were thermally activated; the specific heat will be explained by 180° cyanide reorientations mediated by quantum tunneling. We will get the magnitude of this time-dependent specific heat within a factor of two. To give some context for this result, we can compare it to a naive extrapolation from low concentrations: if the cyanides did not interact, and each contributed the Schottky anomaly characteristic of dilute impurities, the specific heat at 0.1 K would be 10^5 times larger than that observed on a 10-sec time scale.[28] The factor of 10^5 reduction of the specific heat comes out of the calculation in two ways. First, interactions between cyanides raise the barrier to reorientation. Unlike dilute cyanides, most cyanides in the glassy regime will not relax by quantum tunneling on a human time scale, and will not contribute to the specific heat. Second, interactions between cyanides produce an asymmetry between the two preferred orientations. Only those cyanides whose two orientations are degenerate within $k_B T$ will contribute to the specific heat; most cyanides are frozen into their lower state.

As mentioned in the introduction, the low-temperature properties of glasses have traditionally been explained using tunneling centers.[3,4] These centers are formed by atoms or groups of atoms of unknown microscopic structure, with two metastable configurations. Just as for the cyanide impurities, these centers are characterized by a barrier height V and an asymmetry ϵ (FIG. 7). Because the microscopic structure of the centers is unknown in glasses, little can be said about the distribution of tunneling center parameters $P(\epsilon, V)$. Significant conclusions, however, can be drawn on very general grounds. For example, the distribution must be even in the energy asymmetry $P(\epsilon, V) = P(-\epsilon, V)$ (each center has two wells, one higher in energy, one lower), and smooth on scales of tenths of eV. Thus, assuming $P(0, V) \neq 0$, the density of tunneling states is roughly constant in the energy range of interest. Roughly speaking (see below), because only those centers with $\epsilon \lesssim kT$ contribute, this leads to a specific heat linear in temperature.

By using our microscopic model of the tunneling centers in $(KBr)_{1-x}(KCN)_x$, we will be able to make more detailed predictions about the low-temperature properties. In particular, we have definite information about the distribution of tunneling centers, gleaned from our understanding of the isolated cyanide impurity and the pure alkali cyanides. The traditional model *assumes* a flat distribution in the tunneling parameter, which corresponds to a distribution $P(V, \epsilon) \propto V^{-1/2}$. From the dielectric loss experiments, we know the distribution of barrier heights is the Gaussian of equation 1. We do not have a direct experimental probe of the distribution of asymmetries ϵ; however, the inverse density of centers near zero asymmetry should be roughly given by the asymmetry energy ~340 K between the two orientations in the pure cyanide antiferroelectric phase.

The rate Γ of tunneling by an angle θ through a barrier of height V is given by a simple WKB calculation

$$\Gamma(V) = \Gamma_0 \exp\left(-A\theta\sqrt{2IV}/\hbar\right) \tag{23}$$

where A is a numerical constant; for a sinusoidal $V(\theta)$, $A = 4/\pi$. Because our cyanides are reorienting by 180°, $\theta = \pi$. The magnitude of the attempt frequency Γ_0 has been estimated in the pure cyanide to be 8.3×10^{13} Hz;[7] our results are quite insensitive to the value chosen for this prefactor. We use $I = 2.65 \times 10^{-39}\ \mathrm{g \cdot cm^2} \approx 16\ \mathrm{amu \cdot Å^2}$. The probability that a cyanide molecule with barrier height V will flip over within t seconds is thus $1 - \exp(-\Gamma(V)t)$. In calculating the specific heat, we will weight the contribution of each cyanide by this factor. To get a rough idea of the number of active cyanides at a measuring time t, let us define a maximum crossable barrier $V_{max}(t)$, for which $\Gamma(V_{max}(t)) = t^{-1}$. For $t = 1$ sec, $V_{max}(t) = 96$ K (FIG. 5).[d] This implies (evaluating

[d] At temperatures above about 2 K, the dominant relaxation process on a 1-sec time scale will be thermal barrier crossing. This will lead to a temperature dependence of the number of active tunneling centers, which gives a small contribution to the specific heat that goes roughly as T^3. In addition, there should be modes within the individual wells (librational modes) that give a similar contribution. Perhaps the anomalous T^3 specific heat in real glasses will be explained eventually by these effects. The experimental value for this term in $(KBr)_{0.5}(KCN)_{0.5}$ is compatible with zero, however, and the theoretical value for the effects of thermal activation and libration are swamped by the Debye phonon contribution. There appears to be a substantial anomalous T^3 term for $x = 0.25$; we await measurements of the dielectric loss.

an error function) that 0.35% of the cyanide molecules have a flip rate higher than 1 Hz. Thus roughly one cyanide in 300 can tunnel in 1 sec, and one in 340 has asymmetry small enough to contribute to the specific heat below 1 K; crudely, these together explain why the specific heat is down by 10^5.

More explicitly, at low temperatures, the cyanide double-well center can for most purposes be considered a two-level system. The lowest two states of the double well can be nearly degenerate, and higher states are not accessible at low temperatures.[e] We can truncate, keeping only these states; we write the Hamiltonian $\mathcal{H}_{2L}$ in a "position" eigenstate basis (where (1, 0) and (0, 1) correspond to the two stable orientations of the cyanide). The two-level system is then defined by the asymmetry ϵ and a tunneling matrix element Δ

$$\mathcal{H}_{2L} = \begin{pmatrix} -\epsilon/2 & -\Delta \\ -\Delta & \epsilon/2 \end{pmatrix} \tag{24}$$

The energy splitting between the two levels is $2[(\epsilon^2/4)^2 + \Delta^2]^{1/2}$. Because $\Delta \ll \epsilon$ for most centers contributing to the specific heat, we approximate the energy splitting by ϵ.[f] As in the traditional models, we make the approximation that $P(\epsilon, V) = n_0 P(V)$ (that is, is independent of the asymmetry ϵ) for $\epsilon \lesssim kT$. With these approximations, the specific heat of the distribution of two-level systems is[3,4]

$$A(t) = \frac{\pi^2}{6} k_B^2 n_0 \int_0^\infty dV P(V)[1 - \exp(-\Gamma(V)t)] \tag{25}$$

Fitting the experimental specific heat at $t = 1$ sec gives the value $n_0 = 1/(k_B \times 292\text{ K})$. This agrees well with the inverse of the asymmetry (340 K) found for the pure cyanide (L. Wu, Y. H. Jeong, and S. R. Nagel, personal communication).[g] We then use equation 25 to predict the time dependence of the linear term in the specific heat (FIG. 8). This one parameter fit is in good agreement with the experimental data, and is almost as good as the two-parameter fit given by the traditional models (straight line in FIG. 8).

CONCLUSIONS ABOUT GLASSES

We have explained that the time-dependent specific heat and the dielectric loss peak in $(KBr)_{0.5}(KCN)_{0.5}$ are due to 180° cyanide reorientations. These properties are universal in glasses; all glasses show a broad distribution of relaxation times, and all show the same time-dependent linear term in the specific heat. We have deliberately chosen to study these universal properties in a single well-characterized material. Many rather general and abstract theories of glassy behavior have been motivated by

[e] They can be accessed virtually, however, and are important in mediating tunneling in the presence of interactions.[12,13]

[f] We have estimated the importance of Δ in the specific heat of $(KBr)_{1-x}(KCN)_x$; it produces slight deviations from the linear form at low temperatures.

[g] This agreement is probably accidental; the two numbers should agree maybe to within a factor of two.

the idea that universal properties cannot depend on details of the particular glass. The converse, however, is also true—an explanation for a universal property must apply to each individual glass, and in particular to $(KBr)_{0.5}(KCN)_{0.5}$.

Although the microscopic structure of $(KBr)_{1-x}(KCN)_x$ is well understood, we emphasize that our model is not microscopically derived: we ignore the frustration completely. The elastic interactions that freeze the quadrupoles do not typically favor parallel relative orientations; they compete, and this competition is responsible both for the variety of phases between $x = 0.6$ and $x = 1$ and for the orientational glass state for $x < 0.6$. Although the results of Kanter and Sompolinsky[25] suggest that completely random interactions do not form a sensible model for this material, they do demonstrate that new physical effects—particularly level repulsion causing a reduced density of low barriers—should very generally follow from incorporating frustration. We may suggest that the dilution substantially reduces the effects of frustration and "level repulsion," but we should certainly agree that a model calculation with more realistic interactions ought to be done.

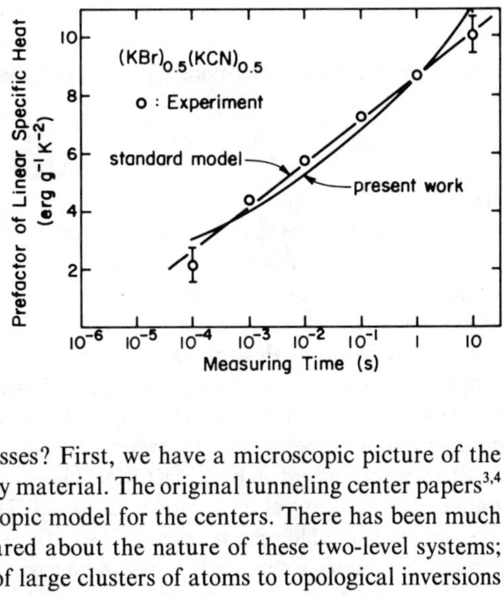

FIGURE 8. The open circles are the prefactors $A(t)$ of the measured linear specific-heat anomaly (FIG. 2). The curve is the result of our calculation; it has one free parameter n_0 that we can independently estimate within a factor of two. The straight line is the prediction of the standard model; both the slope and the intercept are free parameters.

What have we learned about glasses? First, we have a microscopic picture of the tunneling centers in at least one glassy material. The original tunneling center papers[3,4] did not attempt to provide a microscopic model for the centers. There has been much speculation since these papers appeared about the nature of these two-level systems; proposals run from rearrangements of large clusters of atoms to topological inversions of defect lines.[39] Although these exotic centers may occur in some materials, in this case ordinary tunneling defects of the form well known in alkali halides[10] determine the glassy behavior.

Second, we have started with roughly one tunneling center per molecular unit, and explained the magnitudes of the universal properties. The form of the specific heat is naturally explained by the traditional tunneling center models, but the small variation found in the magnitude is surprising: why should all glasses have the same density of active tunneling centers (within a factor of 10)? At least in this case, the density arises naturally from the statistical origin of the distribution of barrier heights.

Third, we have connected the low-temperature and low-frequency properties, explaining both in terms of the same centers. Our discussion cannot be directly transferred to glasses in general. Dielectric loss peaks are not usually of the log-normal form found in $(KBr)_{0.5}(KCN)_{0.5}$. Apart from another glassy crystal,[40] most low-frequency β-relaxation peaks found in glasses are asymmetric, often fit by the Williams–Watts function mentioned in the introduction. It is possible that the glassy crystals are in fact atypical. On the other hand, it is intriguing to speculate that they may be ideal, clean systems, where the low-frequency physics is not obscured by dirt effects so far unavoidable in normal glasses. What we can surely conclude from the cyanide system is that the low-frequency and low-temperature properties are at times linked; we should look for a common origin in glasses as well.

Finally, the $(KBr)_{1-x}(KCN)_x$ system may provide answers to many of the remaining puzzles about glasses. We have already mentioned a possible role for thermal reorientations and librational modes of the cyanides in explaining the anomalous T^3 specific heat; this would presumably carry over directly to normal glasses. The plateau in the thermal conductivity between 1 K and 10 K has also provoked much theoretical speculation over the years, without a satisfactory resolution. A rough calculation of the low-temperature T^2 thermal conductivity (due to resonant scattering from the cyanide centers) has been presented;[6] we are currently working to understand the thermal conductivity in more detail.[41]

We believe that more insight is yet to come from the glassy crystals, as new experiments are done and more sophisticated theoretical methods are used.

SUMMARY

Glasses are amorphous like liquids, but are rigid like crystals. Unlike a crystal, in which each atom has a set position, a glass will have a completely different configuration of atoms each time it is formed. Sometimes atoms or local groups of atoms will have two low-energy configurations and will be able to shift between them. Some of these atoms shift much faster than others, leading to a broad distribution of relaxation times. These two-level systems are also responsible for the universal low-temperature properties shared by all configurational glasses. The microscopic configurations of the two-level systems, however, have remained a mystery.

The mixed crystal $(KBr)_{1-x}(KCN)_x$ shares many properties with glasses; however, unlike glasses its structure is well understood. The two-level systems in this material are cyanide ions which can reorient by 180°. Because each cyanide has a different set of cyanide neighbors, some can flip over more easily than others: there is a distribution of barrier heights hindering the 180° rotations. The broad distribution of relaxation times exhibited by the dielectric loss measurements comes from random thermal rotations of cyanide ions over these barriers. The time-dependent specific heat (one of the universal low-temperature properties of glasses) comes from quantum tunneling of these ions through the same barriers.

ACKNOWLEDGMENTS

This work was originally done in collaboration with K. Chow, M. Randeria, S. R. Nagel, T. V. Ramakrishnan, J. J. DeYoreo, and R. O. Pohl.

REFERENCES

1. ANDERSON, P. W. 1984. Basic Notions of Condensed Matter Physics. Benjamin/Cummings. Menlo Park, CA.
2. ZELLER, R. C. & R. O. POHL. 1971. Phys. Rev. B **4:** 2029.
3. ANDERSON, P. W., B. I. HALPERIN & C. M. VARMA. 1972. Philos. Mag. **25:** 1.
4. PHILLIPS, W. A. 1972. J. Low Temp. Phys. **7:** 351.
5. SETHNA, J. P. & K. S. CHOW. 1985. Phase Transitions **5:** 317.
6. MEISSNER, M., W. KNAAK, J. P. SETHNA, K. S. CHOW, J. J. DEYOREO & R. O. POHL. 1985. Phys. Rev. B (Rapid Communications) **32:** 6091.
7. SETHNA, J. P., S. R. NAGEL & T. V. RAMAKRISHNAN. 1984. Phys. Rev. Lett. **53:** 2489.
8. LOIDL, A. & K. KNORR. 1986. Ann. N.Y. Acad. Sci. This volume.
9. BEYELER, H. U. 1975. Phys. Rev. B **11:** 3078.
10. NARAYANAMURTI, V. & R. O. POHL. 1970. Rev. Mod. Phys. **42:** 201.
11. LÜTY, F. & J. ORTIZ-LOPEZ. 1983. Phys. Rev. Lett. **50:** 1289.
12. SETHNA, J. P. 1981. Phys. Rev. B **24:** 698.
13. SETHNA, J. P. 1982. Phys. Rev. B **25:** 5050.
14. LÜTY, F. 1982. *In* Defects in Insulating Crystals. V. M. Turkevich & K. K. Swartz, Eds.: 69. Springer–Verlag. Berlin.
15. MICHEL, K. H. & J. NAUDTS. 1977. Phys. Rev. Lett. **39:** 212.
16. MICHEL, K. H. & J. NAUDTS. 1977. J. Chem. Phys. **67:** 547.
17. MICHEL, K. H. & J. NAUDTS. 1978. J. Chem. Phys. **68:** 216.
18. MAHANTI, S. D. & D. SAHU. 1982. Phys. Rev. Lett. **48:** 936.
19. SAHU, D. & S. D. MAHANTI. 1982. Phys. Rev. B **26:** 2981.
20. HOLDEN, A. J., V. HEINE, J. C. INKSON, C. M. VARMA & M. A. BOSCH. 1979. J. Phys. C **12:** 1035.
21. MOKROSS, B. J. & R. PIRC. 1978. J. Chem. Phys. **68:** 4823.
22. BOUNDS, D. G., M. L. KLEIN & I. R. MCDONALD. 1981. Phys. Rev. Lett. **46:** 1682.
23. ROWE, J. M., J. J. RUSH & S. SUSMAN. Phys. Rev. B **28:** 3506.
24. KNORR, K. & A. LOIDL. Phys. Rev. B **31:** 5387.
25. KANTER, I. & H. SOMPOLINSKY. Submitted for publication.
26. KNORR, K., A. LOIDL & J. K. KJEMS. 1985. Phys. Rev. Lett. **55:** 2445.
27. DE YOREO, J. J., M. MEISSNER, R. O. POHL, J. M. ROWE, J. J. RUSH & S. SUSMAN. 1983. Phys. Rev. Lett. **51:** 1050.
28. MOY, D., J. N. DOBBS & A. C. ANDERSON. 1984. Phys. Rev. B **29:** 2160.
29. BERRET, J. F., M. MEISSNER, P. DOUSSINEAU, A. LEVELUT & W. SCHÖN. 1985. Phys. Rev. Lett. **55:** 2013.
30. BHATTACHARYA, S., S. R. NAGEL, L. FLEISHMAN & S. SUSMAN. 1982. Phys. Rev. Lett. **48:** 1267.
31. LOIDL, A., R. FEILE & K. KNORR. 1982. Phys. Rev. Lett. **48:** 1263.
32. VOLKMANN, U. G., R. BÖHMER, A. LOIDL, K. KNORR, U. T. HÖCHLI & S. HAUSSÜHL. 1986. Phys. Rev. Lett. **56:** 1716.
33. BIRGE, N. O., Y. H. JEONG, S. R. NAGEL, S. BHATTACHARYA & S. SUSMAN. 1984. Phys. Rev. B **30:** 2306.
34. FISCHER, B. & M. W. KLEIN. 1979. Phys. Rev. Lett. **43:** 289.
35. MICHEL, K. H., J. NAUDTS & B. DE RAEDT. 1978. Phys. Rev. B **18:** 648.
36. KLENIN, M. A. 1983. Phys. Rev. B **28:** 5199.
37. IHM, J. 1985. Phys. Rev. B (Rapid Communications) **31:** 1674.
38. WALTON, D., J. J. VANDERWAL & Z. HU. 1985. Phys. Rev. Lett. **55:** 1307.
39. DUFFY, D. M. & N. RIVIER. 1981. Physica. **108B:** 1261.
40. HÖCHLI, U. T. 1982. Phys. Rev. Lett. **48:** 1494.
41. RANDERIA, M. & J. P. SETHNA. Work in progress.

DISCUSSION OF THE PAPER

A. NAVROTSKY (*Princeton University, Princeton, NJ*): The model for a range of activation energies rests on the binomial distribution of next-nearest neighbor occupancies, that is, a random CN-Br distribution. Do experiments detect either clustering (positive ΔH_{mix}) or ordering (negative ΔH_{mix})? If there is local order, how seriously are the theoretical predictions affected? What would be the stable low-temperature structure, as opposed to glass? [NOTE: This same question is answered in the discussion of the paper by A. Loidl and K. Knorr in this volume—*Ed.*]

SETHNA: The mixed crystal $(KBr)_{1-x}(KCN)_x$ forms good single crystals, which cleave nicely and are optically transparent, throughout the concentration range. This indicates that there is no large-scale inhomogeneity. I naturally will defer to my experimental colleagues for further details, but I know of no measurements that rule out clustering or ordering on a local level. I can say that, although the qualitative features of the theory would survive some clustering, the Gaussian form for the barrier height distribution assumes independent contributions to the barrier height—clustering even on a very local level would distort the distribution. (In some ways this may be evidence against clustering because the Gaussian is observed experimentally.)

Let me also emphasize here that our model is not an explanation from first principles. In order to produce the observed behavior, we assumed that the interactions were ferroelastic, tending to align the cyanides. Although it is true that the high-temperature phase in the pure cyanide is ferroelastic, the microscopic interactions are surely more complex, and frustrated. I have hopes that dilution may reduce the frustration, but certainly a numerical simulation with honest pair interactions is necessary before the Gaussian distribution of barrier heights can be considered understood.

J. T. BENDLER (*General Electric Research and Development Center, Schenectady, NY*): I wish to agree with Dr. Sethna's comment that any nonexponential time-decay law may be represented in terms of a superposition of exponential decay processes, and in the case of the Kohlrausch function, this leads directly to the L'evy stable probability densities

$$e^{-(t/\tau)^{\alpha}} = \int_0^{\infty} \rho(\mu, \alpha)\, e^{-(t/\tau)^{\mu}}\, d\mu$$

with $\mu = C/T$ a dimensionless relaxation rate. The inverse transform is an asymmetric stable density of characteristic exponent α (Montroll, E. W. & J. T. Bendler, *J. Stat. Phys.* **34:** 129, 1984). These solutions to the Chapman–Kolmogorov equation were studied by L'evy as generalizations of (or exceptions to) the central limit theorem, and were termed stable because, like the Gaussian, distributions of sums of stable random variables are invariant to addition of further variables. Asymptotically they behave as power laws, and hence have interesting scaling properties, which Mandelbrot has exploited in modeling phenomena with self-similarity (Mandelbrot, B. B., The Fractal Geometry of Nature, W. H. Freeman, San Francisco, CA, 1982).

If one interprets the Kohlrausch law as resulting directly from jumps over barrier distributions, it is hard to see why a stable density appears as a natural choice. On the other hand, if the dipole relaxation is dependent on another process, such as defect diffusion, the defects may hop over a Poisson barrier distribution (Bendler, J. T. & M. F. Schlesinger, *Macromolecules* **18:** 591, 1985), and this, folded into the first-passage time distribution to reach the dipole, generates the Kohlrausch law (Schlesinger, M. F. & E. W. Montroll, *Proc. Natl. Acad. Sci. USA* **81:** 1280, 1984). In polymers, bond isomerizations display activation barriers even in dilute solution so that a Gaussian barrier distribution did not seem appropriate.

Let me emphasize that I agree that the Kohlrausch function does not fit the KBr-KCN dispersion, although the Wagner function does. In this case of low-temperature relaxation, the direct model described by Dr. Sethna is reasonable. It is for the high-temperature (Kohlrausch) behavior that something else is needed.

SETHNA: The fact that the binomial distribution converges to a Gaussian is no accident. It is a special case of the central limit theorem; the sum of many uncorrelated random quantities, if they are "well behaved," will have a Gaussian distribution. In particular, the sum of any number of uncorrelated quantities, each of which has a Gaussian distribution, automatically is Gaussian: the Gaussian distribution is stable. Dr. Bendler points out that there are other stable distributions with longer tails, which have been discussed in the context of $1/f$ noise and the Kohlrausch form for the α-relaxation in glasses. There are two relevant issues I should mention, however, to avoid confusion.

First, as I understand it, these other distributions are sums of random times, whereas the log-normal frequency distribution in $(KBr)_{1-x}(KCN)_x$ comes from sums of random contributions to the barrier height.

Second, at least within our model, there is no question that the contributions to the barrier height are well behaved. The contributions may or may not be correlated (that is, clustering might be important, as Dr. Navrotsky points out), and they may or may not sum to form the total barrier height (that is, the quadrupolar fields add but the barrier, as an eigenvalue difference, may not, as Kanter and Sompolinsky point out). But surely each contribution has a finite range of possible values (no long tail), and the contributions have roughly equal magnitudes; they are well-enough behaved.

Three Coupled Relations for Relaxations in Complex Systems[a]

K. L. NGAI, R. W. RENDELL,[b] A. K. RAJAGOPAL, AND S. TEITLER

Naval Research Laboratory
Washington, DC 20375

INTRODUCTION AND MODEL DESCRIPTIONS

Some time ago, one of us[1,2] introduced a model of relaxation in complex systems in which the relaxation rate conventionally considered as time independent in simpler systems may become time dependent. Relaxations are manifested by the decay (or saturating growth) of some macroscopic variable such as polarization, stress, electric field, volume, or enthalpy, in appropriate complex systems such as glasses, amorphous polymers, polymer melts, viscous liquids, ionic conductors, and amorphous semiconductors.[3] The mechanism for the macroscopic relaxation can often be traced on the microscopic level, however, to a fundamental relaxation mode with simple relaxation rate $W_0 = \tau_0^{-1}$ of the primitive species (PS) that is isolated except for its interaction with the heat bath. The PS can be a molecular moiety or a structural unit in structural relaxations; a dipolar group or a mobile charged entity (ionic or electronic) in dielectric relaxations; a polymer chain in polymer melt terminal viscoelastic relaxation; etc.

In many complex systems, the fundamental relaxation mode often occurs within the domain of short-range order so that all PSs have essentially the same W_0. For example, most glasses, metallic glasses, and viscous liquids have some short-range order. Even in a linear polymer melt, monodispersive samples can be prepared such that all chains have almost the same length (or molecular weight, M). Here the PS can be identified as the individual polymer chains. An example of a fundamental mode is the terminal ($p = 1$) Rouse mode[4] which describes the longest wavelength relaxation of the chain. The simple relaxation rate W_0 is the same for all chains.

It is the nature of complex systems, however, that an individual PS cannot be considered to be completely isolated from other PSs or other neighboring species. Rather, a PS is coupled via molecular, ionic, electronic, or other interactions with its complex environment, an environment that may include other PSs. Relaxation of a PS then involves sequential cooperative adjustments by both its complex environment and the PS itself. For the example of entangled monodisperse linear polymer melts, the terminal Rouse mode of any chain cannot relax without coupling via entanglements with other chains. Such cooperative adjustments act to slow the simple relaxation rate W_0 at times longer than a time $t_c = \omega_c^{-1}$ that is characteristic of the cooperative adjustments. The time t_c is determined by the complex environment, which has a

[a]This work was supported in part by Contracts N00014-85-C-2315 and N00014-86-WR-24016 with the Office of Naval Research.

[b]Sachs/Freeman Associates.

larger spatial scale than that of the residual order. The interaction between the PS and the heat bath, which is responsible for the simple relaxation rate W_0, occurs at times short compared to t_c so that any frequency ν characteristic of the heat bath such as lattice and molecular vibrations must be much larger than $\omega_c/2\pi$. The result is a time-dependent rate $W(t) = W_0 f(t)$ where $f(t) = 1$ for $\omega_c t \ll 1$ and $f(t) < 1$ for $\omega_c t \gg 1$. The function $f(t)$ will depend on the strength of coupling between the PS and the environment as well as on the degree of complexity of the environment. In this physical picture, concepts for PSs, fundamental modes, a simple relaxation rate W_0 with microscopic meaning, cooperative adjustments roughly after $t_c = \omega_c^{-1}$, and a time-dependent transition rate $W(t)$ have all emerged. A schematic representation of these features is given in FIGURE 1.

The next step is a calculation or deduction of $f(t)$. We have approached this problem in several different, yet essentially compatible, ways.[1,2,5–7] In each we have attempted to account for the rather "universal" properties of various relaxations in

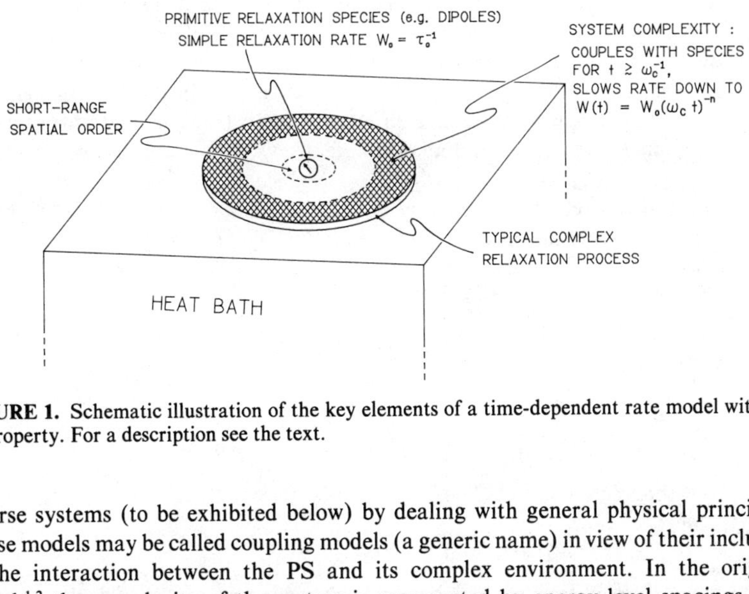

FIGURE 1. Schematic illustration of the key elements of a time-dependent rate model with the W property. For a description see the text.

diverse systems (to be exhibited below) by dealing with general physical principles. These models may be called coupling models (a generic name) in view of their inclusion of the interaction between the PS and its complex environment. In the original model,[1,2] the complexity of the system is represented by energy level spacings. The distribution of spacings for a very large class of complex condensed matter systems has a universal character of being linearly proportional to the magnitude of the spacing itself for sufficiently small spacings. This remarkable property is a consequence of the fundamental physical principle of level repulsion in Hamiltonian systems (the Wigner–von Neumann theorem), and an invariant result of random matrix theory. Recent calculations based on specific Hamiltonians appropriate for complex nonintegrable systems[8,9] have substantiated this result. Thus there are strong indications that, in the absence of imposed special symmetries, the linearity of the distribution of spacings at small spacings is a generic property of complex systems. It follows that one should incorporate this property into any Hamiltonian description of relaxation in complex systems.

In the original model,[1,2] the PS perturbs the spacing structure and, as a consequence its simple relaxation rate, W_0 is modulated to become $W(t)$. A direct calculation in the semiclassical limit $kT \gg \hbar\omega_c$ of the rate $W(t)$ led to

$$W(t) = \begin{cases} W_0 & (\omega_c t \ll 1) \\ W_0(\hat{\omega}_c t)^{-n} & (\omega_c t \gg 1) \end{cases}$$

because the spacing property leads to $f(t) = (\hat{\omega}_c t)^{-n}$, $\omega_c t \gg 1$. Note that neither $\hbar$ nor $\hbar\omega_c/kT$ enters directly into this relation. We refer to this equation hereafter as the W property. The W property involves an intimate entwinement of long-time behavior with its short-time precedent; the quantum mechanical methods used in the derivation may be appropriate even though the problem is nominally judged to be "manifestly" classical. Here ω_c, also identified as the reciprocal of the characteristic time for cooperative adjustments, is the upper cut-off angular frequency above which the linearity of the distribution of the spacings ceases, and $\hat{\omega}_c$ is related to ω_c by constant factors of order unity. An alternative approach[5] that also leads to the W property is based on reaction-rate theory. The sequential time-dependent decrease in the entropy of the spacing structure caused by its displacement from thermal equilibrium by coupling to the relaxation PS provides a time-dependent modulation of W_0. The devolution in the entropy of the spacing structure offers some insight into the source of the slowing of W_0. Of course, the overall entropy of the relaxation process is increasing.

The normalized relaxation function $\phi(t)$ for a relaxing quantity Q defined by $\phi(t) = Q(t)/Q(0)$ is obtained by solving the equation $dQ/dt = W(t)Q$. For $\omega_c t > 1$, the solution has the Kohlrausch fractional exponential[10] form (hereafter referred to as the K form)

$$\phi(t) = \exp -(t/\tau^*)^{1-n} \qquad (\omega_c t > 1) \tag{1}$$

Here

$$\tau^* = [(1-n)\omega_c^n \tau_0]^{1/(1-n)} \qquad (\tau_0 = W_0^{-1}) \tag{2}$$

For $\omega_c t < 1$, the solution has the linear exponential form

$$\phi(t) = \exp -(t/\tau_0) \qquad (\omega_c t < 1) \tag{3}$$

Furthermore, coupled relations 1–3 have the interesting properties of dominance and crossover.[6] If $\omega_c\tau_0 \ll 1$ (a condition satisfied at high temperatures for thermally activated τ_0), the relaxation function follows the linear exponential of relation 3 from unity to almost zero. We call such behavior linear dominance because the predominance of the decay occurs in the form of a linear exponential independent of the value of n. On the other hand, if $\omega_c\tau_0 \gg 1$, $\phi(t)$ has hardly decayed from unity by the linear exponential function of relation 3 before it assumes the K form corresponding to $W(t) = W_0(\hat{\omega}_c t)^{-n}$ when $\omega_c t \gg 1$. It is then dominated by the fractional exponential. A crossover of dominance occurs in a neighborhood defined by $\omega_c\tau_0 \simeq 1$. Thus, three predictions follow from the W property demonstrated quantitatively in FIGURE 2 for a choice of $n = 0.70$ and $\log(\omega_c/\text{sec}^{-1}) = 9$. In the left panel τ_0 is taken to be 10^{-5} sec; in the right panel it is taken to be 10^{-11} sec.

The three coupled predictions may be summarized as follows:

[1] Relaxation follows the K form $\exp -(t/\tau^*)^{1-n}$ for $\omega_c t \gg 1$ and is dominated by it if $\omega_c \tau_0 \gg 1$. (See relation 1.)
[2] The effective τ^* is related to τ_o by relation 2.
[3] Relaxation follows $\exp -(t/\tau_0)$ for $\omega_c t \ll 1$ and is dominated by it if $\omega_c \tau_0 \ll 1$. (See relation 3.)

By coupled predictions we mean the same experimentally accessible parameters n, τ^*, τ_0, and ω_c occur simultaneously in all three predictions and in the crossover condition $\omega_c \tau_0 \simeq 1$. These three predictions are directly applicable to a relaxation process in a

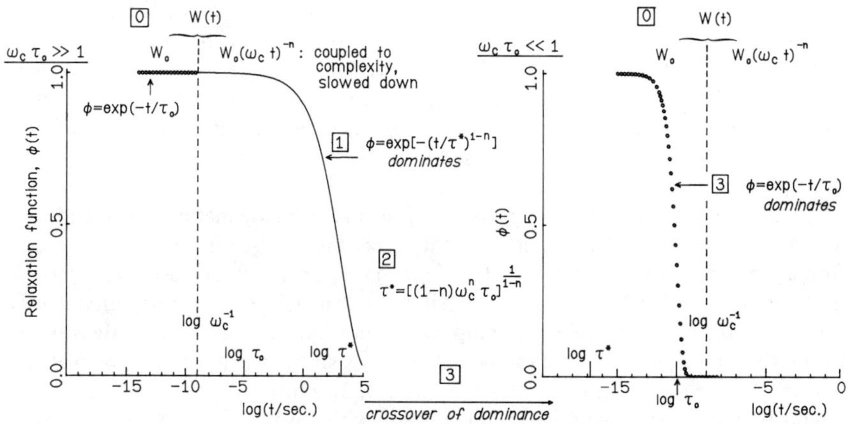

FIGURE 2. Demonstration of how the W property, [0], leads to coupled predictions [1], [2], and [3]. The crossover frequency ω_c is chosen to be 10^9 sec^{-1}. On the left we have $\tau_0 = 10^{-5}$ sec and the regime of $\omega_c\tau_0 \gg 1$. The relaxation function $\phi(t)$ obtained as a solution of the master equation with a time-dependent relaxation rate with the W property is dominated by the Kohlrausch function, relation 1, where the complex or effective τ^* is related to the simple τ_0 of the primitive species by relation 2. On the right we have chosen $\tau_0 = 10^{-11}$ sec and have the regime of $\omega_c\tau_0 \ll 1$. The relaxation function $\phi(t)$ as shown is now dominated by the linear exponential $\exp(-t/\tau_0)$. The horizontal arrow at the bottom pointing from left to right further illustrates the crossover of dominance property, [3].

system with sufficient residual order to have essentially a single simple or primitive relaxation rate $W_0 = \tau_0^{-1}$. In this way the predictions can be individually and independently checked by experimental measurements. Validity of the predictions is only established if the parameters n, τ^*, τ_0, and ω_c determined separately in each of the three predictions are compatible. As will be seen below, this compatibility is a stringent condition and thereby provides convincing experimental evidence for the validity of the W property.

Throughout the tables and figures used to exhibit this compatibility, we use the symbols [1], [2], and [3] to indicate the realizations of the three predictions. Our goal is, by means of the pattern revealed by the symbols, to guide the reader in acquiring the feel for the use of the three coupled relations in correlating experimental data. One cannot help but be struck by the wide range of disciplines and complex systems in which the three predictions are verified. The conclusion that there is more to relaxation in complex systems than compatibility with the K form at long times seems inescapable.

Elsewhere in this volume, we have used the extensive verifications of the three predictions to view them as empirical relations that must be satisfied by models of relaxation. There we indicate the properties of some prototypical models[11-14] that are aimed at describing the physical basis for the K form. By their construction, these alternatives do not incorporate the W property. They cannot arrive at relations 2 and 3 or their equivalents, and some obtain only a tail portion of the K form. Even if the coupling models had simply taken the W property as an initial assumption (TABLE 1) instead of attempting to derive it from generic properties of Hamiltonians as described above, the predictions 1, 2, and 3 would still follow. Note than in TABLE 1 each of the other models[11-14] begin with different initial assumptions. As we shall see, the experimental results provide convincing evidence for the existence of the W property. It follows that modeling activities that ignore it are not applicable for these experiments. Indeed, the appropriate goal for present-day modeling should be a combination of provision of further insight into the W property, and the development of its physical basis.

It may be noted that general phenomenological and mathematical constraints can also be used to obtain information about the relaxation function. For example, we have formulated an "all-purpose" relaxation function $\phi_A(t)$, which has been described in detail elsewhere,[6] that is 1) the characteristic function of a spectral probability density that is insensitive to the size of the complex system (that is, the probability density has the mathematical property of stability, but not Lévy stability in its usual form, which is not appropriate for physical relaxation functions); 2) compatible with requirements on a relaxation function around $t = 0$; 3) consistent with the W property for intermediate and long times. The function $\phi_A(t)$ consists of an initial quadratic exponential exp $-(t/\tau_2)^2$ for $0 \leqslant t < T_Q$, an intermediate linear exponential exp $-(t/\tau_1)$ for $T_Q < t < T_L$ $(T_Q < T_L)$ and a final fractional exponential exp $-(t/\tau_\alpha)^\alpha$ for $t > T_L$. Continuity at T_Q and T_L provides relations among the values for τ_k $(k = \alpha, 1, 2)$ so that there is only one independent τ_k. For example, $\tau_\alpha = (T_L^{\alpha-1}\tau_1)^{1/\alpha}$ so that with the identification $\alpha \equiv 1 - n$, $T_L^{-n} \equiv (1 - n)\omega_c^n$, $\tau_\alpha \equiv \tau^*$, and $\tau_1 \equiv \tau_0$ (the relation between τ_α and τ_1 is just equation 2). This $\phi_A(t)$ is all-purpose in the sense that the observed relaxation can appear to have any one of the quadratic, linear, or fractional exponential behaviors depending, respectively, on whether τ_k $(k = \alpha, 1, 2)$ falls well within $(0, T_Q)$, (T_Q, T_L), or (T_L, ∞). We call this property of $\phi_A(t)$ k dominance, and we say its corresponding probability density has k dominance stability. The latter can be viewed as a physical principle so that the k dominance of $\phi_A(t)$ and hence also the W property follow from k dominance stability.

We have also initiated an effort to develop a purely classical model of the W property involving the irregular trajectories of a complex conservative system described by a classical Hamiltonian. This classical approach shares with our other approaches the occurrence of an $f(t)$ for $\omega_c t > 1$ that has the scaleless property $\delta f(t)/f(t) = -n\delta t/t$. Because of space limitations here, we will describe this effort elsewhere. We turn now to the experimental verifications of the three coupled predictions.

TABLE 1. Some Representative Kohlrausch Relaxation Models

	Model			
	Coupling	Hierarchically Constrained Dynamics	Free Volume	Defect Diffusion Continuous Time Random Walk
Assumptions	Time-dependent relaxation rate[a] and the W property	Serial distribution of relaxation times[b] or matrix elements of master equations in ultrametric space	Free volume, parallel distribution of relaxation times[c]	Defects as random walkers with pausing time distribution that has an algebraic tail[d]
Predictions	[1], [2], and [3]	[1] for special choice of $\{\tau_i\}$ and $\{W_i\}$ or matrix elements, but not [2]	[1] for asymptotically large t, but not [2]	A tail portion of [1], but not [2]
References	1, 2, 5–7	11, 12	13	14

[a]Actually derived in the original coupling model.
[b]$G(\tau) = \Sigma_i W_i \delta(\tau - \tau_i)$ for suitable choices of series $\{\tau_i\}$ and $\{W_i\}$.
[c]$\tau(\nu) = \tau_0 \nu^x$ $(x > 0)$ and the probability distribution $g(\nu) = A \exp(-c\nu^y)$ $(y > 0)$. These forms are suggested by percolation calculations.
[d]$\psi(t) \sim t^{-1-\alpha}$ $(0 < \alpha < 1$ at large $t)$.

THREE COUPLED PREDICTIONS: EXPERIMENTAL VERIFICATIONS

At the time of the first statement of the coupled relations 1 and 2 in 1979,[1] the K form represented a least among equals as a means of describing relaxation phenomena.[15] In a standard text of viscoelasticity of polymers,[4] the K form had hardly been mentioned. For mechanical relaxations, the interest in the K form involved mainly silicate and borate glasses and had been advocated intermittently by workers such as Pierce, deBast and Gilard, Douglas, Kurkjian, Hopkins, and others. Later, in 1971, Williams and Watts[16] revisited the K form in dipolar relaxations of several viscous liquids and polymers. A new application was found in 1973 by Moynihan *et al.*[16] for electrical relaxations in vitreous ionic conductors and fused salts.

Since the 1978 model implied some degree of universality, the first step was to determine the breadth of applicability of the K form. Thus we examined the above relaxations and others including spin lattice, luminescence, electronic hopping, magnetization, dipolar secondary, entangled polymer melt terminal, and nonlinear viscoelastic relaxations as well as creep compliance and other related fluctuation and noise phenomena.[1,2,67] The relevance of the K form to these observed phenomena was established. In mechanical and dielectric relaxations, we enhanced the credibility of the K form by an exhaustive examination of available data. Having been convinced of at least some limited universality of the first relation a few years ago, we were ready for the next step[1,2,67] of testing the two other relations, that is, relations 2 and 3. This more recent task requires a much more penetrating understanding in each field of the material, its structure, the relaxation process, the experimental measurements, the treatment of data, and the jargon as well as of the existing physical ideas and models. It is a nontrivial task that is often time consuming, but one that was and continues to be important.

In this section we give examples of the effort in simultaneous testing and verification of the three coupled relations. The minimum requirement for an example to be presented here is that enough experimental data exists such that at least two of the three relations can be tested. As will be seen, there are several ways of testing the coupled relations.

One utilization of the predictions of the model is to apply simultaneously coupled relations 1 and 2 to relaxation phenomena in complex systems that exhibit puzzling behavior. Examples include $M^{3.4}$ dependence of the viscosity η and terminal relaxation time of entangled monodisperse linear polymer melts, and the anomalous isotope mass dependence m^{γ} of lithium ion conductivity in borate glasses. In these systems we first identify the primitive species and the relevant dependence $\tau_0(x)$ of the simple relaxation time $\tau_0 = W_0^{-1}$ on a physical quantity x such as M^2 dependence of the terminal Rouse mode, $m^{1/2}$ dependence for a Li ion. When the preliminary requirements of identical and simple τ_0 are satisfied, the first prediction is that the relaxation (that is, the terminal relaxation of the polymer melt and the conductivity relaxation of the Li ions) is well described by the K form. The fit of the K form to the time, frequency, or temperature (at constant frequency) dependences should be satisfactory for some n, and this determines n, the coupling parameter. After n is determined, we use the coupled relation 2 to predict with the same n such dependences of τ^* as

$$[\tau_0(x)]^{1(1-n)},\ M^{2/(1-n)},\ m^{1/2(1-n)},\ \text{and}\ \exp\,[E_a/(1-n)kT]$$

and hence the dependencies of the viscosity, the resistivity, etc. An acid test of the model is whether these two coupled relations can predict or account for the observed anomalous dependences. The remarkable thing is that they do![1,2,67] In this way the two coupled relations have been verified, and, in the process, the model resolves apparent anomalies in various fields.

In a number of cases either the magnitude of τ_0 or its dependence $\tau_0(x)$ can be measured directly or deduced indirectly. For example, we can follow the same relaxation process to higher temperatures, check the crossover of dominance from relation 1 to relation 3, and directly obtain τ_0 and its dependences. The crossover also offers an estimate of ω_c. We can indirectly monitor the same relaxation process in a dilute solution or a crystalline version (if available or possible) of the same material where there is lack of complexity and zero coupling ($n = 0$) so that the measured τ^* reduces to τ_0. If the dependence of τ_0 on a parameter x (for example, T or P) is deduced and the dependence is extrapolated to the regime where relation 1 dominates for a nonzero n, then relation 2 can be used to predict the dependence of the measured τ^* on x. If the magnitude of τ_0 is known, then relation 2 can determine ω_c. The value of ω_c obtained must be physically reasonable and consistent with the estimate inferred from the crossover of dominance to relation 3. Other ways of testing the three relations are best described when considering actual data and hence will be discussed throughout the text. The symbols ⊡1, ⊡2, and ⊡3 found repeatedly in tables and graphs remind the reader of either the application or confirmation of the three coupled relations. Unfortunately, limitations on space has forced us to exclude some examples and to abbreviate our descriptions of those examples that remained. We have designed almost self-explanatory figures and organized material in tables in order to include as much information as possible.

Monodisperse Polymer Melt Viscoelasticity, Rheology, and Self-diffusion

For $M < M_c$, the critical molecular weight for entanglement, the linear chains can be considered uncoupled (FIG. 3a). The fundamental mode we consider is the terminal Rouse mode $\exp(-t/\tau_0)$ with $\tau_0 \sim \zeta_0(T)M^2$ where $\zeta_0(T)$ is the simple friction factor. The Rouse theory gives a very good description of the viscosity η for $M < M_c$ (FIG. 3d, left panel). If $M > M_c$, the chains are entangled and coupled (FIG. 3b). The loss modulus G'' data for hydrogenated polybutadiene (HPB) are well fit by relation 1 with $n_\eta = 0.41$ (FIG. 3c). Then relation 2 predicts $\tau^* = [(1 - n)\omega_c^n\tau_0]^{1/(1-n)} \sim \zeta_0(T)^{1/(1-n)} M^{2/(1-n)}$. For $n = 0.41$, the zero shear viscosity $\eta = G_N\tau^*$ (G_N is the plateau modulus) will have the $M^{3.4}$ dependence in accord with the observed $M^{3.4}$ or $M^{3.5}$ dependence observed in linear polymer systems (FIG. 3d, right panel). The measured T dependences of η for entangled linear HPB and polyethylene (PE) are both Arrhenius. The former has an activation energy $E^*_{a_\eta}$ of 7.2 kcal/mol; the latter, an $E^*_{a_\eta}$ of 6.35 kcal/mol. Relation 2 predicts $E^*_{a_\eta}$ is related to the fundamental activation energy E_a of ζ_0 by the relation $E^*_a = E_a/(1 - n)$. In TABLE 2 (left panel) we have summarized, for the same n used to predict the MW dependence, the steps for the deduction of E_a. For PE it is 3.6 kcal/mol, which turns out to be almost the same as the internal rotational barrier of the lower molecular weight alkanes (TABLE 3 and FIG. 3e, left panel) and a high-molecular-weight polyethylene oxide chain in dilute solution. The internal

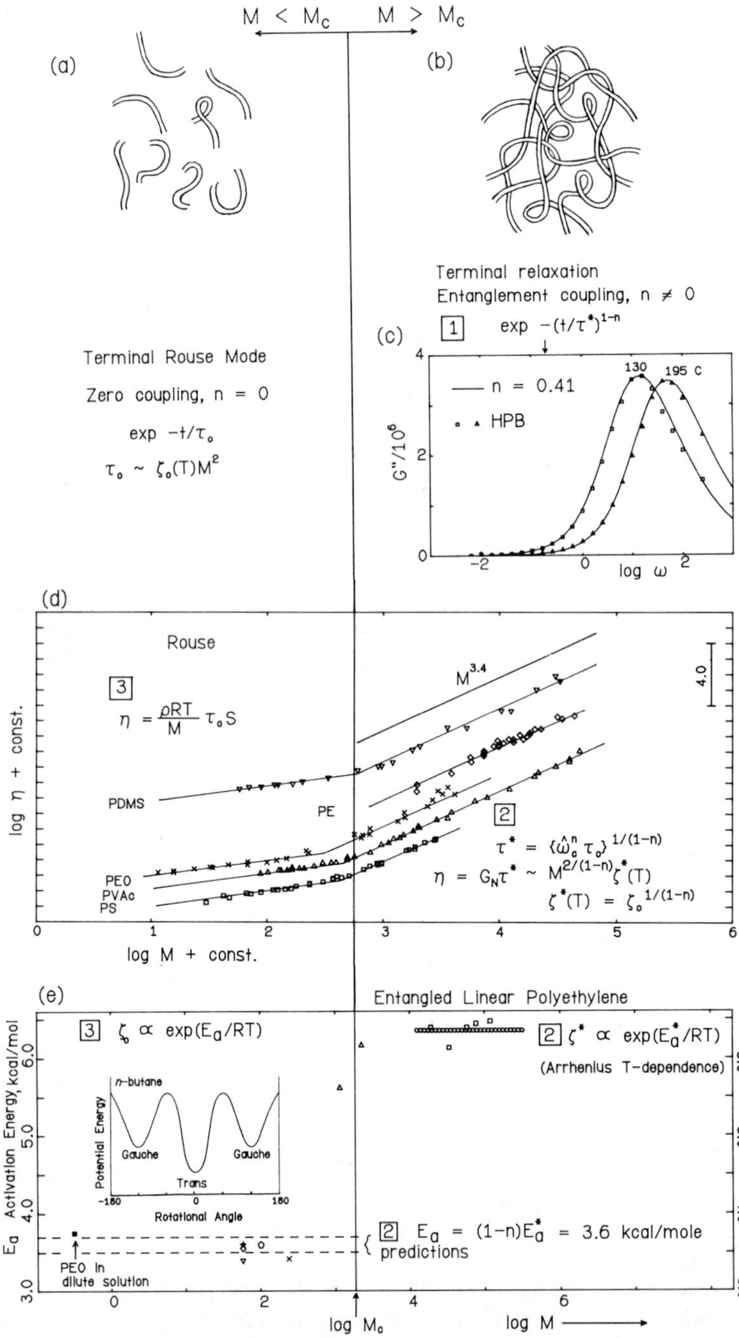
M < Mc
M > Mc
(a)
(b)
Terminal relaxation
Entanglement coupling, n ≠ 0
1 exp −(t/τ*)^(1−n)
(c)
Terminal Rouse Mode
Zero coupling, n = 0
exp −t/τo
τo ~ ζo(T)M²
n = 0.41
HPB
130
195 C
G"/10⁶
log ω
(d)
Rouse
3 η = ρRT/M τoS
M^3.4
4.0
log η + const.
PDMS
PE
PEO
PVAc
PS
2 τ* = {ω̂o^n τo}^(1/(1−n))
η = GNτ* ~ M^(2/(1−n)) ζ*(T)
ζ*(T) = ζo^(1/(1−n))
log M + const.
(e)
Entangled Linear Polyethylene
3 ζo ∝ exp(Ea/RT)
2 ζ* ∝ exp(Ea*/RT)
(Arrhenius T-dependence)
n-butane
Potential Energy
Gauche
Trans
Gauche
Rotational Angle
Ea Activation Energy, kcal/mol
Ea* Flow Activation Energy, kcal/mol
2 Ea = (1−n)Ea* = 3.6 kcal/mole
predictions
PEO in dilute solution
log Mc
log M

TABLE 2. Parallel Comparison of the Zero Shear Viscosity and the Self-diffusion Coefficient: Experimental Data and Coupling Model Predictions

	Zero Shear Viscosity[a] (η)	Self-diffusion Coefficient[b] (D_s)
		Polyethylene[c]
Determination	$n_\eta = 0.43$	$n_D = 1/3$
Prediction	$\eta \propto M_w^{3.5}$ [2] [d]	$D_s = M^{-2}$ [2] [e]
Determination	$E^*_{a\eta} = 6.35$ kcal/mole	$E^*_{aD} = 5.47$ kcal/mole[f]
Prediction	$E_a = 3.6$ kcal/mole [2] [g]	$E_a = 3.7$ kcal/mole [2] [e]
		Hydrogenated Polybutadiene[h]
Determination	$n_\eta = 0.41$	$n_D = 1/3$
Prediction	$\eta \propto M_w^{3.4}$ [2] [d]	$D \propto M^{-2}$ [2] [e]
Determination	$E^*_{a\eta} = 7.2$ kcal/mole	$E^*_{aD} = 6.19$ kcal/mole[f]
Prediction	$E_a = 4.2$ kcal/mole [2] [g]	$E_a = 4.1$ kcal/mole [2] [e]

[a]The value for n_η was determined, from [2], with the following equations:

$$\tau_{\text{eff}} \propto (\omega_c^{n_\eta}\tau_o)^{1/(1-n_\eta)}$$

$$\eta \propto G_N^o \tau_{\text{eff}} \propto M^{2/(1-n_\eta)} \exp(E_a/(1-n_\eta)RT)$$

[b]The value for n_D was determined, from [2], with the following equations:

$$\tau'_{\text{eff}} \propto (\omega_c^{n_D}\tau_o)^{1/(1-n_D)}$$

$$D_s \propto R_g^2/\tau'_{\text{eff}} \propto M^{1-2/(1-n_D)} \exp(-E_a/(1-n_D)RT)$$

[c]The value of n_η was determined independently from viscoelasticity data; that for n_D was determined from M^{-2} dependence of D_s.

[d]After Ngai and Rendell.[17]

[e]After McKenna *et al.*[19]

[f]After Tirrell.[20]

[g]After Ngai and Plazek.[21]

[h]The value for n_η was determined independently from $G''(\omega)$ (data after Raju *et al.*[18]); that for n_D was determined from M^{-2} dependence of D_s.

FIGURE 3. (**a**) Unentangled ($M < M_c$) monodisperse linear chains. (**b**) Entangled ($M > M_c$) monodisperse linear chains. (**c**) Loss modulus data of HPB obtained by Raju *et al.*[18] The solid lines are fits of the data with $n = 0.41$. (**d**) Log η, plus a constant, plotted against log M, plus a constant, for four different linear polymers. The Rouse theory works remarkably well for $M < M_c$ ($\eta = (\rho RT/M)\tau_0 S$ where $S = 1.645$). The slope calculated using the coupling model is 3.4 for $M > M_c$. (**e**) *Left side:* internal rotational energy barriers E_a for several hydrocarbons, for long-chain PEO in dilute solution, and for lower molecular weight PEs obtained by various experiments. The inset shows the potential energy of n-butane as a function of rotational angle. *Right side:* flow activation energy E_a data for high-molecular-weight ($M > M_c$) entangled linear polyethylene.

rotational barrier is the fundamental quantity that limits chain flexibility (FIG. 3e, inset), and it is readily identifiable with our fundamental E_a for the simple friction ζ_0. It is gratifying to see this confirmed by relation 2. FIGURE 4b heuristically suggests that a star-branched molecule will have a stronger coupling to the complexities when the arms are sufficiently long. In fact, for branched PE in a limited molecular weight range, a coupling parameter $n_b = 0.70$ is required to explain the $M^{6.56}$ dependence of η (FIG. 4c). With the same n_b, relation 2 reduces the observed E^*_{ab} of 11.66 kcal/mol (FIG. 4d, right panel) to a fundamental E_a of 3.5 kcal/mol. The latter is in good agreement with the fundamental E_a of 3.6 kcal/mol deduced for linear PE (as expected) because, independent of polymer architecture, the simple ζ_0 has the same energy barrier. A similar success is attained in using relation 2 to consider the molecular weight and temperature dependence of a star-branched HPB. Shear viscous deformation involves orientation whereas self-diffusion does not.[19] We intuitively expect $n_D < n_\eta$, and indeed we have found $n_D = 1/3$. Both the M^{-2} dependence of the self-diffusion coefficient D_s and its activation energy E^*_{aD} can be explained by relation

TABLE 3. Internal Rotational Isomerism Barrier of Alkanes

Material	U_0 (kcal/mole)	Method
n-Butane	3.55	Thermodynamic data[a]
	3.4 ± 0.4	Ultrasonic data[a]
	3.6	Calculation[a]
Heptane	3.6	Excimer/fluorescence data[a]
Polyethylene oxide chain (in dilute solution)	3.7–3.8	Excimer/fluorescence data[a]
Linear polyethylene melt	3.6	Rheological data[b] (coupling model)
	3.7	Self-diffusion data[c] (coupling model)
Branched polyethylene melt	3.5	Rheological data[b] (coupling model)

[a]For source see reference 21.
[b]After Ngai and Plazek.[21]
[c]After McKenna *et al.*[19]

2 with $n_D = 1/3$ and the same fundamental E_a as determined from η for both PE and HPB (TABLE 2, right panel). One can consider the use of relation 2 in relating E^*_{aD} with E_a to determine $n_D \simeq 1/3$. Then use of relation 2 for a second time predicts the celebrated $D_s \sim M^{-2}$ dependence.[4]

FIGURE 5 presents recoverable creep compliance data of a polydisperse[23] and a monodisperse polystyrene melt.[24] The recoverable compliance of the terminal mode for the monodisperse sample can be well fit by the recoverable compliance function calculated from the K form with $n_\eta = 0.45$ (solid curve),[25] which corresponds to an acceptable $M^{3.6}$ dependence. On the other hand, the terminal zone of the polydisperse sample cannot be well fit by a compliance function calculated from a single K form for any n_η. Both these opposite situations are expected by the coupling model. The monodisperse sample satisifies the requirement of the model for approximately identical and simple expressions for τ_0 whereas the polydisperse sample obviously does not. For the latter an inhomogeneous distribution of the simple τ_0 must be considered in

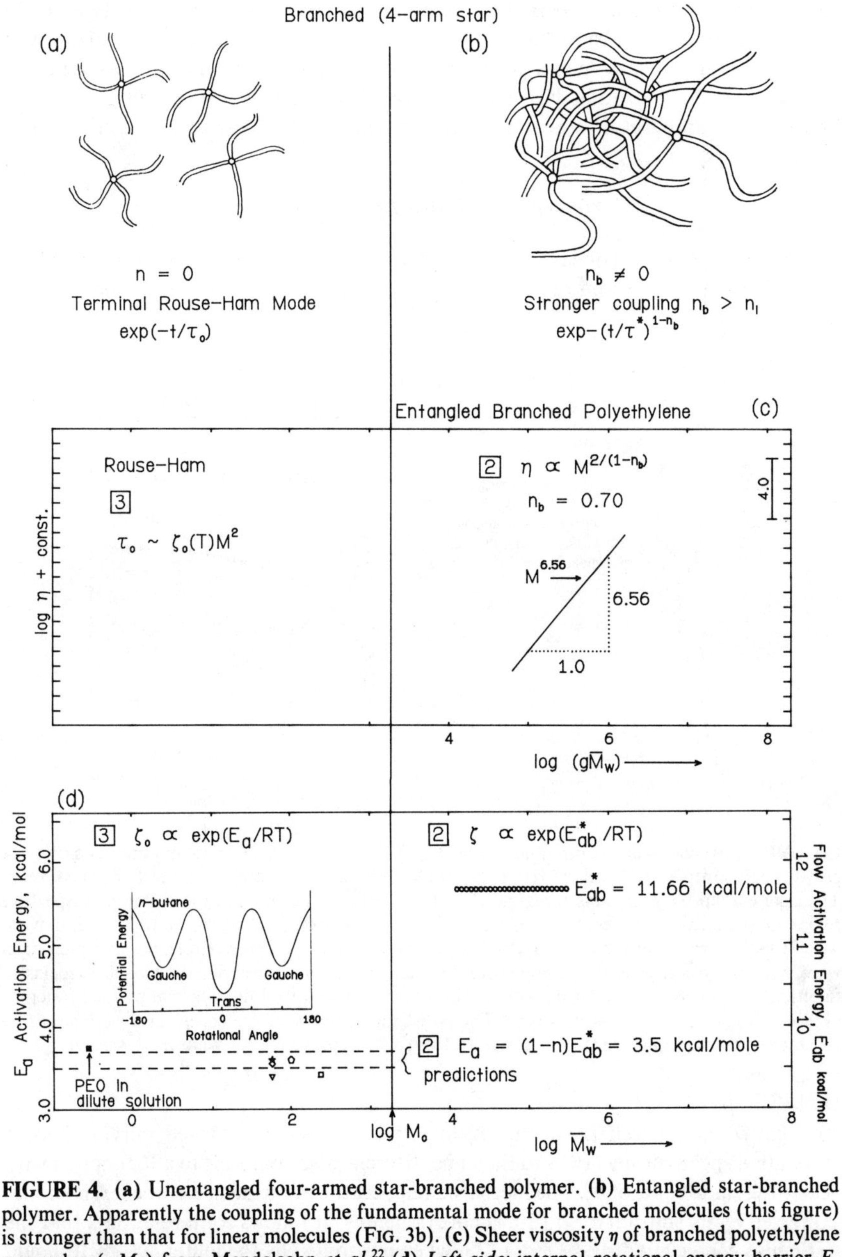

FIGURE 4. (**a**) Unentangled four-armed star-branched polymer. (**b**) Entangled star-branched polymer. Apparently the coupling of the fundamental mode for branched molecules (this figure) is stronger than that for linear molecules (FIG. 3b). (**c**) Sheer viscosity η of branched polyethylene versus log (gM_w) from Mendelsohn *et al.*[22] (**d**) *Left side:* internal rotational energy barrier E_a (same as in FIG. 3e). *Right side:* flow average activation energy of the entangled branched polyethylene samples studied by Mendelsohn *et al.*[22]

the coupling model, and the relaxation function may no longer have the K form. In this case it makes no physical sense to fit the data with an overall K form. Even if, by accident, a particular $\tilde{n}$ can be found, it should not be used as the coupling parameter n in relation 2. This example of a polydisperse melt warns against the careless use of the K form to fit any relaxation data without a criterion that tells us when this procedure is physically meaningful. Fortunately, in the coupling model we have such a criterion.

Primary, α, and Structural Relaxations

Let us start with polystyrene and the compliance J_r data of FIGURE 5. The α relaxation in the softening dispersion, where J_r changes approximately from 10^{-10} to

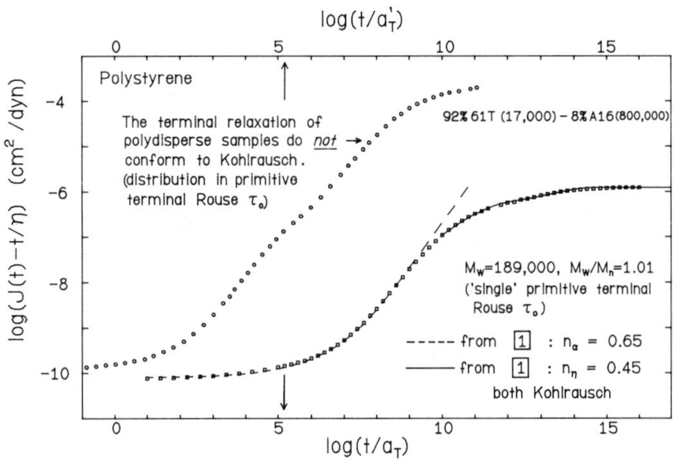

FIGURE 5. Recoverable compliance data ($J_r(t) = J(t) - t/\eta$) from experiments with a polydisperse sample[23] (○) and experiments with a monodisperse sample[24] (□) of L-2 polystyrene. The data extend from the glassy region ($J_r \sim 10^{-10}$ cm^2/dyne), through the softening dispersion region, to the terminal region ($J_r \sim 10^{-6}$ cm^2/dyne). As we expected, the monodisperse data were well fit by K forms and the polydisperse data were not. The monodisperse terminal region data were well fit by a recoverable compliance function obtained via relations between viscoelastic functions from a K form with $n_\eta = 0.45$; the relaxation was well fit by a compliance function obtained from a K form with $n_\alpha = 0.65$. The polydisperse terminal dispersion data were not fit by any recoverable compliance function derived from a K form, no matter what n_η was used.

10^{-8} cm^2/dyne, is well fit by the K form with $n_\alpha = 0.65$ (dashed curve). Photon correlation spectroscopy (PCS) data on polystyrene also conforms to a K function with nearly the same value of n_α. The measured temperature dependence of the α relaxation, expressed by its shift factor $a_\alpha(T)$, is not Arrhenius. It follows from relation 2 that the Arrhenius coefficient $d\ln a_\alpha(T)/d(1/T)$, when multiplied by the factor $(1 - n_\alpha)$, should be the Arrhenius coefficient $d\ln\zeta_0(T)/d(1/T)$ of the simple friction factor $\zeta_0(T)$. The simple $\zeta_0(T)$ is the same for both the α and the terminal η relaxations. Hence, after the Arrhenius coefficient of the shift factor for η is calculated from experimental data, a

TABLE 4. Explanation of the Differences between the Temperature Dependence of the Viscoelastic Softening and the Terminal Dispersions of Linear Amorphous Polymers[a]

Material	α Relaxation[b] (n_α)	Terminal Relaxation[c] (n_η)	Relation 2[d] [2]
PVAc	0.57	(0.40, 0.45)	Verified[a]
PS	(0.62, 0.65)	(0.40, 0.45)	Verified[a]
Amorphous PP	0.66 ± 0.02	(0.40, 0.45)	Verified

[a]After Ngai and Plazek.[25]
[b]$\exp[-(t/\tau_\alpha^*)^{1-n_\alpha}]$ [1].
[c]$\exp[-(t/\tau_\eta^*)^{1-n_\eta}]$ [1].
[d]$(1-n_\alpha)\mathrm{d}\ln a_\alpha(T)/\mathrm{d}(1/T) = \mathrm{d}\ln \zeta_0(T)/\mathrm{d}(1/T) = (1-n_\eta)\mathrm{d}\ln a_\eta(T)/\mathrm{d}(1/T)$ [2].

double use of relation 2 gives the equality $(1 - n_\alpha)\mathrm{d}\ \ln a_\alpha(T)/\mathrm{d}(1/T) = (1 - n_\eta)\mathrm{d}\ \ln a_\eta(T)/\mathrm{d}(1/T)$. This is verified for three amorphous polymers (TABLE 4), and a very fundamental property of amorphous polymers is explained at the same time.[4,24,25]

The complete sets of data available for polyphenylmethyl siloxane (PPMS) allow a thorough test of the three relations. The α relaxation follows a K form with $n_\alpha = 0.56$ (FIG. 6). Both T and P dependences of τ^* are obtained from PCS data (TABLE 5). The PCS samples studied are of low molecular weight and are not entangled. Hence $n_D = 0$ for the diffusion and local concentration fluctuation (caused by diffusion) data directly gives the T and P dependences of the primitive ζ_0. A direct test of relation 2 can be made. As demonstrated in TABLE 5, both predictions, $(1 - n_\alpha)\ B^* = B_0$ and $(1 - n_\alpha)a^* = a_0$, are verified. Relation 3 is supported by Brillouin scattering data that allow a determination of the simple prefactor τ_∞. Another application of relation 2 fixes the values of ω_c at $10^{9.3}\ \mathrm{sec}^{-1}$, which is physically acceptable and consistent with the crossover of dominance property. In TABLES 6–8 we present similar demonstrations of how well the three predictions correlate and explain various data sets. Fundamental activation barrier energies directly determined by neutron scattering and by relation 3 agree well with the prediction of relation 2.

Similar results and analyses are presented in TABLES 9–11 for the nonpolymeric

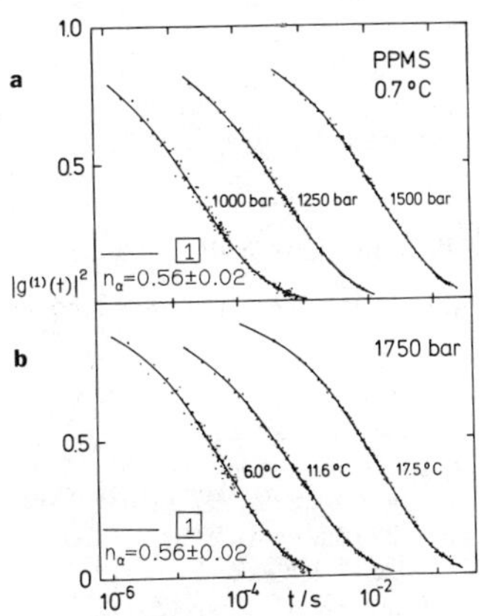

FIGURE 6. Normalized composite net photon correlation functions (data) of the primary relaxation process measured for PPMS at (**a**) constant pressure (1750 bar) and different temperatures (lower half) and at (**b**) constant temperature (0.7 °C) and different pressures (upper half). Each solid line is a fit of the K form (relation 1) to the experimental data. The data are after Fytas *et al.*[26]

TABLE 5. Explanation of the Differences between the Temperature and Pressure Dependencies of the Density and Concentration Fluctuations in Amorphous Polyphenylmethyl Siloxane[a]

From [1][b]			From [2][c] and [3][d]					
			B_o/R (K)		a_o (cm^3/mole)			
τ_∞^* (sec)	B^*/R (K)	a^* (cm^3/mole)	[2][e]	[3][f]	[2][g]	[3][f]	τ_∞ (sec) [3][h]	ω_c (sec^{-1}) [2][i]
3×10^{-19}	1659	67.6	730	712	30.0	32.4	7×10^{-14}	$10^{9.3}$

[a]After Ngai and Fytas.[27]

[b]Here $n_\alpha = 0.56 \pm 0.03$ and $\tau^* = \tau_\infty^* \exp[(B^* + a^*P)/(T - T_0)]$ from photon correlation spectroscopy data. After Fytas *et al.*[26]

[c]From [2]: $\tau^* = [(1 - n_\alpha)\omega_c^{n_\alpha}\tau_o]^{1/(1-n_\alpha)}$.

[d]The primitive relaxation time τ_o is written in the form shown below:

$$\tau_o = \tau_\infty \exp[(B_o + a_oP)/(T - T_o)].$$

[e]From [2]: $(1 - n_\alpha)B^* = B_o$.

[f]From local concentration fluctuation data After Fytas *et al.*[26,27]

[g]From [2]: $(1 - n_\alpha)a^* = a_o$.

[h]This value of τ_∞ is physically reasonable and identifiable with the pre-exponential of the primitive τ_o (from Brillouin scattering data). After Fytas *et al.*[28]

[i]The crossover frequency ω_c is obtained from [2]; $\tau_\infty^* = (\hat{\omega}_c^{n_\alpha}\tau_\infty)^{1/(1-n_\alpha)}$ with n_α, τ_∞^*, and τ_∞ as given. The magnitude of $\omega_c \simeq 10^{9.3}$ sec^{-1} so determined is consistent with relation 3 of crossover and dominance crossover as revealed by Brillouin scattering and photon correlation spectroscopy.

materials *o*-terphenyl and 1, 2, 6-hexanetriol and for the glass-forming molten salt 0.4 $Ca(NO_3)_2$-0.6 KNO_3 to illustrate the general nature of the three relations. For the molten salt we have analyzed ultrasonic mechanical relaxation[45] and found the dominance of the K form. The n is T dependent and increases with decreasing T (FIG. 7). With the simple τ_0 as given in TABLE 11 and these n values, relation 2 predicts

TABLE 6. Photon Correlation Spectroscopy, Brillouin Scattering, Neutron Scattering, and Shear Viscosity Data[a]

Material	Method	E_a (kcal/mole)
Polypropylene glycol	Photon correlation spectroscopy [1][b]	
	↓ Brillouin scattering [3][c]	4.3
Polypropylene oxide	Neutron scattering [3][d]	4.2
Entangled linear polypropylene melt	Shear viscosity [2][e]	4.2

NOTE: The arrow denotes a crossover from $\exp[(-t/\tau^*)^{1-n}]$ dominance to $\exp(-t/\tau_0)$ dominance for $\tau_o \le 10^{-10}$ sec.

[a]After Ngai *et al.*[29]

[b]$0.37 < (1 - n) < 0.41$ for $\tau^* > 10^{-6}$ sec. After Wang *et al.*[30]

[c]$\tau_0 = \tau_\infty \exp(E_a/RT) \le 10^{-10}$ sec; $\tau_\infty \simeq 2.5 \times 10^{-14}$ sec. This value of τ_∞ is physically reasonable and identifiable with the pre-exponential of the primitive τ_0. After Wang and Huang.[31,32]

[d]After Allen *et al.*[33]

[e]From $\eta \propto M^{3.4} \exp(E_a^*/RT)$, $E_a^* = 7.2$ kcal/mole for $T > 500$ K. It follows from [2] that $3.4 = 2/(1 - n_\eta)$ or $n_\eta = 0.41$ and that $E_a^* = E_a/(1 - n_\eta)$.

TABLE 7. Brillouin Scattering, Shear Viscosity, and Excimer Emission Data

Material	Method	E_a (kcal/mole)
Polyethylene glycol	Brillouin scattering[a]	3.85 $\boxed{3}$
Polyethylene oxide chain in dilute solution	Excimer emission[b]	3.7–3.8 $\boxed{3}$
Entangled polyethylene oxide melt[c]	Shear viscosity	>3.7; <4.1 $\boxed{2}$ [d,e]
Entangled linear polyethylene melt	Shear viscosity	3.8 $\boxed{2}$ [d,f]

[a]exp $(-t/\tau_0)$ dominance; $\tau_0 = \tau_\infty \exp(E_a/RT)$; $\tau_\infty = 2.2 \times 10^{-14}$ sec. The value for τ_∞ is physically reasonable and identifiable with the pre-exponential of the primitive τ_0. After Wang *et al.*[34]

[b]Dilute solution implies zero coupling, that is, $n = 0$. After Liao and Morawetz.[35]

[c]$7400 < M_v < 42\,000$.

[d]From $\eta \propto M^{2/(1-n_\eta)} \exp(E_a^*/RT)$, $E_a^* = E_a(1 - n_\eta)$, and $n_\eta = 0.41$.

[e]Here $6.2 < E_a^* < 7$ kcal/mole. After Teramoto and Fujita.[36]

[f]Here $E_a^* = 6.4$ kcal/mole. After Mendelsohn *et al.*[22]

rather well the observed τ^* with an ω_c that is consistent with relation 3 of the crossover.[45] If the expected departure of τ_0 from the Arrhenius T dependence is incorporated into relation 2, the overall agreements can be improved.

Secondary Relaxation

We discuss only the γ relaxation of polycarbonate (BPA and chloral), the NMR and dynamical mechanical relaxation data of which have been extensively studied.[49] Also available are dilute solution[51] NMR data, which give the fundamental activation energy E_a directly. As revealed by NMR line-shape studies,[48] there is additional randomness affecting the γ relaxation that gives an inhomogeneous distribution in τ_0. Hence the NMR T_1 data and the loss modulus G'' data should not be used to deduce the coupling parameter n by fitting to a K form. In fact, before the results of NMR line-shape studies were available, an attempt to do just that showed a large departure[49] of data from the K form for any n. We reverse the order and apply relation 2 before relation 1. The activation energy E_a^* in the bulk of the solution (50 kJ/mol) and the E_a in dilute solution (13 to 15 kJ/mol) are related by relation 2. The relations $E_a^* = E_a/(1 - n)$ can be used to determine n (TABLE 12). A repeated use of relation 2 ($\tau_\infty^* =$

TABLE 8. Shear Viscosity and Neutron Scattering Data for Linear Polydimethyl Siloxane

Method	E_a (kcal/mole)
Shear viscosity $\boxed{2}$ [a]	2.2
Neutron scattering $\boxed{3}$ [b]	2.0

[a]For high-molecular-weight, entangled samples, $\eta \propto M^{3.4} \exp(E_a^*/RT)$, $E_a^* = 3.65$ kcal/mole. Data after Plazek *et al.*[37] It follows from $\boxed{2}$ that $3.4 = 2/(1 - n_\eta)$ or $n_\eta = 0.41$ and that $E_a^* = E_a/(1 - n_\eta)$.

[b]Internal conformational motions of polymer chain. Data after Allen *et al.*[33]

TABLE 9. Photon Correlation Spectroscopy, Brillouin Scattering, NMR Self-diffusion, and Viscosity Data for *o*-Terphenyl

Method	E_a (kcal/mole)
Photon correlation spectroscopy [1] [a]	
↓	
Brillouin scattering [3] [b]	6
NMR self-diffusion[c]	6
Viscosity[d]	6

NOTE: The arrow denotes the crossover from $\exp[-(t/\tau^*)^{1-n}]$ dominance ($T < 271$ K) to $\exp[-t/\tau_0(T)]$ dominance ($T > 370$ K).

[a] $T < 271$ K; $\tau^* > 5 \times 10^{-5}$ sec. After Fytas *et al.*[38]

[b] $T > 370$ K and $\tau_0 = \tau_\infty \exp(E_a/RT) < 5 \times 10^{-10}$ sec where $\tau_\infty = 2.9 \times 10^{-13}$ sec. This value of τ_∞ is physically reasonable and identifiable with the pre-exponential of the primitive τ_0. After Higashigaki and Wang.[39]

[c] $T > 370$ K. After McCall *et al.*[40]

[d] $T > 370$ K. After Greet and Turnbull.[41]

$[(1 - n)\omega_c^n\tau_\infty]^{1/(1-n)}$) for the prefactors with the same n gives $\omega_c \approx 10^{10}$ to 10^{11} sec^{-1}, which is physically acceptable. Note also the physically reasonable magnitude of τ_∞ from dilute solution data. After $n = 0.70$ is determined by using relation 2, we compare the dynamical loss modulus G'' data of chloral with that predicted by using relation 1 with the same n. This comparison (FIG. 8) reveals the excess contribution from the additional inhomogeneous distribution of the simple τ_0 as confirmed by NMR line-shape analyses.

TABLE 10. Dielectric, Brillouin Scattering, and Depolarized Light Scattering Data for 1,2,6-Hexanetriol

Method	Results
Dielectric relaxation[a]	$n \simeq 0.40$ [1]
	$\tau^* = \tau_\infty^* \exp(E_a^*/T)$ [1]
	$E_a^* = 7800$ K
	$\tau_\infty^* = 7.3 \times 10^{-20}$ sec
Brillouin scattering and depolarized light scattering[b]	$n = 0$ [3]
	$\tau_0 = \tau_\infty \exp(E_a/T)$ [3]
	$E_a = 4129$ K
	$\tau_\infty = 4.6 \times 10^{-15}$ sec
Deduction based on [2]	$E_a = 4700$ K [2] [c]
	$\omega_c = 2.5 \times 10^8$ sec^{-1} [2] [d]

[a] After McDuffie and Litovitz.[42]

[b] After Dorfmüller *et al.*[43]

[c] $E_a = (1 - n)\,E_a^*$.

[d] $\tau_\infty^* = [(1 - n)\,\omega_c^n\,\tau_\infty]^{1/(1-n)}$ using n, τ_∞^*, and τ_∞ from the experimental data as given. The value for ω_c is reasonable for the characteristic frequency of crossover of dominance.

TABLE 11. Ultrasonic Relaxation and Brillouin Scattering Data for a Glass-forming $0.4Ca(NO_3)_2$–$0.6KNO_3$ Molten Salt

Method	Dominance	Data
Ultrasonic relaxation[a]	$\exp[-(t/\tau^*)^{1-n}]$ [1]	
($f < 10^8$ Hz)	where $n \neq 0$ ↓	
Brillouin scattering[b]	$\exp[-(t/\tau_0)]$ [3]	$\tau_\infty = 4 \times 10^{-14}$ sec[d]
	augmented by shear viscosity[c]	$E_a = 6$ kcal/mole

NOTE: The arrow denotes the crossover.
[a]The data are after Weiler *et al.*[44] The analysis is after Ngai *et al.*[45]
[b]After Torell.[46]
[c]$\tau_0 = \eta/G_\infty = \tau_\infty \exp(E_a/RT)$. After Angell and Torell.[47]
[d]This value for τ_∞ is physically reasonable and identifiable with the pre-exponential of the primitive τ_0. Moreover, it is in agreement with the far-infrared data of Angel and Torell.[47]

Ionic Electrical Conductivity Relaxations

Moynihan *et al.*[16] were the first to discover that the K form gives an excellent description of the conductivity relaxation due to the diffusion of mobile ions in vitreous ionic conductors. The relaxation of diffusing ions has the same K function time dependence as dipolar and mechanical relaxations, as suggested by the coupling model. Any model that proposes a different time dependence for the diffusion of ions or other defects[14] contradicts these experimental facts. We have demonstrated the general applicability of the K form in conductivity relaxation to a larger number of ionic conductors; one of these conductors is the fast ion conductor Na β-alumina, which is not vitreous. The example given in FIGURES 9 & 10 for lithium borate glass shows how well the K form can describe the electric modulus data. TABLE 13 summarizes the relevant information and indicates the applicability of the other two relations. The

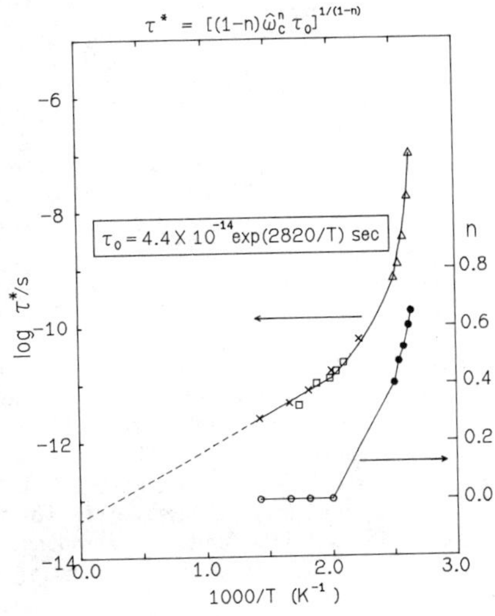

FIGURE 7. Ultrasonic data after Weiler *et al.*[44] (△), Brillouin scattering data (□), and shear viscosity data for $0.4Ca(NO_3)_2$-$0.60KNO_3$ (x) were combined on a single plot by Angell and Torell.[47] Extrapolation to infinite temperature gives the primitive τ_0 shown in the figure. The pre-exponential factor $\tau_\infty = 4.2 \times 10^{-14}$ sec is in good agreement with the far-infrared absorption peak frequency. The values of n for the open and filled circles are shown on the far right. For all of the open circles, $n = 0$, which is consistent with Angell and Torell's finding, from their analysis of the Brillouin scattering data,[47] that the relaxation function becomes a single exponential (relation 3). For each of the filled circles, n is different; these values of n were determined by fits of the K form (relation 1) to the ultrasonic data. If the open and filled circles are interpolated by the continuous line as shown, relation 2 with τ_0 as given predicts τ^* as shown by the solid curve.

TABLE 12. Activation Energies and Pre-exponential Factors of γ Relaxation (π Flips of the Phenyl Group) in Polycarbonate

	BPA	Chloral
Bulk data[a]		
E_a^* (kJ/mole)	50	50
τ_∞^* (sec)[b]	2.29×10^{-16}	2.29×10^{-16}
Dilute solution data[c]		
E_a (kJ/mole)	13	15
τ_∞ (sec)[d]	8×10^{-13}	8×10^{-13}
Deduction based on [2]		
n (where $E_a^* = E_a/(1-n)$)	0.74	0.70
ω_c ($\sec^{-1}$)[e]	2.8×10^{10}	1.1×10^{11}

[a] $\int d\tau_0 g(\tau_0) \exp[-(\tau/\tau^*)^{1-n}]$. After Jones *et al.*[49] and Ngai *et al.*[50]

[b] The value for BPA was assumed to be the same as that for chloral.

[c] $\exp(-t/\tau_0)$. After O'Gara *et al.*[51]

[d] The values for τ_∞ are physically reasonable and identifiable with the pre-exponential of the primitive τ_0.

[e] $\tau_\infty^* = [(1-n)\,\omega_c^n\,\tau_\kappa]^{1/(1-n)}$ using n, τ_∞^*, and τ_∞ as given here. The magnitude for the ω_c deduced from relation 2 is consistent with that expected for the crossover frequency.

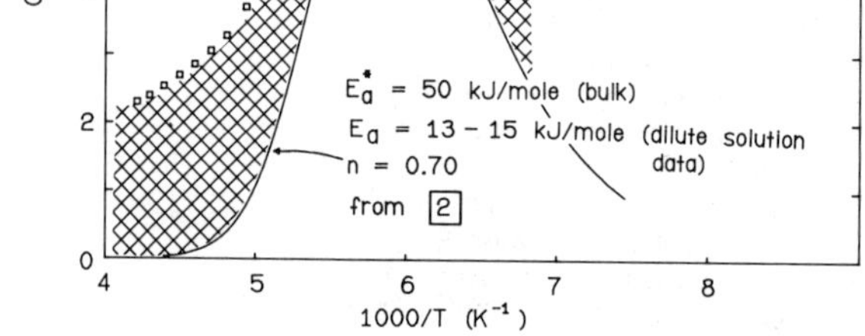

FIGURE 8. The open squares are for the dynamical loss modulus G'' data plotted versus $1000/T$ obtained for polycarbonate (chloral) in the γ relaxation region (data after Yee and Smith[52]). The solid curve is for values of G'' predicted by the K form with $n = 0.70$ without accounting for any possible inhomogeneous distribution of the simple τ_0. The value $n = 0.70$ is obtained from using relation 2 to relate the two activation energies E_a^* and E_a measured for the same γ relaxation in the bulk and in dilute solution, respectively. The shaded area shows the large discrepancy between theory and experiment and provides evidence for additional inhomogeneous distribution of the fundamental τ_0. This is confirmed by the NMR line shape studies by Spiess[48] and by Jones *et al.*[49,50]

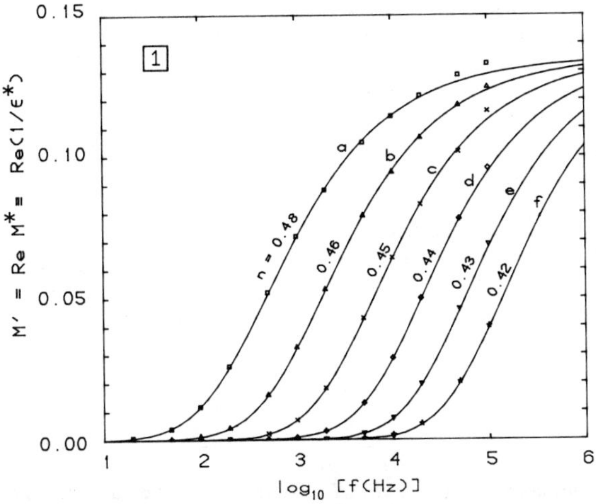

FIGURE 9. Real part of the electric modulus M^* plotted against frequency at a range of temperatures for a $Li_2O \cdot 3B_2O_3$ glass containing predominantly ^6Li alkali ions. The experimental data are after Jain *et al.*[53] The solid curves are theoretical fits with $M'(\omega)$ calculated from the Kohlrausch function; the n for each fit is indicated. **(a)** $T = 108.9°$ C; **(b)** $T = 131.0°$ C; **(c)** $T = 153.8°$ C; **(d)** $T = 177.5°$ C; **(e)** $T = 200.5°$ C; **(f)** $T = 222.2°$ C.

mixed molten salt discussed earlier, $0.4Ca(NO_3)_2$-$0.6KNO_3$, exhibits very interesting relaxation behavior. The fractional exponent for the structural relaxation decreases while that for the conductivity relaxation increases as T approaches T_g from above. Such behavior is difficult to accommodate in some models, but not in the coupling model. Relation 2 gives predictions that are consistent with both of the opposing trends.

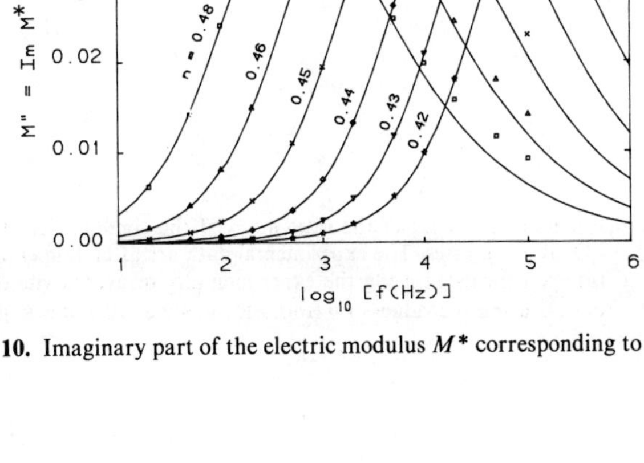

FIGURE 10. Imaginary part of the electric modulus M^* corresponding to FIGURE 9.

TABLE 13. Lithium Ion Conductivity Data, NMR Spin-Lattice Relaxation Data, and Far-infrared Data for Lithium Borate Glasses

Method	Results
Conductivity relaxation	
Data[a]	$E_a^* = 84$ kJ/mole
Analysis [1] [b]	$n = 0.48$
NMR spin-lattice relaxation	
Data[c]	$E_a = 41.9$ kJ/mole
Analysis [2] [d]	$E_a = 43.7$ kJ/mole
Far-infrared	
Data [3] [e]	$\tau_\infty = 1.3 \times 10^{-14}$ sec
Analysis [2] [f]	$\omega_c = 5 \times 10^{12}$ sec^{-1}

[a]After Jain *et al.*[53]
[b]Analysis of Kohlrausch-dominated ($T < T_g$) electric modulus data, which showed that n has a slight temperature dependence, after Ngai *et al.*[54]
[c]After Jain *et al.*[55]
[d]$E_a = (1 - n)\, E_a^*$. After Ngai and Jain.[56]
[e]This is a direct measurement of τ_∞, the pre-exponential of the primitive τ_0. After Exarhos and Risen.[57]
[f]This magnitude of ω_c determined from relation 2 is consistent with the crossover behavior.[56] This is evidenced by the observed approach of the apparent activation energy of conductivity toward the primitive E_a value at $T \simeq 1000°C$, when $\langle\tau\rangle < 10^{-11}$ sec. Data after Shartsis *et al.*[58]

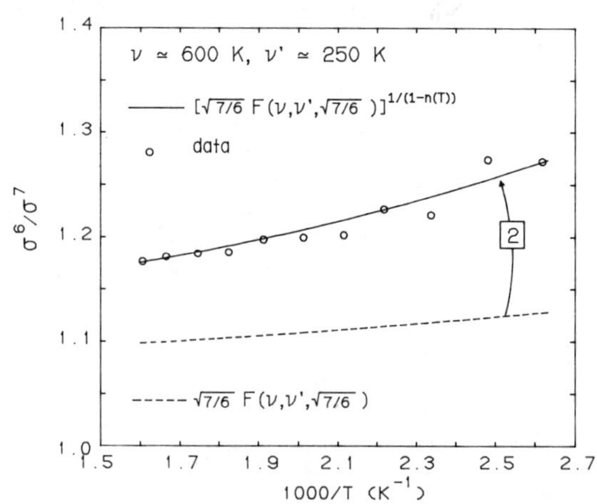

FIGURE 11. Isotope mass dependence of the ratio σ_6/σ_7 of the conductivity of ^{6}Li and ^{7}Li in $^6Li_2O \cdot 3B_2O_3$ and $^7Li_2O \cdot 3B_2O_3$ glasses. The experimental data are after Jain *et al.*[53] The dashed line is the classical isotope expression using the experimentally measured vibration frequency. The solid curve is from [2] using the values of n from FIGURES 9 & 10 and is in good agreement with the data.

Features of conductivity relaxation seen in Na β-alumina are also well explained by relation 2.

Next, let us consider the isotope mass dependence[53–55] of the ratio σ_6/σ_7 of the conductivity of ^{6}Li and ^{7}Li in lithium-6 borate and lithium-7 borate glasses. If the ion jump rate W_0 is proportional to $m^{1/2}$, where m is the mass of the ion, then we would expect $\sigma_6/\sigma_7 = \sqrt{7/6} \simeq 1.08$. The data deviates considerably from this result. When entropy change is incorporated into absolute-rate theory, the σ_6/σ_7 ratio is given by $\sqrt{7/6}\ F(\nu, \nu', \sqrt{7/6})$, the ν and ν' of which are ion vibrational frequencies at the potential energy minimum and the saddle point, respectively. The vibrational frequency $\nu \simeq 600$ K is taken from far-infrared measurements in lithium borate glasses. Even with this correction, the expected σ_6/σ_7 ratio (FIG. 11, dashed curve) still deviates significantly from the data. From $\sigma_6/\sigma_7 = (\tau^*)_7/(\tau^*)_6$, however, relation 2

TABLE 14. Wet and Dry a-SiO_2: Relationship between the Hole Transit Time Activation Energy and the Actual Activation Energy

	Wet a-SiO_2 (Sandia Laboratory Sample)	Dry a-SiO_2 (Harry Diamond Laboratory Sample)
Transit time activation energy		
E_a^* (eV)[a]	0.37	0.60
Observed value for n from [1][b]	0.70	0.82
Primitive activation energy E_a (eV)		
Derived from [2][c]	0.11	0.11
Direct measurement[d]	0.14	0.14

[a]Obtained from dispersive transient transport data for two samples with different water content: wet a-SiO_2 (after Hughes[59]) and dry a-SiO_2 (after McLean *et al.*[60]).

[b] If [1] describes the hole-hopping relaxation function, then the transient current $I(t)$ will have[61] the form $\sim t^{-n}$ at times less than the transit time.

[c]$E_a = (1 - n)E_a^*$. Note that the primitive activation energies E_a so deduced are the same for both samples, as they should for small polaron hopping.

[d]Obtained directly from small polaron prompt mobility measurements. After Hughes and Emin.[62]

predicts $\sigma_6/\sigma_7 = [\sqrt{7/6}F(\nu, \nu', \sqrt{7/6})]^{1/(1-n(T))}$ where the coupling parameter $n(T)$ is as determined in FIGURES 9 & 10. The predicted σ_6/σ_7 is in good agreement with actual measurements (FIG. 11). In our opinion, this ability to account quantitatively for the anomalous lithium isotope mass dependence observed in $(^6Li/^7Li)_2O \cdot 3B_2O_3$ glasses provides the most convincing evidence for the relevance of relation 2 for conductivity relaxation.

Electronic Hopping Conductivity Relaxations

It has been generally agreed that holes of the $2p$ oxygen lone pair orbital in a-SiO_2 form small polarons. Short-range order in a-SiO_2 ensures that the simple small polaron hopping rate W_0 will be the same for every hop. Thus the requirements for immediate applicability of the three relations are satisfied. We consider transient current measurements in a-SiO_2 that show a characteristic t^{-n} dependence for $t \ll t_T$ and a more rapid decay for $t \gg t_T$ where t_T is called the transit time. Starting from the K form for hopping relaxation we can arrive at the observed time dependencies with the same n. The dependence of the transit time on the ratio of sample thickness L to applied electric field E and on T is $[(L/E) \exp (E_a/kT)]^{1/(1-n)}$. Here E_a is the fundamental activation energy for small polaron hopping. The dependencies on n arise as a consequence of relation 2. TABLE 14, which summarizes the data for two different oxides, and FIGURE 12 demonstrate how well relation 2 works in relating the observed transit time to the simple small polaron mobility, which has been directly measured by a separate prompt mobility experiment.[62] The "wet" and "dry" oxides exhibit only a

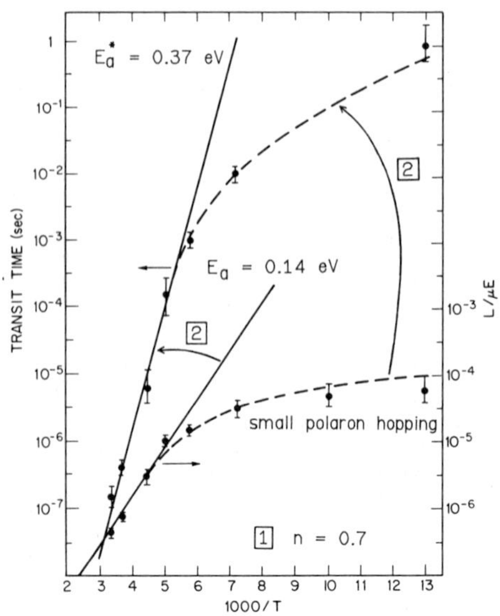

FIGURE 12. Arrhenius plot of the hole transit time data of Hughes[59] and the prompt mobility data of Hughes and Emin[62] for g-SiO_2. The solid lines indicate Arrhenius behavior for $T > 200$ K with an E_a^* of 0.37 eV and an E_a of 0.14 eV. The relation between E_a and E_a^* is in accordance with 2. For $T > 200$, the temperature dependence of the prompt mobility data is that expected for a small polaron. These $L/\mu E$ data points are interpolated to give a continuous dashed curve. Using this dashed curve and relation 2, we predict the upper dashed curve for the transit time temperature dependence for $T < 200$ K. The experimental data are in good agreement with the calculations. The value of n taken from the experimental t^{-n} dependence of the transient current is 0.70.

slight difference in the coupling parameter n, but this difference is large enough to explain the large difference between the observed activation energies by means of relation 2. Small polaron formation and transport, that is, small polaron hopping, in chalcogenide glasses has been a subject of controversy. Evidence recently reported by Nagel[63] seems to reaffirm its existence in at least the archetypal As_2Se_3 glasses. If small polaron hopping is responsible for dispersive transient transport, then the criterion for applicability of our model is satisfied. TABLE 15 and FIGURE 13 summarize the data from the tests of the predictions for virgin and doped a-As_2Se_3. The coupling parameter n may or may not change depending on the nature and the quantity of the dopant. The fundamental E_a values obtained from relation 2, however, are always close to 0.32 eV, the small polaron hopping activation energy as deduced from the difference between the activation energies of dark conductivity and thermoelectric power.

TABLE 15. Pure and Impurity-doped Chalcogenide Glasses: Relationship between the Transit Time Activation Energy and the Actual Activation Energy[a]

	a-As_2Se_3			
		Doped		
	Pure	Tl ($10^{17}-10^{19}$ cm^{-3})	Tl (10^{20} cm^{-3})	I ($10^{17}-10^{20}$ cm^{-3})
Transit time activation energy E_a^* (eV)[b]	0.6	0.62	0.5	0.62
Observed value for n from [1][c]	0.46	0.46	0.31	0.46
Primitive activation energy E_a (eV)				
Derived from [2][d]	0.33	0.33	0.34	0.33
Direct measurement[e]	0.32			

[a]After Taylor and Ngai.[64]
[b]After Pfister *et al.*[65]
[c]If [1] describes the carrier-hopping relaxation function, then the transient current $I(t)$ will have[61] the form $\sim t^{-n}$ at times less than the transit time. Data after Pfister *et al.*[65]
[d]$E_a = (1-n)E_a^*$. Note that the primitive activation energies E_a so deduced are the same for all samples, as they should be.
[e]After the dark conductivity and thermoelectric power measurements of Emin *et al.*[66]

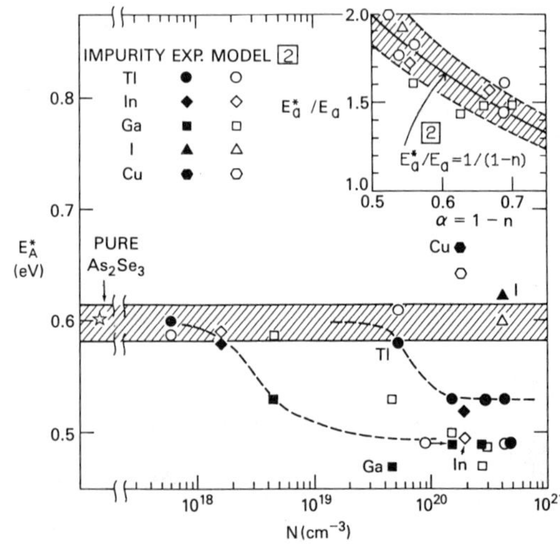

FIGURE 13. Dependence of the experimentally measured activation energy for the transit time of transient hole transport at 300 K (solid symbols) in pure a-As_2Se_3 and impurity-doped a-As_2Se_3 as a function of impurity concentration. The width of the shaded region indicates the uncertainty of the measured activation energy of a-As_2Se_3. The coupling parameter n obtained from the t^{-n} time dependence of the transient current is given in the inset. The value of n may vary with impurity concentration and impurity content. The solid line in the inset is the predicted variation of E_a^*/E_a with $1 - n$ from relation 2. Within the uncertainty in E_a^*/E_a (shaded area in inset) there is good agreement with the coupling model analysis of the experimental data. The data are after Pfister and Morgan.[65]

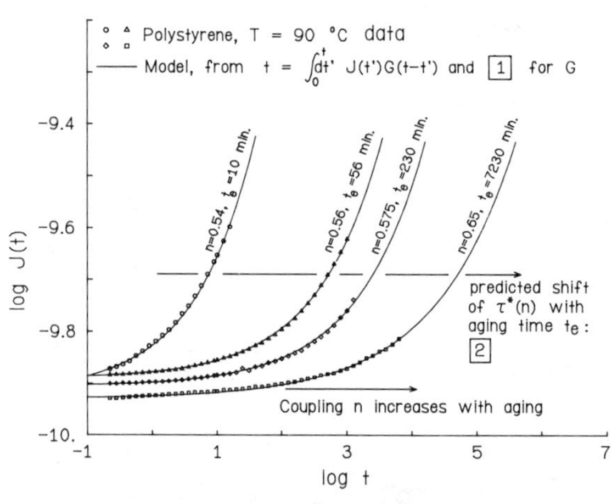

FIGURE 14. Physical aging in glassy polystyrene. The coupling model fits of the creep compliance data (after Plazek *et al.*[68]) at each aging time t_e. The coupling parameter n increases with aging and correctly predicts the shift of τ^* via relation 2.

Physical Aging, Enthalpy, and Volume Recovery

Liquids and polymers cooled below T_g are found to be out of structural equilibrium when the rate of cooling exceeds the rate of structural (molecular) rearrangements. The subsequent structural recovery of the glass has been discussed extensively by Moynihan, Kovacs, and others in previous volumes in the *Annals* series of the New York Academy of Sciences. The observed recovery shows 1) nonlinearity with respect to the magnitude of departure from equilibrium, 2) asymmetry with respect to the sign of the departure, and 3) memory effects that are sensitive to the thermal and physical aging history. All of these features are consistent with the coupling model, and, in several cases, sufficient data are available to provide quantitative verification.

The coupling parameter n, the simple relaxation time τ_0, and the crossover frequency ω_c may change their values during the course of a structural recovery experiment. An example of this is shown in polystyrene samples that have been cooled from equilibrium to $T < T_g$ (FIG. 14). The creep compliance $J(t)$ has been measured

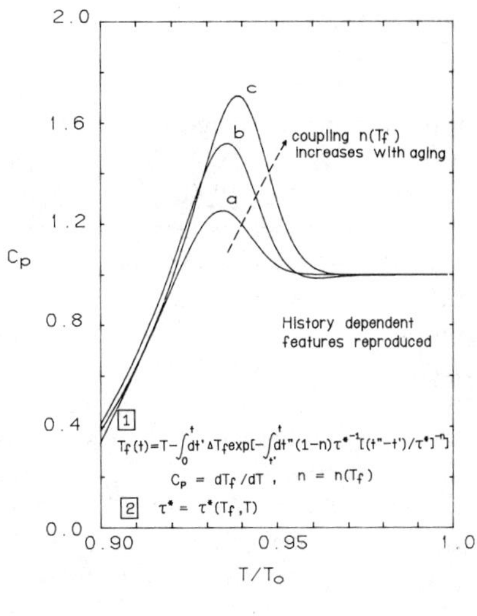

FIGURE 15. Self-consistent numerical calculation from the coupling model that reproduces shifts of the diffraction scanning calorimetry peak after aging for (**a**) 18, (**b**) 180, and (**c**) 1800 sec. In this example, $n(T_f)$ decreases linearly with T_f and the Narayanaswamy form is used for τ_0. After Rendell *et al.*[69]

(in the linear viscoelastic regime) at different aging times t_e after the quench.[68] Although the relaxation modulus $G(t)$ is described by the K form, the compliance is related to it by the well-known convolution expression $t = \int_0^t dt' J(t') G(t - t')$. FIGURE 14 shows that this describes the compliance well at each t_e but that n increases as aging proceeds and the glass densifies. Relation 2 then predicts that τ^* should increase with t_e. The predicted shifts are verified (FIG. 14), although there may also be contributions from changes in τ_0 and ω_c. Similar results are found in other polymers and glasses.

A description of how n actually evolves with aging requires us to follow the structural state of the glass, which we denote by a fictive temperature $T_f(t)$. During the course of structural recovery, the time-dependent rate equation cannot be integrated to the K form. Rather, we must use the more general decay function $\phi = (T_f - T)/\Delta T = \exp\left[-\int_0^t dt' \, \tau_0^{-1} (\omega_c t')^{-n}\right] = \exp\left[-\int_0^t dt' \, (1 - n)\tau^{*-1}(t'/\tau^*)^{-n}\right]$ where relation 2 has

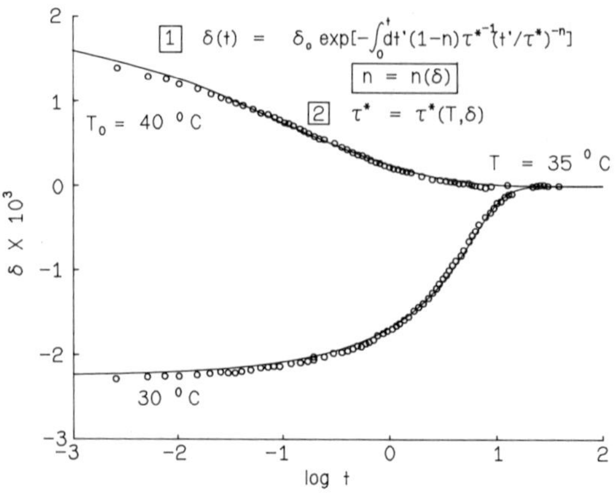

FIGURE 16. Volume recovery in PVAc and coupling model fits[72] of $\delta = \Delta\alpha(T_f - T) \equiv (v - v_\infty)/v_\infty$ (solid line) corresponding to experimental data (O). The data are after Kovacs.[70] The decrease in $n(\delta)$ used ($n = 0.58$ at $T_f = 30°$ C to $n = 0.53$ at $T_f = 35°$ C) is consistent with shear and dielectric data. The $\tau^*(T, \delta)$ agrees with the shear measurements of Kovacs *et al.*[71] (FIG. 18). Here $\Delta\alpha = 4.65 \times 10^{-4}$ K^{-1}.

been used to eliminate τ_0 and ω_c. Here the $n = n(T_f)$ and $\tau^* = \tau^*(T_f, T)$ dependences on T_f self-consistently determine the evolution of T_f, and this produces the observed nonlinearity, asymmetry, and memory effects. A numerical calculation[69] of the self-consistent evolution (FIG. 15) exhibits sharpening and shifting of the $C_p = \mathrm{d}T_f/\mathrm{d}T$ peaks with increased aging as commonly observed in differential scanning calorimetry experiments.

For the case of PVAc, information about the $n(T_f)$ and $\tau^*(T_f, T)$ dependencies is available from independent measurements so that the coupling model predictions can be quantitatively checked during structural recovery. The volume deviation $\delta(t) = \Delta\alpha(T_f - T) \equiv (v - v_\infty)/v_\infty$ following a temperature jump, T_0 to T, has been isothermally measured in the classic work of Kovacs (FIGS. 16 & 17). Under the same conditions, Kovacs, Stratton, and Ferry (KSF) measured a_T, $\delta = \tau^*(T, \delta)/\tau^*(35, 0)$ from dynamic mechanical shear out of volume equilibrium (FIG. 18). The KSF study concluded that the same molecular motions were responsible for both the shear and the volume deformations. KSF also found that the relaxation spectrum changes shape as a function of T and as δ recovers, indicating that $n(\delta)$ is changing during volume recovery (FIG. 19). Similar n variations in PVAc are found in dielectric measurements as a function of T. A calculation of the self-consistent evolution of the volume describes

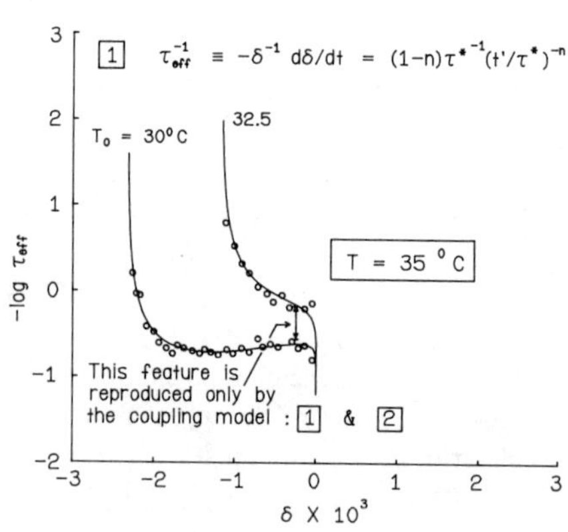

FIGURE 17. Volume recovery in PVAc and coupling model fits of $\tau_{\mathrm{eff}}^{-1} = -\delta^{-1}\mathrm{d}\delta/\mathrm{d}t$ (solid line) corresponding to the expansion data (O) of FIGURE 16. The difference in τ_{eff} near $\delta = 0$ is reproduced; this difference is only reproduced by the coupling model (1 and 2). After Rendell *et al.*[72]

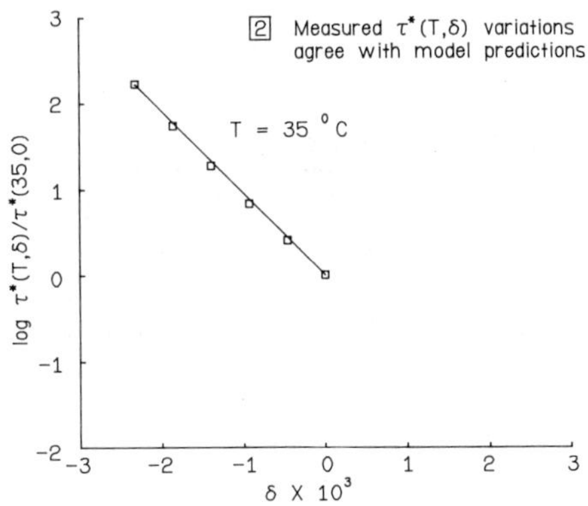

FIGURE 18. Coupling model prediction[72] of $\tau^*(T = 35, \delta)$ (solid line), which is based on the fits in FIGURES 16 & 17, compared to dynamic mechanical shear measurements out of volume equilibrium (□). The data are after Kovacs *et al.*[71]

very well the $\delta(t)$ measurements (FIG. 16) as well as the more sensitive $\tau_{\text{eff}}^{-1} = -\delta^{-1}\text{d}\delta/\text{d}t$ plots (FIG. 17). In particular, as emphasized by Aklonis, the significant difference observed in τ_{eff} near $\delta = 0$ for initial temperatures differing by only 2.5° C ($T_0 = 30°$ and 32.5° C) is a real effect and has been accurately established. In collaboration with Aklonis,[72] we have reproduced this difference for $\delta < 0$ using the coupling model (FIG. 17). No other approach is able to describe this feature. Finally, the $\tau^*(T, \delta)$ dependencies required to describe $\delta(t)$ and τ_{eff} (FIGS. 16 & 17) agree with the independent KSF measurements (FIG. 18). Thus the W property may be verified even when near T_g in those transitions in which the rate parameters are changing during the experiment and the relaxation function does not integrate to the simple K form. Descriptions of phenomena that are seen near T_g and future theories of the glass transition must incorporate the W property.

SUMMARY

We started with a general physical description of the way the relaxation rate of a fundamental mode for a PS in a simple environment is slowed in the presence of a

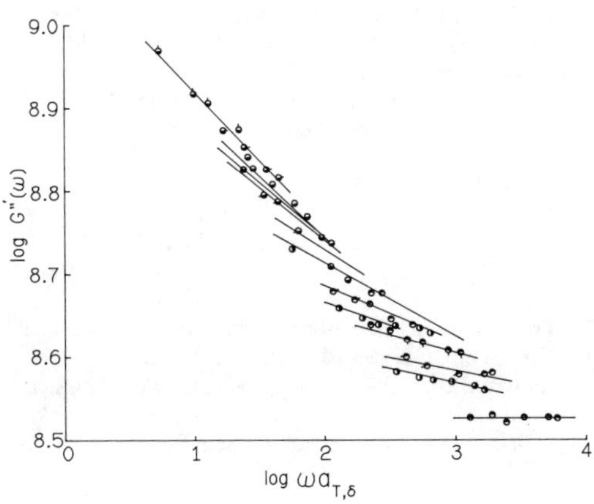

FIGURE 19. Composite plot of G'' shear data out of equilibrium with $a_{T,\delta} = \tau^*(35, 0)$. The data are after Kovacs *et al.*[71] The relaxation spectrum changes shape as a function of T and as δ recovers, indicating that n changes with δ (1).

complex environment. This description was followed by an outline of the general components of the original coupling model, a model that leads to a parameterized but quantitative description of the relaxation rate in a complex environment, that is, of the W property. The W property consists of a time-independent simple relaxation rate W_0 for times before a crossover time $t_c = \omega_c^{-1}$ followed by a time-dependent complex relaxation rate $W(t) = W_0(\hat{\omega}_c t)^{-n}$. The expression for the complex relaxation rate reflects the coupling to the complex environment, and $\hat{\omega}_c$ is related to ω_c by constant factors.

Three coupled predictions, which have been designated [1], [2], and [3] and which correspond to relations 1–3, follow from the W property. We have shown, in our figures and tables, that the three predictions are repeatedly verified in a wide range of experiments. In the process of verification, we have also provided explanations for many puzzling physical phenomena. We have concluded that modeling activities aimed at describing such experimental results should be focused on gaining insight into, or providing a physical basis for, the W property. We have also outlined some of our more recent efforts to do just that by expanding the viewpoint of the original coupling model.

ACKNOWLEDGMENTS

We would like to thank M. H. Cohen for his helpful comments in our conversations.

REFERENCES

1. NGAI, K. L. 1979. Comments Solid State Phys. **9:** 127.
2. NGAI, K. L. 1980. Comments Solid State Phys. **9:** 141.
3. NGAI, K. L. & G. B. WRIGHT, EDS. 1985. Relaxations in Complex Systems. U.S. Government Printing Office. Washington, DC. (Available from the National Technical Information Service, 5285 Port Royal Road, Springfield, VA 22161.)
4. FERRY, J. D. 1980. Viscoelastic Properties of Polymers. John Wiley & Sons. New York, NY.
5. RAJAGOPAL, A. K., S. TEITLER & K. L. NGAI. 1984. J. Phys. C **17:** 6611.
6. NGAI, K. L., A. K. RAJAGOPAL & S. TEITLER. 1985. Physica **133A:** 213.
7. NGAI, K. L., A. K. RAJAGOPAL, R. W. RENDELL & S. TEITLER. 1986. IEEE Trans. Electr. Insul. **EI21:** 313.
8. CASATI, G., ED. 1985. Chaotic Behavior in Quantum Systems. Plenum. New York, NY.
9. NGAI, K. L. & R. W. RENDELL. 1985. *In* Handbook of Conjugated Polymeric Conductors. T. Skotheim, Ed.: 967–1039. Marcel Dekker. New York, NY.
10. KOHLRAUSCH, R. 1847. Pogg Ann. Phys. **12**(3): 393; 1854. Pogg Ann. Phys. IV-91: 56, 79.
11. PALMER, R. G., D. L. STEIN, E. ABRAHAMS & P. W. ANDERSON. 1984. Phys. Rev. Lett. **53:** 958.
12. OGIELSKI, A. T. & D. L. STEIN. 1985. Phys. Rev. Lett. **55:** 1634.
13. COHEN, M. H. & G. S. GREST. 1981. Phys. Rev. B **24:** 4091.
14. SHLESINGER, M. F. & E. W. MONTROLL. 1984. Proc. Natl. Acad. Sci. USA **81:** 1280.
15. BOTTCHER, C. J. F. & P. BORDEWIJK. 1978. Theory of Electric Polarizations. Vol. 2. Elsevier. Amsterdam.
16. WILLIAMS, G. & D. C. WATTS. 1971. Trans. Faraday Soc. **66:** 80; MOYNIHAN, C. T., L. P. BOESCH & N. L. LABERGE. 1973. Phys. Chem. Glasses **14:** 122.
17. NGAI, K. L. & R. W. RENDELL. 1982. Polym. Prepr. Am. Chem. Soc., Div. Polym. Chem. **23:** 46.
18. RAJU, V. R., H. RACHAPUDY & W. W. GRAESSLEY. 1979. J. Polym. Sci., Polym. Phys. Ed. **17:** 1223.

19. McKENNA, G. B., K. L. NGAI & D. J. PLAZEK. 1985. Polymer **26:** 1651.
20. TIRRELL, M. 1984. Rubber Chem. Technol. **57:** 523.
21. NGAI, K. L. & D. J. PLAZEK. 1985. J. Polym. Sci., Polym. Phys. Ed. **23:** 2159.
22. MENDELSON, R. A., W. A. BOWLES & G. L. FINGERS. 1970. J. Polym. Sci., Part A **8:** 105.
23. ORBON, S. J. & D. J. PLAZEK. 1979. J. Polym. Sci., Polym. Phys. Ed. **17:** 1871.
24. PLAZEK, D. J. 1965. J. Phys. Chem. **69:** 3480.
25. NGAI, K. L. & D. J. PLAZEK. 1986. J. Polym. Sci., Part B **24:** 619.
26. FYTAS, G., T. DORFMÜLLER & B. CHU. 1984. J. Polym. Sci., Polym. Phys. Ed. **22:** 1471.
27. NGAI, K. L. & G. FYTAS. 1986. J. Polym. Sci., Part B **24:** 1683.
28. FYTAS, G., G. MEIER & T. DORFMÜLLER. 1985. Macromolecules **18:** 993.
29. NGAI, K. L., C. H. WANG, G. FYTAS, D. L. PLAZEK & D. J. PLAZEK. 1986. Preprint.
30. WANG, C. H., G. FYTAS, D. LILGE & T. DORFMÜLLER. 1981. Macromolecules **14:** 1363.
31. WANG, C. H. & Y. Y. HUANG. 1975. J. Chem. Phys. **62:** 120.
32. WANG, C. H. & Y. Y. HUANG. 1976. J. Chem. Phys. **64:** 4847.
33. ALLEN, G., P. N. BRIER, G. GOODYEAR & J. S. HIGGINS. 1972. Faraday Symp. Chem. Soc. **6:** 169.
34. WANG, C. H., Y. -H. LIN & D. R. JONES. 1979. Mol. Phys. **37:** 287.
35. LIAO, T. -P. & H. MORAWETZ. 1980. Macromolecules **13:** 1228.
36. TERAMOTO, A. & H. FUJITA. 1965. Makromol. Chem. **85:** 261.
37. PLAZEK, D. J., W. DANNHAUSER & J. D. FERRY. 1961. J. Colloid Sci. **16:** 101.
38. FYTAS, G., C. H. WANG, D. LILGE & T. DORFMÜLLER. 1981. J. Chem. Phys. **75:** 4247.
39. HIGASHIGAKI, Y. & C. H. WANG. 1981. J. Chem. Phys. **74:** 3175.
40. McCALL, D. M., D. C. DOUGLAS & V. R. FALCONE. 1969. J. Chem. Phys. **50:** 3839.
41. GREET, R. J. & O. TURNBULL. 1967. J. Chem. Phys. **46:** 1243.
42. McDUFFIE, G. E. & T. A. LITOVITZ. 1962. J. Chem. Phys. **37:** 1699.
43. DORFMÜLLER, T., H. DUX, G. FYTAS & W. MERSCH. 1979. J. Chem. Phys. **71:** 366.
44. WEILER, R., R. BOSE & P. B. MACEDO. 1970. J. Chem. Phys. **53:** 1258.
45. NGAI, K. L., A. K. RAJAGOPAL & C. Y. HUANG. 1984. J. Appl. Phys. **55:** 1714.
46. TORELL, L. M. 1982. J. Chem. Phys. **76:** 3467.
47. ANGELL, C. A. & L. M. TORRELL. 1983. J. Chem. Phys. **78:** 937.
48. SPIESS, H. W. 1985. Adv. Polym. Sci. **66:** 24.
49. JONES, A. A., J. F. O'GARA, P. T. INGLEFIELD, J. T. BENDLER, A. F. YEE & K. L. NGAI. 1983. Macromolecules **16:** 658.
50. NGAI, K. L., R. W. RENDELL & A. F. YEE. Submitted for publication.
51. O'GARA, J. F., DESJARDINS & A. A. JONES. 1981. Macromolecules **14:** 64.
52. YEE, A. E. & S. A. SMITH. 1981. Macromolecules **14:** 54.
53. JAIN, H., H. L. DOWNING & N. L. PETERSON. 1984. J. Non-Cryst. Solids **64:** 335.
54. NGAI, K. L., R. W. RENDELL & H. JAIN. 1984. Phys. Rev. B **30:** 2133.
55. JAIN, H., G. BALZER-JÖLLENBECK & O. KANERT. 1985. J. Am. Ceram. Soc. **68:** C24.
56. NGAI, K. L. & H. JAIN. 1986. Solid State Ionics **18** & **19:** 362.
57. EXARHOS, G. J. & W. M. RISEN. 1972. Solid State Commun. **11:** 755.
58. SHARTSIS, L., W. CAPPS & S. SPINNER. 1953. J. Am. Ceram. Soc. **36:** 319.
59. HUGHES, R. C. 1977. Phys. Rev. B **15:** 2012.
60. McLEAN, F. B., H. E. BOESCH & J. M. McGARRITY. 1976. IEEE Trans. Nucl. Sci. **NS23:** 1506.
61. NGAI, K. L. 1984. *In* Polarons and Excitons in Polar Semiconductors and Ionic Crystals. J. T. Devreese & F. Peeters, Eds.: 383. Plenum. New York, NY.
62. HUGHES, R. C. & D. EMIN. 1978. *In* Proceedings of the International Conference on the Physics of SiO_2 and Its Interfaces. S. Pantelides, Ed.: 14. Pergamon Press. New York, NY.
63. NAGEL, S. P. 1985. Paper presented at the Seventh Trieste International Symposium on "Hopping Transport," August 27–30, Trieste.
64. TAYLOR, P. C. & K. L. NGAI. 1981. Solid State Commun. **40:** 525.
65. PFISTER, G. & M. MORGAN. 1980. Philos. Mag. **B41:** 209.
66. EMIN, D., C. H. SEAGER & R. K. QUINN. 1972. Phys. Rev. Lett. **28:** 813.
67. RENDELL, R. W. & K. L. NGAI, Eds. 1985. *In* Relaxations in Complex Systems: 309–345. U.S. Government Printing Office. Washington, DC. (Available from the National Technical Information Service, 5285 Port Royal Road, Springfield, VA 22161.)
68. PLAZEK, D. J., K. L. NGAI & R. W. RENDELL. 1984. Polym. Eng. Sci. **24**(10): 1111.

69. RENDELL, R. W., T. K. LEE & K. L. NGAI. 1984. Polym. Eng. Sci. **24**(10): 1104.
70. KOVAKS, A. J. 1963. Fortschr. Hochpolym.-Forsch. **3:** 394.
71. KOVAKS, A. J., R. A. STRATTON & J. D. FERRY. 1963. J. Phys. Chem. **67:** 152.
72. RENDELL, R. W., J. J. AKLONIS, K. L. NGAI & G. R. FONG. 1986. Macromolecules. In press.

DISCUSSION OF THE PAPER

G. H. FREDRICKSON (*AT&T Bell Laboratories, Murray Hill, NJ*): In the polymer physics literature the 3.4 exponent that describes the relationship between the longest relaxation time and the molecular weight of a polymer melt is believed to arise from the de Gennes reptation mechanism for chain motion. The de Gennes theory takes careful account of the peculiar topological constraints that exist in a dense polymeric system. How can your theory, which clearly does not attempt to realistically describe these constraints, provide the correct exponent?

NGAI: This question raises two issues that should be addressed: 1) the contention that the de Gennes reptation theory takes careful account of the topological constraints and properly describes the molecular weight dependence and other aspects of polymer melts; and 2) the way in which our coupling model incorporates the entanglement effects and provides the correct exponent for the molecular weight and other aspects of polymer melts. We will respond to each of these in turn.

1) In 1971 de Gennes proposed reptation by long unattached linear chains in a medium of *permanent* topological obstacles. In this case, the topological constraints are clear and reptation, that is, curvilinear diffusion along the linear chain's own contour, is the dominant mechanism for chain motion. This should not be generalized without question to polymer melts of flexible linear chains because there are no fixed obstacles in this case. The topological constraints are entanglements. Not much is known about entanglements on the molecular level except that chains cannot cross each other. The entanglements, however, are not permanent, but fluctuate in space and time. Motion of the chain is coupled via entanglements to other chains, but the motion need not be reduced drastically to reptation.

Nevertheless, Doi and Edwards developed a theory of entanglement effects in polymer melts with the *assumption* that reptation is the dominant motion. They obtain a $\eta_0 \propto M^{3.0}$ dependence of the viscosity on the chain length that differs from the experimentally observed dependence of $\eta_0 \propto M^{3.4}$. Doi and Edwards obtained an expression for the product of the plateau modulus G_N^0 and the recoverable compliance J_e^0, $G_N^0 J_e^0 = 1.2$, which deviates from the experimental value of ≈ 3. Though nonexperts on polymer rheology, we point out that these deviations, among others, are serious problems that have plagued reptation. Several suggestions have been made to the effect that pure reptation in a fixed tube is not correct for a polymer melt (W. W. Graessley, *Adv. Polym. Sci.* **47:** 68, 1982). Notably tube leakage and constraint release have been proposed in conjunction with reptation. These modifications somewhat improve the comparison between theory and experiment. In the limit of very long chain lengths, however, the viscosity is predicted to return to the $\eta_0 \propto M^{3.0}$ dependence. So far, all experimental data reported conform to the $\eta_0 \propto M^{3.4}$ dependence, in disagreement with the modified reptation theory. Shear viscosity of cyclic polystyrene systems has been measured by McKenna *et al.* (*Macromolecules.* In press.), and it has nearly the same $M^{3.4}$ dependence on molecular weight as in linear systems. This result poses a problem for the reptation model because obviously cyclic polymers cannot reptate.

We, as well as several others, believe it is unrealistic to assume that chains reptate in flexible polymer melts. Let us quote from a very recent paper by H. Fujita and Y. Einaga (*Polym. J.* **17:** 1131, 1985): "We suspect that the chains in flexible polymer concentrates do not suffer as strong topological constraints of entanglements as those assumed in the reptation theory, but can wriggle rather freely and isotropically as if they were in dilute solution, except for the following important point: in entangled systems, moving chains induce movement of their surrounding chains through interactions at the entangling points . . ."

How, then, is a theory to be developed for the difficult problem of entanglements for realistic systems of polymer melts? Doi and Edwards set up the dynamics of their chain motions using a kinetic equation with constant rate or damping factors but subject to a set of uncrossable spatial constraints in the form of a tube that prevents sustained lateral motions. Kinetic equations with constant damping factors were previously successful in describing noninteracting chains (Rouse model), and they are not unexpected for simple systems. Doi and Edwards took this type of kinetic equation without question over to entangled polymer melts and merely supplemented it with spatial constraints. Such an approach is almost certain to miss completely the deep aspects of the dynamics that are suggested by the relaxation behavior in a wide range of complex systems. Let us quote from a paper by P. G. de Gennes and L. Léger (*Annu. Rev. Phys. Chem.* **33:** 60, 1982): ". . . as always in science, we should always be aware of the limitations of our current scheme and search for more profound approaches to polymer dynamics. Perhaps the difference between $a = 3.0$ and $a = 3.3$ in the exponent . . . conceals a very deep theoretical problem." We believe this to be the case. Any approach that attempts to build in the complex entanglement effects explicitly will not be successful. The coupling model provides a description of dynamics in complex systems based on fundamental properties of dynamical systems. The form of the kinetic equation itself is modified by the entanglement effects, and it is not just a matter of adding spatial constraints to a simple kinetic equation. This brings us to the other issue raised.

2) It is known from a large body of work that the dynamics of Hamiltonian systems exhibit certain generic features (G. Casati, Ed., Chaotic Behavior in Quantum Systems, Plenum, New York, NY, 1985; A. J. Lichtenberg and M. A. Lieberman, Regular and Stochastic Motion, Springer–Verlag, New York, NY, 1983). For instance, quantum Hamiltonians generically possess a linear distribution of level spacings. In the corresponding classical limit, Hamiltonians generically possess a characteristic structure of regular and stochastic orbits in phase space. Only certain simple Hamiltonians can avoid these generic features. The generic features had never before been incorporated into a calculation of relaxation. The coupling model attempts to do just that.

The original version of the coupling model used the linear distribution of level spacings, which is generic to quantum Hamiltonians, to modify a constant relaxation rate τ_0^{-1} to the time-dependent form $\tau_0^{-1}(\omega_c t)^{-n}$. The model can be understood as an extension of the standard treatments of relaxation. The origin of kinetic equations with constant rate or damping factors from the underlying equations of motion was studied thirty years ago by van Hove, Prigogine, and others. They described conditions for which a Hamiltonian system can approximate a kinetic equation with a constant relaxation rate. We have recently been able to formulate the coupling model, both classically and quantum mechanically, directly within the framework of relaxation developed by van Hove and Prigogine by extending the class of Hamiltonians under consideration (R. W. Rendell and K. L. Ngai, unpublished results). Those studied by van Hove and Prigogine are actually quite restrictive and do not possess the generic properties. If Hamiltonians with generic properties are used, then all the elements of the original coupling model can be identified.

In particular, this framework can be used to develop the coupling model completely on the basis of the classical equations of motion (R. W. Rendell and K. L. Ngai, unpublished results). Although we could use a quantum mechanical starting point, the use of the corresponding classical Hamiltonian is quite natural for application to relaxations in complex systems. This development is an example of our continued effort to improve our earlier models and to seek new avenues for derivation of the W property. Even the classical and very simple coupled chain models we have constructed exhibit the generic properties. If it were possible to construct a realistic Hamiltonian that incorporates the chain entanglements in detail, we should expect it to possess the generic properties. Then when relaxation of a chain is considered, the accompanying rate will have a time dependence of the form $\tau_0^{-1}(\omega_c t)^{-n}$. This intrinsically time-dependent rate arises from fundamental properties of complex dynamical systems and should be incorporated into any model of entangled polymer melts. These physical features are not considered in the reptation model, and, from our viewpoint, this omission must lead to deviations from experimentally observed behavior.

As we have shown in the past two years, the application of the coupling model to flexible polymer melts does lead to a description of the experimentally observed behavior. In particular, the value of n obtained from the shape of the terminal relaxation does lead to the correct 3.4 exponent for the viscosity of both linear and cyclic polymers.

D. MONROE (*AT&T Bell Laboratories, Murray Hill, NJ*): I'm concerned about the validity of your relationship for activation energies when the exponent is temperature dependent. For example, hole transport in glassy As_2Se_3 exhibits a weak linear temperature dependence of the exponent, but this is sufficient to give a very substantial apparent activation energy E_a^* that has no relation to a microscopic activation energy. This warning is relevant when considering activation energies.

NGAI: In the text we have discussed transient hole transport in g-SiO_2 and in g-As_2Se_3. The situation in g-SiO_2 is clear. The exponent n is observed to be temperature independent. The hole in the $2p$ lone pair orbital of oxygen forms a small polaron, and the hole transport mechanism is small polaron hopping. There is general agreement on this between Mott and Emin. Our relaxation model, when applied to transient transport in g-SiO_2, has verified the relaxation $E_a^* = E_a/(1 - n)$ between the observed activation energy E_a^* and a microscopic activation energy E_a. The situation in chalcogenide glasses, including g-As_2Se_3, is not as clear. The transport mechanism is still controversial. Emin has maintained that small polarons are formed in chalcogenide glasses. Mott has disagreed with this, although he has never dismissed it. The main evidence for small polarons put forward by Emin is the inequality of the activation energies deduced from the conductivity and thermopower. An alternative interpretation of this inequality by Nagel, Callearts, and Denayer has been quoted to counter the small polaron picture. Nagel, however, has recently given a lecture (at the International Symposium on Electron Hopping, Trieste, Italy, August 1985) in which he reexamined the experimental data and came out in favor of small polaron formation in chalcogenide glasses. Recent work by Cohen, Soukoulis, Economou has revitalized the role of small polarons. Hence, the small polaron and small-polaron-hopping conduction remain viable possibilities in chalcogenide glasses.

Transient hole transport was studied extensively by Pfister. He generated his free carriers by a 5-nsec light flash from a nitrogen laser directed onto one surface of the sample. The nitrogen laser light has a wavelength of 337 nm, which corresponds to an energy of 3.68 eV, which is much larger than the 2.48-eV band gap of g-As_2Se_3. He found the shape of the transient and thereby found n to be remarkably stable with respect to temperature. The sources of the experimental data of hole transport in g-As_2Se_3 you quoted that exhibit a weak linear temperature dependence of the exponent are from transient photoconductivity studies by Orenstein and Kastner and transient transport measurements by Khan, Kastner, and Adler. Orenstein and Kastner used exciting light with a sub-band-gap energy of about 1.7 eV. The latter used a 483 nm wavelength or light with an energy of 2.57 eV. Your recent work (*Phys. Rev. B* **33:** 8881, 1986) has used pulsed light having a photon energy as low as 1.4 eV. There are several luminescence processes in g-As_2Se_3, and their intensities depend on the energy of the exciting light (Murayama and co-workers). This fact together with the drastic difference in the exciting light energy used by Pfister as compared with those used by the MIT group, means care must be exercised in comparing data. Under his conditions, Pfister may produce small polarons for transient transport. On the other hand, the near-band-gap and sub-band-gap light used by others may involve defect states in the gap that give rise to the temperature dependence of the exponent. In brief, one cannot dismiss the findings of Pfister in view of the recent data because the energy of the exciting light used is very different. At the very least, the MIT group would have to repeat Pfister's experiment under the same conditions before a definitive conclusion could be drawn. Details of points discussed can be found in a recently published review (K. L. Ngai, *In* Polarons and Excitons in Polar Semiconductors and Ionic Crystals, J. T. Devreese and F. Peeters, Eds., Plenum, New York, NY, 1984).

We have applied our model only to those data of Pfister's for which n is independent of temperature. Since the MIT data may involve defect states, they are not appropriate candidates for analysis by our model without some generalization because the model requires a homogeneous τ_0.

J. R. STEVENS (*University of Guelph, Guelph, Ontario, Canada*): Is it correct that you have three adjustable parameters, τ_0, ω_c, and n?

NGAI: The model provides a framework with which to analyze and interpret data in a self-consistent way. The data are not from a single experiment but from a coupled set of independent experiments, all of whose results must be consistent with the parameters in the framework. With input about the form of the fundamental relaxation time τ_0 from direct measurements or molecular models (for example, mass dependence, molecular weight dependence, and temperature dependence), the possible experiments and relationships among parameters are sufficient in number that the model yields predictions with no arbitrary (unfixed) parameters. In effect, some of the experiments are used to "compute" the parameters so that the parameters are known for other experiments. There are nominally four parameters that enter the framework. These are the exponent parameter n ($0 < n < 1$) and the relaxation time τ^*, which enter into $\exp[-(t/\tau^*)^{1-n}]$, the predicted K form; the fundamental time τ_0, which enters in the short-time regime of the complex relaxation or in the relaxation of the related simple system (both of which have linear exponential behavior, $\exp(-t/\tau_0)$); and the parameter essentially defining the boundary between long and short times ω_c such that $\omega_c\tau_0 < 1$ represents the short-time regime and $\omega_c\tau_0 > 1$ represents the long-time regime for relaxations in complex systems. From general considerations of the model, ω_c, although dependent on the specific complex system, is expected to be 10^9–10^{12} $\sec^{-1}$. A single long-time relaxation experiment under the condition $\omega_c\tau_0 > 1$ that yields a K form serves to determine both n and τ^*. The fundamental relaxation time τ_0 can be taken from molecular models or obtained by measurement in the corresponding simple system, or by a measurement in the complex system under conditions $\omega_c\tau_0 < 1$ for which crossover has occurred to dominant linear exponential relaxation. The occurrence of crossover also provides an estimate of ω_c. An alternate determination of ω is provided by the relation $\tau^* = [(1 - n)\omega_c\tau_0]^{1/(1-n)}$ once n, τ^*, and τ_0 have been determined. This determination must be consistent with any experiments that show crossover. Further, the relation between τ^* and τ_0 means that all dependencies that enter into τ_0 must be renormalized by being raised to the power $(1 - n)^{-1}$. For example, in considering entanglement effects in monodisperse polymer melts, τ_0 taken from the Rouse model has the chain-length dependence of M^2. The corresponding τ^* is predicted to have the dependence of $M^{2/(1-n)}$. If comparison of the predicted τ^* with measurement is restricted to only its power law dependence on M, then only a single parameter n is involved. As mentioned before, n can be determined by a single relaxation experiment. Many examples of the success of this renormalization procedure in resolving existing anomalies as well as in leading to new predictions of behavior (predictions ultimately verified) have been given in the body of the paper. The important point is that once the parameters have been fixed by the related experiments, these renormalizations are carried out with no arbitrary parameters and are tested by further independent experiments. In brief, the procedure in using the model is to first make a consistent determination of pertinent parameters and then to use them to explain or predict behavior in additional independent experiments.

Linear and Nonlinear Experiments for a Spin Model with Cooperative Dynamics

GLENN H. FREDRICKSON
AT&T Bell Laboratories
Murray Hill, New Jersey 07974

INTRODUCTION

In this paper we discuss a simple spin model for the description of dynamical properties of viscous liquids in the vicinity of their glass transition. The model was constructed to reflect the belief that structural relaxation phenomena in dense fluids are very highly cooperative. Monte Carlo simulations that probe the linear and nonlinear dynamical response of the spin system are presented.

The equilibrium relaxation properties of the model were investigated in collaboration with Steven A. Brawer and Hans C. Andersen. Analytical work on the model has been described in two publications,[1,2] and a detailed Monte Carlo study of its equilibrium properties and linear response has recently been completed.[3] The present paper contains the first quantitative analysis of nonlinear relaxation in the spin model.

Viscous (supercooled) liquids exhibit a variety of anomalous kinetic properties in the vicinity of their glass transition.[4,5] These anomalous properties are "universal" in the sense that many different classes of glass-forming substances exhibit the same qualitative behavior. For example, amorphous polymers such as polystyrene show relaxation spectra and temperature dependence of viscosity that are very similar to the corresponding properties of low-molecular-weight glasses such as glycerol.[4–6] We need to understand why such diverse types of materials behave similarly if the glass transition and relaxation phenomena are to be completely understood.

One of the most characteristic dynamical properties of a supercooled liquid is that the decay of fluctuations about the (metastable) equilibrium liquidus line is nonexponential. A variety of experiments show that the decay of density or enthalpy fluctuations can be accurately fit over many decades in time to the Kohlrausch–Williams–Watts (KWW) empirical expression

$$\psi(t) = \exp\left[-(t/\tau)^{\beta}\right] \tag{1}$$

where τ and β are adjustable parameters. Although β is usually only weakly temperature dependent in the glass transition region (where typically $0 < \beta \leq 0.5$), the average relaxation time τ has an extremely strong temperature dependence. For most materials it is found that τ has non-Arrhenius behavior with an activation energy that increases as the temperature is lowered. Experimental data for the average relaxation time (or viscosity) can often be fit with the Vogel–Fulcher (VF) expression

$$\tau = \tau_f \exp\left[E/(T-T_0)\right] \tag{2}$$

where τ_f, E, and T_0 are adjustable constants. Another commonly used expression to describe the temperature dependence of the average relaxation time is that suggested by Adam and Gibbs[7]

$$\tau = \tau_0 \exp [A/(TS_c)] \tag{3}$$

where S_c is the configurational entropy of the liquid (loosely defined as the entropy of the metastable equilibrium liquid minus the entropy of the stable crystalline phase at the same temperature). Often it is found that extrapolations of entropy data to temperatures below the laboratory glass transition temperature, T_g, predict that the configurational entropy vanishes at a temperature T_2. If such extrapolations are to be believed, one must resolve how at T_2, which is sometimes hundreds of degrees above absolute zero, the entropy of an amorphous phase can become smaller than the entropy of a stable crystalline phase.[8] Empirically, T_0 and T_2 often roughly coincide.

A successful theory of structural relaxation in dense fluids and the glass transition must be able to explain some, or preferably all, of the above experimental observations. Two commonly cited theories are the Cohen–Grest free volume theory[9,10] and the Adam–Gibbs entropy theory.[7] In both it is suggested that the dramatic increase in structural relaxation times of fluids at low temperature is caused by a *thermodynamic* singularity below the T_g observed in the laboratory. Although the two theories have been successful in many of their predictions, they are not based on a precisely defined statistical mechanical model with consistent thermodynamic and kinetic properties. Furthermore, it is difficult (or impossible) to relax the severe approximations in the theories in order to assess their significance.

In the past few years several theories of relaxation have been proposed that suggest the sluggish kinetic behavior near a glass transition is caused by a purely *dynamical* singularity at a lower temperature.[11–15] In general, these theories are based on the predictions of a particular mode-coupling approximation[16] for nonlinear equations of compressible hydrodynamics. A recent paper by Siggia,[17] however, suggests that the singularity predicted by these theories is an artifact of the mode-coupling approximation and is not actually present in the exact solution to the model equations. This question is apparently still open.[18]

In the present paper we discuss the so-called two-spin-facilitated kinetic Ising model (2SFM) introduced in reference 1. Related spin models have proved extremely useful for the description of critical phenomena and phase transitions. The basic idea is to replace the many degrees of freedom in a region of dense fluid (roughly the size of an equilibrium correlation length) by a spin that can take only one of two states: up or down. As an example, an up spin might represent a region of fluid with lower than average density, while a down spin could represent a region with higher than average density. This spin model has a precisely defined Hamiltonian and thus has thermodynamic properties, but it was also given kinetic properties by defining spin-flip transition probabilities that are consistent with the form of the Hamiltonian. The key feature of the model is that the transition probability for the flipping of a particular spin depends crucially on the states of the surrounding spins. In particular, the probability is zero unless the environment of a spin exceeds a certain level of excitation. A consequence of this choice of spin-flip rates is that relaxation in the 2SFM becomes a highly cooperative sequence of events at low temperature, involving the participation of large numbers of spins. Furthermore, the model was constructed so that it has no thermodynamic singularities for the temperatures of interest.

The initial analytical work on the 2SFM[1,2] indicated that the spin system might undergo a purely dynamical transition of the type proposed by Leutheusser[11] at a nonzero temperature. A renormalized pertubation theory was developed for the model that led to approximations very similar to the mode-coupling approximations for the hydrodynamic theories described above. Subsequent Monte Carlo simulations on the spin model, however, demonstrated that the kinetic singularity predicted by the analytical work was an artifact of the mode-coupling approximation.[3] These results seem to support Siggia's conjecture[17] that the singularities in the mode-coupling approximations for the hydrodynamic models are also unphysical.

In spite of the fact that the 2SFM has no actual kinetic or thermodynamic singularity at a nonzero temperature, the Monte Carlo results of reference 3 and those in the present paper indicate that it has extremely rich linear and nonlinear dynamical responses in qualitative agreement with experiments on viscous liquids. Furthermore, because the spin model can be simulated by computer in a straightforward and efficient manner (with moderate computational power one can calculate relaxation functions over seven decades in time), it is ideal for the study of cooperative relaxation phenomena. There seems to be no other existing model that provides such extensive qualitative agreement with experiment and that can be investigated in such a complete and rigorous manner.

THE n-SPIN FACILITATED MODELS

In order to introduce the 2SFM it is helpful to discuss the more general n-spin-facilitated model. We consider a collection of N spins on a lattice with, for the purpose of simulation, periodic boundary conditions. The state of the ith spin, σ_i, can take the values $+1$ (spin up) or -1 (spin down). The energy of a spin configuration σ is defined by the Ising Hamiltonian

$$H(\sigma) = h\sum_{i=1}^{N} \sigma_i - J\sum_{i<j} \sigma_i\sigma_j \tag{4}$$

where h is an external magnetic field that acts uniformly on each spin, J describes the strength of interactions between spins, and the second sum is taken over all near-neighbor spins. The sign of h is chosen to make the down-spin state energetically more favorable. Thus, with the interpretation of a spin down as a high-density region of fluid, increasing h or lowering the temperature, T, densifies the system. There is a formal analogy between the Ising model and the lattice gas model of a fluid that allows one to interpret the magnetic field h as an external pressure or chemical potential in a fluid system.[19]

For simplicity in the present paper we consider the special case of equation 4 in which $J = 0$, that is, in which all spins are independent at thermal equilibrium. (This is reasonable if the spins are to represent fluid regions whose size is comparable with the equilibrium correlation length of the fluid at the temperature of interest. Computer simulations with interactions between spins are in principle no more difficult than for $J = 0$ and are left for future work.) Because of the simplicity of equation 4 for $J = 0$, we can calculate exact analytical expressions for all thermodynamic properties of the model. Of particular interest are the entropy

$$S(T) = \ln [2 \cosh (1/T)] - T^{-1} \tanh (1/T) \tag{5}$$

the average fraction of up spins

$$c = \frac{1}{1 + \exp (2/T)} \tag{6}$$

and the magnetization

$$M = \frac{1}{N}\sum_{i=1}^{N} \langle \sigma_i \rangle = 2c - 1 \tag{7}$$

In equations 5–7, as in the remainder of the paper, the temperature is expressed in units of h/k_B, where k_B is Boltzmann's constant, and the entropy is expressed in units of k_B.

The kinetic properties of the n-spin facilitated models are defined in a manner similar to that of Glauber.[20] The dynamics are generated from a master equation for the probability distribution function of spin configuration σ at time t, $P(\sigma,t)$

$$\frac{d}{dt}P(\sigma,t) = -\sum_{i=1}^{N} W_i(\sigma)P(\sigma,t) + \sum_{i=1}^{N} W_i(\sigma')P(\sigma',t) \tag{8}$$

where σ' is the configuration obtained from σ by flipping the ith spin. $W_i(\sigma)$ is the probability per unit time that the ith spin changes its state and for the n-spin-facilitated models is defined by

$$W_i(\sigma) = \frac{\alpha}{n!} \exp [(\sigma_i - 1)/T] \prod_{k=0}^{n-1} (m_i - k) \tag{9}$$

where m_i is the number of nearest neighbors of the ith spin that are in the spin-up state in configuration σ and α is a constant with the dimensions of inverse time. In the subsequent discussion all times are expressed in units of α^{-1}.

From equation 9 it is apparent that the rate at which a spin flips in the n-spin-facilitated model is zero unless the spin has n or more spin-up neighbors. Hence the relaxation of a spin is facilitated by neighboring up spins. Because the concentration of up spins, c, decreases in the model according to equation 6 as the temperature is lowered, relaxation becomes impeded and highly cooperative at low temperature.

The flip rates in equation 9 were constructed to satisfy detailed balance in order to insure that equation 8 relaxes to the canonical distribution consistent with equation 4 in the limit of long times. This is only guaranteed, however, if the Markov process defined by equations 8 and 9 is *irreducible,* that is, if all states can be reached by the dynamics. In reference 2 we presented a strong argument that the Markov processes for the models with $n = 1$ and $n = 2$ (the 2SFM) on hypercubic lattices in dimensions greater than one are irreducible in the thermodynamic limit (except for a set of states with measure zero). The n-spin-facilitated models, however, are *reducible* for a finite system of spins. In reference 3 we demonstrated that, at least for the 2SFM on a square lattice, a system with 50 spins on a side is large enough that the effect of the reducibility on the static and dynamic properties is inconsequential for the temperatures we are able to simulate ($T > 0.8$). This demonstration is sufficient to ensure that the simulations in the present paper are also not affected. A much more detailed discussion of the ergodicity of the 2SFM can be found in references 2 and 3.

The one-spin-facilitated model (1SFM) is obtained by choosing $n = 1$ in equation 9. The mechanism for relaxation in the 1SFM at low temperatures is easily established. At such temperatures, where $c \ll 1$, there is a small population of isolated up spins, spins that will be referred to as defects. As illustrated in FIGURE 1 for a square lattice in two dimensions, a defect can facilitate one of its neighbors to flip up, following which the original defect spin can flip down. The net result is that the original defect has effectively hopped to an adjacent site on the lattice. Thus, isolated defects can "diffuse" about the lattice at any nonzero temperature in the 1SFM and cause relaxation. This diffusion mechanism can be argued to lead to Arrhenius temperature dependence of the average relaxation time,[2] and this has been verified by Monte Carlo simulations (G. H. Fredrickson, unpublished results). Because most glass formers exhibit non-Arrhenius behavior over some range in temperature, it is likely that the 1SFM is not sufficiently cooperative to describe relaxation in real materials near their glass transition. The model has some obvious similarities to the defect diffusion model of Shlesinger and Montroll.[21]

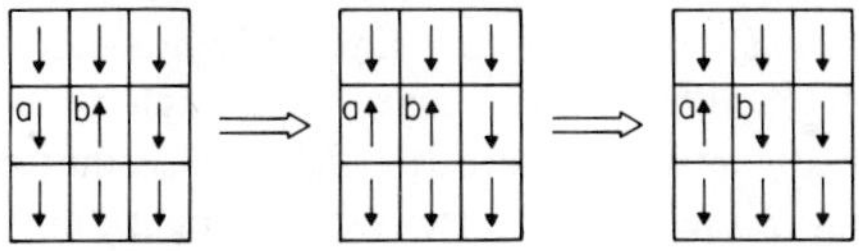

FIGURE 1. Mechanism of low-temperature relaxation in the 1SFM. The isolated defect designated as spin *b* can facilitate spin *a* to flip up. Following this, spin *b* can flip down with the net result that the defect has hopped from *b* to *a*.

The 2SFM is obtained by setting $n = 2$ in equation 9. For the purpose of this paper we consider a square lattice in two dimensions, but the behavior of the model is expected to be qualitatively similar on hypercubic lattices in higher dimension. The 2SFM is much more cooperative at low temperature than the 1SFM. The reason for this is that, unlike the 1SFM, the 2SFM has no isolated defects involving a small number of spins that can propagate in a sea of down spins. For example, in FIGURE 2 the boxed cluster of spins can undergo internal transitions, but cannot by itself cause the down spins on the exterior of the box to flip up. The only way such spins can flip up is for some surface or structure composed of up spins, possibly very far away, to flip up intermediate spins and engulf the original down spins.

That such a mechanism can lead to a temperature-dependent activation energy for relaxation is easily seen. As the temperature is lowered, the concentration of up spins decreases, and hence the number of down spins that some spin up structure must flip to relax a perturbation is increased. The energy barrier for relaxation then increases because each down spin must be flipped up against the magnetic field. As a result, the activation energy for relaxation is likely to be related to the number of spins that must be flipped by some characteristic spin-up structure and hence to the size of some region that must undergo cooperative rearrangement. This idea is essentially the same as that

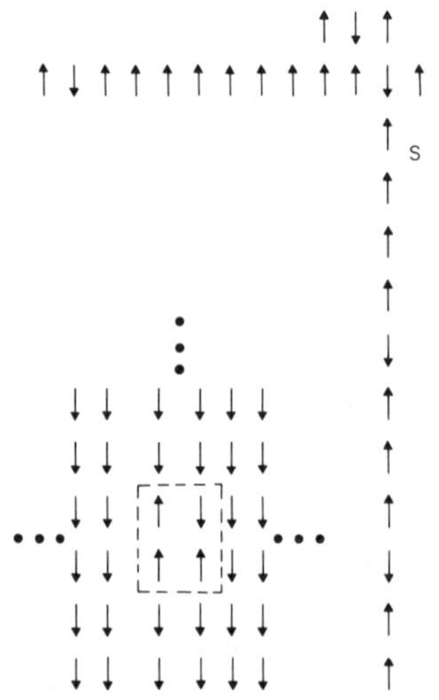

FIGURE 2. A possible mechanism of low-temperature relaxation in the 2SFM. The high-energy cluster of up spins (highlighted by the dashed box) is embedded in a sea of down spins and cannot propagate or attain thermal equilibrium in its immediate environment. The environment of the cluster can change, however, if some structure S, consisting primarily of up spins and possibly very far away, facilitates the flipping up of the spins surrounding the box.

proposed very elegantly by Adam and Gibbs twenty years ago.[7] Indeed, we demonstrate in the next section that the expression for the average relaxation time derived by Adam and Gibbs on the basis of this picture, equation 3, fits our simulation data for the 2SFM extremely well. A more contemporary approach is to say that there is a characteristic length that controls the rate at which a dense fluid equilibrates following a perturbation, and that this length increases as the temperature is lowered. It is likely that such a length considerably exceeds the equilibrium correlation length in the fluid, which is only weakly temperature dependent at high density. At the present we do not understand the mechanism of relaxation in the 2SFM well enough to identify this length.

LINEAR DYNAMICAL RESPONSE OF THE 2SFM

The linear dynamical response of the 2SFM has been investigated through Monte Carlo simulations on a square (two-dimensional) lattice with periodic boundary conditions. The same qualitative behavior is expected for the n-spin-facilitated model ($n \geq 2$) on a hypercubic lattice in d ($\geq n$) dimensions. However, it is anticipated that for a given n, the cooperativity of the n-spin models decreases as the dimensionality of the lattice is increased. The choice of the 2SFM on a square lattice is convenient because it allows the investigation of large spin systems and long simulation runs. All simulation data in the present paper are for systems with 50 spins on a side. For the temperatures and times relevant to this study it has been shown that there are no significant finite size effects in a 50×50 system.[3] The details of the Monte Carlo algorithm and the equilibration procedure are given in reference 3.

Two different linear relaxation functions for the 2SFM have been investigated. The first, $\phi(t)$, is a single-spin time correlation function

$$\phi(t) = \frac{\langle \sigma_i(t)\sigma_i(0) \rangle - \langle \sigma_i \rangle^2}{1 - \langle \sigma_i \rangle^2} \tag{10}$$

and describes the manner in which the system responds to an infinitesimal *localized* perturbation. The angular brackets in equation 10 denote equilibrium ensemble averages with the Hamiltonian in equation 4 ($J = 0$). The second response function of interest is the magnetization time correlation function

$$\psi(t) = \frac{\langle \mu(t)\mu(0) \rangle - M^2}{\langle \mu^2 \rangle - M^2} \tag{11}$$

where μ is the spin operator $N^{-1}\,\Sigma_i\,\sigma_i$ and $M = \langle \mu \rangle$ is given by equation 7. This function describes the system's response to an infinitesimal *long wavelength* perturbation and is equal to the dynamic structure factor at zero wavevector.

The behavior of $\phi(t)$ has been investigated for dimensionless temperatures in the range $0.83 < T < 5.0$, over which the average relaxation time changes by six orders of magnitude. FIGURE 3 shows the decay of $\phi(t)$ for the 2SFM at the highest temperature studied: 5.0. The solid curve in the figure is the best fit of the simulation data to the KWW expression

$$\phi(t) = \exp\,[-(t/\tau_w)^\gamma] \tag{12}$$

with parameters $\tau_w = 1.4$ and $\gamma = 0.54$. Thus, even at the (high) temperature of 5.0, where the average relaxation time is O(1), structural relaxation in the 2SFM is highly nonexponential. This is in marked contrast with more conventional kinetic Ising models, such as the Glauber model,[20] which have $\gamma = 1$ at infinite temperature. From the discussion of the previous section it is clear that the departure from exponential decay for the 2SFM is due to cooperativity of relaxation that persists at high temperature.

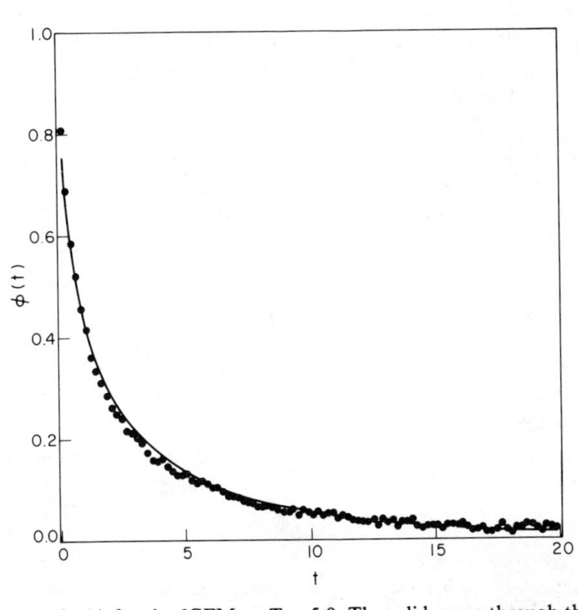

FIGURE 3. Decay of $\phi(t)$ for the 2SFM at $T = 5.0$. The solid curve through the Monte Carlo data is a KWW fit with parameters $\gamma = 0.54$ and $\tau_w = 1.4$.

FIGURE 4 shows a KWW fit to simulation data for $\phi(t)$ at the lower temperature of 0.91. The parameters in this case are $\tau_w = 3700$ and $\gamma = 0.26$. It is apparent that the relaxation of $\phi(t)$ becomes increasingly nonexponential as the temperature is lowered. Indeed, we have found that the temperature dependence of γ can be described reasonably well by the empirical correlation

$$\gamma^{-1} = 1.3 + 2.3T^{-1} \pm 0.1 \tag{13}$$

FIGURE 5 shows this expression plotted with the values of γ obtained from the Monte Carlo simulations for the temperature range $0.83 < T < 5.0$. Fits of experimental data on laboratory glasses to the KWW expression in equation 1 also indicate that the exponent β decreases with decreasing temperature.[4,22] In FIGURE 6 we show that the β^{-1} verses T^{-1} correlation describes the published viscoelastic relaxation data of Matsuoka[22] for polycarbonate extremely well over the complete 145 degree temperature interval of the study. It would be of great interest to see if this correlation can describe the temperature dependence of β for other types of materials.

As discussed in the introduction, non-Arrhenius temperature dependence of average relaxation times is a characteristic property of most viscous liquids. In FIGURE 7 we show that the relaxation time defined by

$$\bar{\tau} = \int_0^\infty \mathrm{d}t\, \phi(t)$$

for the 2SFM exhibits highly non-Arrhenius behavior over the temperature range that was investigated.[a] Furthermore, the activation energy for relaxation is seen to increase as the temperature is lowered, in agreement with experiment.

It is highly instructive to attempt fits of the data shown in FIGURE 7 to the Vogel–Fulcher and Adam–Gibbs expressions (equations 2 and 3). We found that fits of the simulation data for the 2SFM to the Vogel–Fulcher expression were unsatisfactory in that the parameters were very sensitive to the temperature interval used for the fitting. In particular, the parameter T_0 moved to lower temperature as more low T data points were included in the fit. This observation supports our previous conjecture[3] that the 2SFM has no actual kinetic singularity for $T > 0$ and invalidates use of the Vogel–Fulcher relation. In contrast, it was found that the Adam–Gibbs equation could be used to very accurately describe the simulation data for the 2SFM. In FIGURE 8 is shown a fit of equation 3 to the relaxation time data of FIGURE 7, using equation 5 for the temperature dependence of the entropy. The solid line through the data is the result of a linear regression, yielding the relation

$$\bar{\tau} = 1.0 \exp\,[3.4/(TS(T))] \tag{14}$$

This remarkable fit shows that the entropy plays a dominant role in governing structural relaxation in the 2SFM and gives strong support to the Adam–Gibbs picture of cooperative relaxation, at least for the spin model.

Similar results were obtained for the magnetization correlation function of the 2SFM, $\psi(t)$, over the temperature interval that was simulated: $1.0 < T < \infty$. In

[a] The data points at $T^{-1} = 0.4$, 0.6, and 0.9 were obtained with improved statistics and thus differ slightly from those in FIGURE 15 of reference 3.

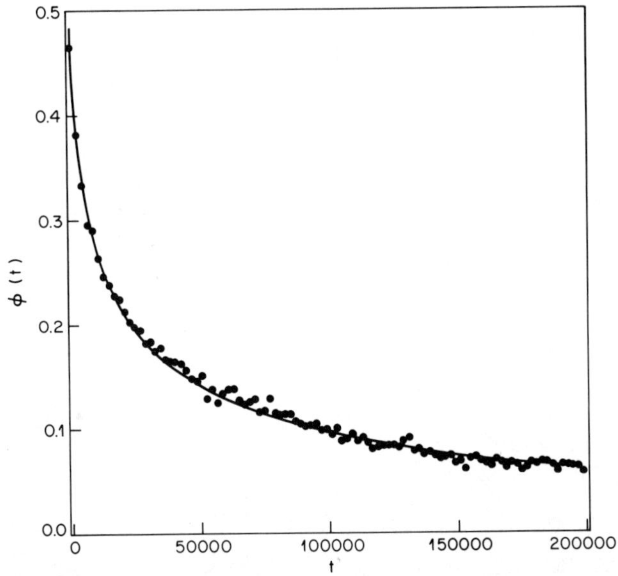

FIGURE 4. Decay of $\phi(t)$ for the 2SFM at $T = 0.91$. The solid curve through the simulation data is a KWW fit with parameters $\gamma = 0.26$ and $\tau_w = 3700$.

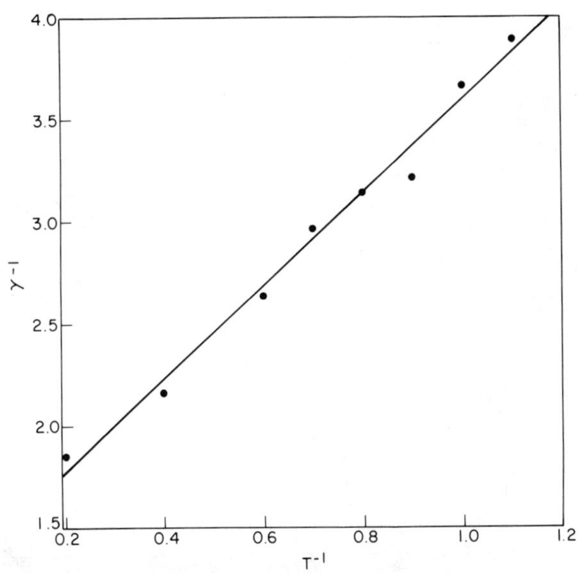

FIGURE 5. Temperature dependence of the KWW exponent γ for the single-spin correlation function $\phi(t)$. The solid line is the result of a least-squares regression yielding the parameters given in equation 13.

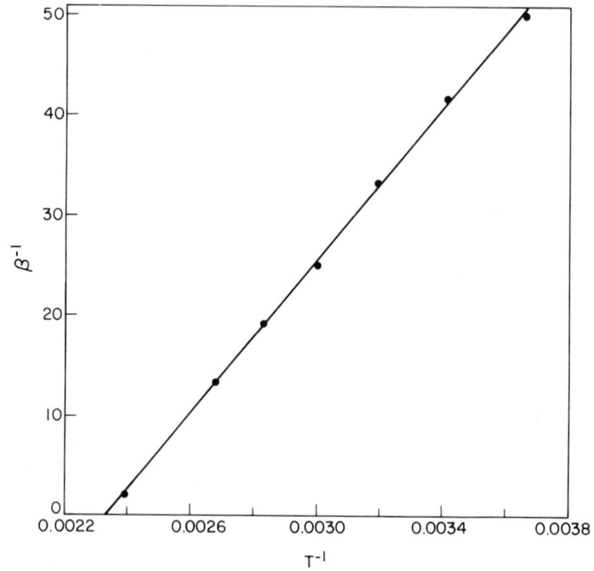

FIGURE 6. Temperature dependence of the KWW exponent β for the polycarbonate mechanical relaxation data given in reference 22. The solid line is the result of a least-squares fit and is described by $\beta^{-1} = -89 + 3.8 \times 10^4\ T^{-1}$.

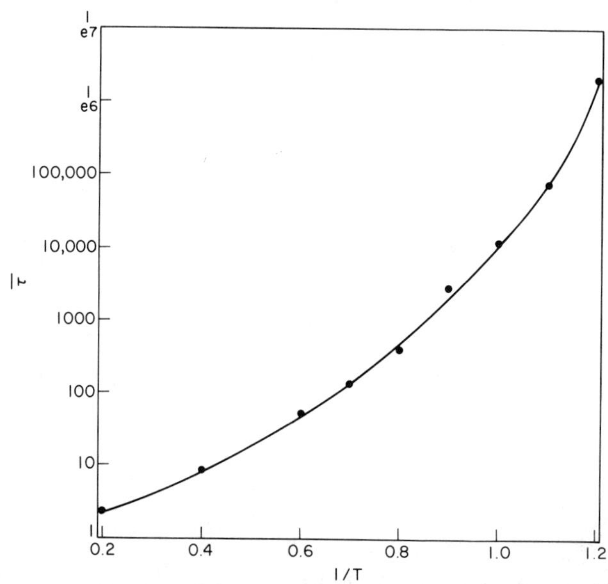

FIGURE 7. Temperature dependence of the average relaxation time $\bar{\tau}$, defined as the area under $\phi(t)$. The 2SFM exhibits highly non-Arrhenius behavior.

particular, we found the following correlation for the KWW exponent

$$\beta^{-1} = B + CT^{-1} \tag{15}$$

with $B = 1.3 \pm 0.1$ and $C = 1.5 \pm 0.1$. Thus, the single-spin correlation function, $\phi(t)$, is somewhat more nonexponential in character than the magnetization correlation function, $\psi(t)$. This observation is in agreement with more conventional kinetic Ising models, but β for the magnetization correlation function of conventional models is temperature independent and almost exactly equal to 1.[23] For the magnetization correlation function of the 2SFM it was also found that the average relaxation time, τ, obtained from KWW

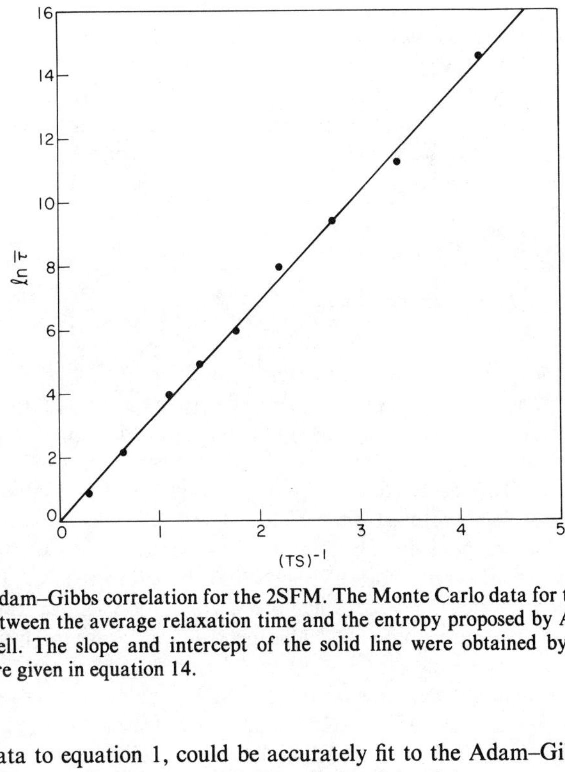

FIGURE 8. Adam–Gibbs correlation for the 2SFM. The Monte Carlo data for the 2SFM follow the relation between the average relaxation time and the entropy proposed by Adam and Gibbs remarkably well. The slope and intercept of the solid line were obtained by a least-squares analysis and are given in equation 14.

fits of $\psi(t)$ data to equation 1, could be accurately fit to the Adam–Gibbs expression, equation 3, with parameters $\tau_0 = 1.3 \pm 0.1$ and $A = 4.1 \pm 0.1$.

NONLINEAR DYNAMICAL RESPONSE OF THE 2SFM

Several extremely successful phenomenological theories have been developed to describe the nonlinear response of viscous liquids and glasses to large perturbations from equilibrium.[24,25] The approach of Moynihan *et al.*[24] is particularly convenient because of its simplicity and ability to provide quantitative predictions of material behavior. Although the Moynihan approach is based on the Narayanaswamy equation[26] for the average relaxation time, a recent theory by Scherer[27] uses the Adam–Gibbs equation and apparently yields equally good results.[27,28] Because Scherer's formalism combines various ideas described by Narayanaswamy, Adam, and Gibbs, we will refer to it as the Narayanaswamy–Adam–Gibbs–Scherer (NAGS) theory. Because of our success in fitting the linear response functions of the 2SFM to the Adam–Gibbs equation, it is natural to apply the NAGS formalism to nonlinear experiments on the spin model. In the present section we compare the predictions of the NAGS theory with Monte Carlo simulation data for finite temperature jumps in the 2SFM.

Temperature-jump simulations were carried out on a 50×50 spin system at a variety of starting and ending temperatures in the range $0.9 < T < 10.0$. Before each temperature jump the system was equilibrated at the starting temperature, T_1, and a sequence of configurations, separated by times longer than the average relaxation time at T_1, were saved. With the system prepared in one of the saved configurations, it was then instantaneously quenched to the final temperature, T_2, and the magnetization was recorded as a function of time. This procedure was repeated for each of the saved configurations and the results were averaged to obtain a smoothed system response. For the simulation data reported here, between 30 and 100 saved configurations were employed. The fictive[24] temperature, T_f, was calculated from the smoothed magnetization data via the relationship

$$M(t) = \frac{2}{1 + \exp\,[2/T_f(t)]} - 1 \qquad \textbf{(16)}$$

which is an obvious nonequilibrium extension of equation 7. Note that for the present case of no interactions between spins ($J = 0$) the fictive temperature for the energy and for the magnetization (for example, density) are both given by equation 16.

FIGURE 9 illustrates the nonlinearity of the 2SFM as it responds to a finite temperature jump. The figure shows Monte Carlo data for the fictive temperature following a jump in temperature from 3.75 to 2.5 and following a jump from 1.25 to 2.5. Although the temperature changes of the two experiments have the same magnitude, they differ in sign. FIGURE 9 shows that the 2SFM requires a significantly longer time to equilibrate at $T = 2.5$ when prepared at a low temperature than when prepared at a temperature above 2.5. This behavior is in accord with laboratory experiments on viscous liquids.

To quantify the nonlinear response of the 2SFM, the simulation results for the fictive temperature were compared with the predictions of the NAGS theory, which we modified slightly for application to the 2SFM. The following equations were used in conjunction with equation 16 to describe the response of the model to a single temperature jump at time t_1

$$M(t) = M_1 + (M_2 - M_1)[1 - \phi(t - t_1, t)] \qquad \textbf{(17)}$$

$$\phi(t - t_1, t) = \exp\left[-\left(\int_{t_1}^{t} \frac{dt'}{\tau(t')}\right)^{\beta(T_f)}\right] \qquad \textbf{(18)}$$

$$\tau = \tau_0 \exp\,(A/[TS(T_f)]) \qquad \textbf{(19)}$$

where M_1 and M_2 are the equilibrium magnetizations at T_1 and T_2, and $S(T_f)$ is defined by equation 5. The parameters appropriate for the magnetization correlation function $\psi(t)$ are to be inserted in equation 19; that is, $\tau_0 = 1.3$ and $A = 4.1$. Furthermore, β is taken to be a constant equal to the average of its values (according to equation 15) at T_1 and T_2.

An important feature of the above equations is that they contain no adjustable parameters. Our independent experiments for the magnetization correlation function $\psi(t)$ were sufficient to determine all the parameters in the NAGS theory. As a result, equations 16–19 can be used to make a definite prediction for the response of the 2SFM to any finite or infinitesimal perturbation.

FIGURES 10 & 11 demonstrate that the above equations are very good at describing the response of the 2SFM to jumps down in temperature. The dotted curves in the

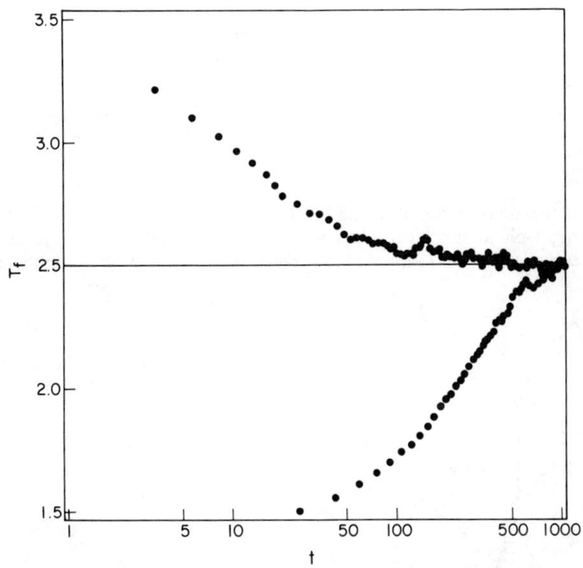

FIGURE 9. Dependence of nonequilibrium relaxation in the 2SFM on the sign of the perturbation. The Monte Carlo data shown were averaged over 30 independent temperature jumps from 3.75 to 2.5 and from 1.25 to 2.5.

figures are the Monte Carlo data, and the solid curves were obtained from numerical solutions of equations 16–19. FIGURE 10 shows the change in fictive temperature following a 50% decrease in temperature from 5.0 to 2.5 Note that because of our temperature scale, the analogous experiment in a real liquid might correspond to a temperature jump of more than 100 degrees. FIGURE 11 shows the results of a similar experiment in which the temperature is jumped from 1.67 to 1.25.

In general it was found that the NAGS formalism is capable of accurately predicting the response of the 2SFM to temperature decreases of 15% or less for the complete temperature range studied: $0.9 < T < 10.0$. At the highest temperatures, however, the above equations are able to predict the behavior of the model for much

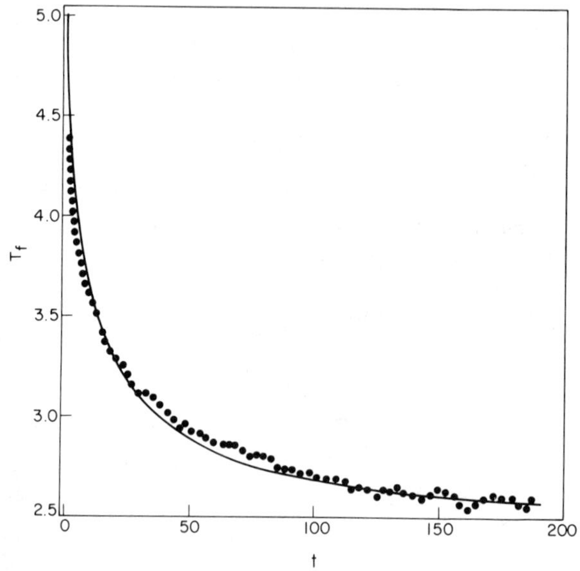

FIGURE 10. Nonlinear response of the 2SFM to a temperature jump from 5.0 to 2.5. The solid curve was obtained from a numerical solution of equations 17–19 with no adjustable parameters, and the dotted curve is the simulation data, averaged over 30 independent temperature jumps.

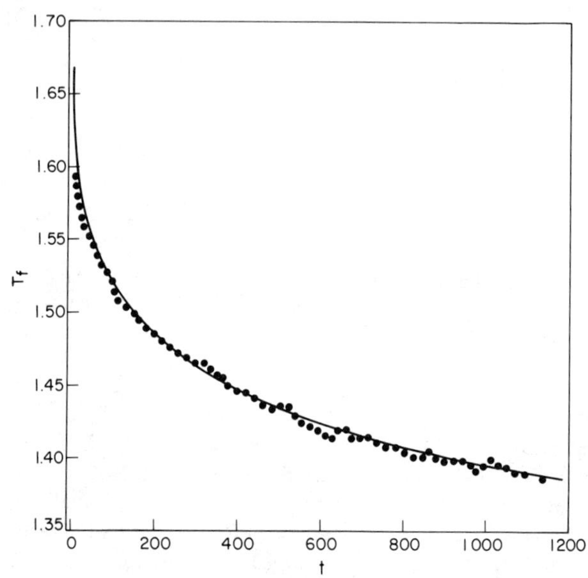

FIGURE 11. Nonlinear response of the 2SFM to a decrease in temperature from 1.67 to 1.25. The curves have the same meaning as in FIGURE 10.

larger temperature decreases. For lower starting temperatures it was found that the theory fails if the temperature decrease exceeds about 15%. FIGURE 12 illustrates this breakdown of the theory for a 50% jump from 2.5 to 1.25. It is interesting that Scherer[29] also finds that the NAGS equations predict volume relaxation functions for plate glass that are too slow for large downward temperature jumps.

FIGURES 13 & 14 demonstrate that equations 16–19 provide a quantitative description of the response of the 2SFM to instantaneous temperature increases from 5.0 to 7.5 and from 2.5 to 3.2, respectively. We found that the theory could accurately predict the behavior of the spin system for temperature increases of less than 10% over the entire temperature range studied, but the theory was accurate for even larger temperature jumps at high temperature. It was necessary to average over more starting configurations for the temperature increase experiments because of larger fluctuations in the dynamical trajectories from the various low-temperature starting configurations. For very large temperature increases, the phenomenological theory cannot quantitatively describe the dynamics of the spin model. FIGURE 15 illustrates the breakdown of the formalism for a 100% increase from 1.25 to 2.5.

Laboratory fluids often exhibit a linear relationship between the equilibrium density (or enthalpy) and the temperature. As a result, the following equations are often used for the fictive temperature instead of equations 16 and 17

$$\frac{T_f(t) - T_2}{T_1 - T_2} = \frac{M(t) - M_2}{M_1 - M_2} \tag{20}$$

$$T_f(t) = T_1 + (T_2 - T_1)[1 - \phi(t - t_1,t)] \tag{21}$$

We have obtained numerical solutions to equations 18–21 and found that they are very similar (indeed, almost identical in the case of shallow temperature jumps) to the results obtained from equations 16–19. Both sets of equations are capable of accurately describing the simulation data for temperature jumps of less than 10%. Although the two approaches are not always quantitative for larger temperature jumps, the predictions of equations 16–19 are somewhat better than those of equations 18–21 in this case.

DISCUSSION AND CONCLUSIONS

It is clear that the 2SFM has many of the dynamical properties of real liquids near their glass transition. In particular, we have found KWW relaxation functions with an exponent that decreases weakly with decreasing temperature, non-Arrhenius behavior of the average relaxation time that follows the Adam–Gibbs equation, and nonlinear response that can be described by the NAGS theory. Furthermore, for temperature jumps of a size consistent with experimental practice, the nonlinear response of the 2SFM gives the appearance of thermorheological simplicity. The remarkable success of the Adam–Gibbs equation for the model suggests that we should take a second look at the various approximations in the formal development of the Adam–Gibbs theory.

There are obvious extensions of the present work. As an example, one could

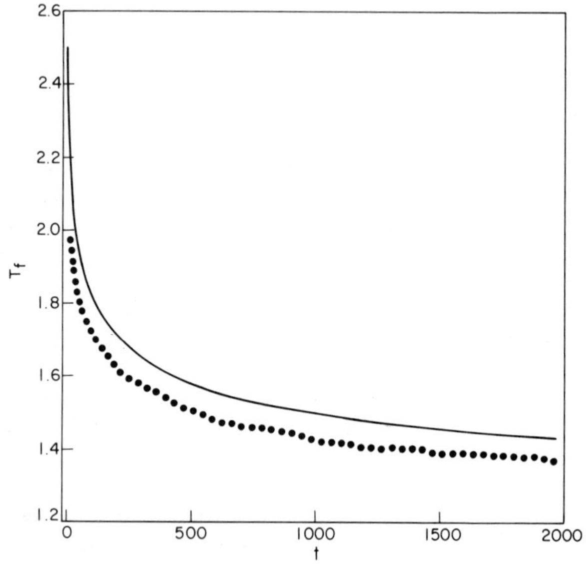

FIGURE 12. Nonlinear response of the 2SFM to a decrease in temperature from 2.5 to 1.25. The NAGS theory is seen to begin to breakdown for this large (50% change in T) departure from equilibrium.

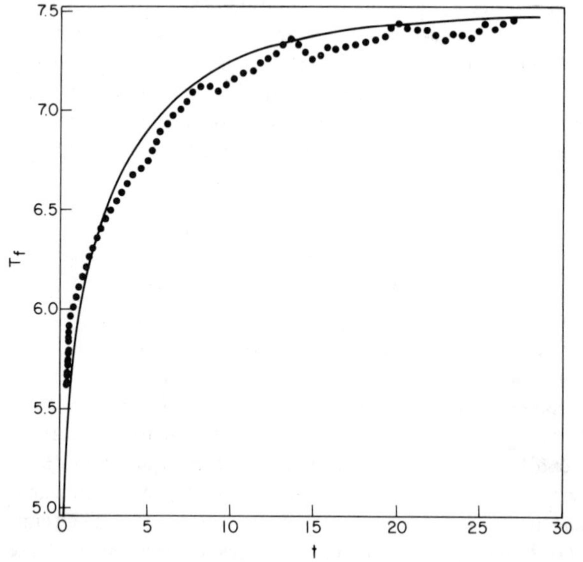

FIGURE 13. Nonlinear response of the 2SFM to an increase in temperature from 5.0 to 7.5. The solid and dotted lines have the same meaning as in FIGURE 10, except that the simulation data were averaged over 100 independent temperature jumps.

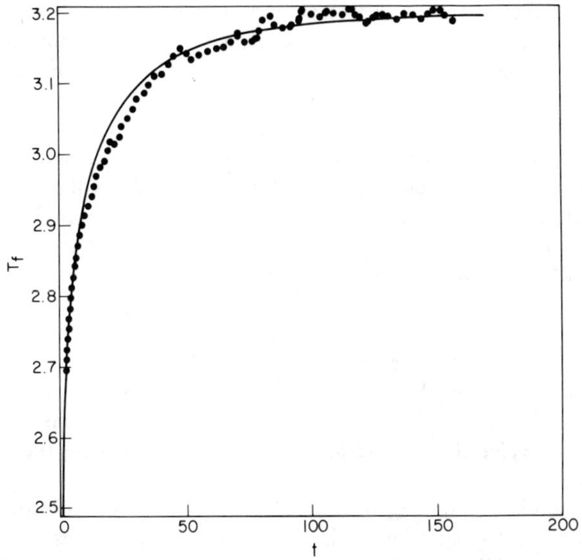

FIGURE 14. Nonlinear response of the 2SFM to an increase in temperature from 2.5 to 3.2. The simulation data shown as the dotted curve were averaged over 100 independent temperature jumps.

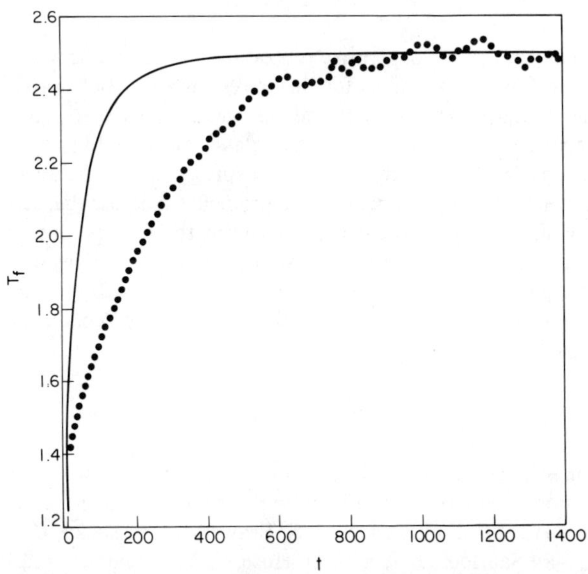

FIGURE 15. Failure of the phenomenological theory for a large (100%) temperature increase from 1.25 to 2.5. The simulation data were averaged over 30 independent jumps.

investigate the effect of including interactions between spins ($J \neq 0$) on the results described here. One could also study the behavior of more restrictive facilitated models, or analyze the 2SFM on lattices of higher coordination number.

In spite of the apparent success of the 2SFM, there are several nagging questions. For example, what is it about the model that provides the universal dynamical behavior that we associate with a viscous liquid? Furthermore, what is the proper interpretation of a spin in terms of the various degrees of freedom in a dense fluid? It is disturbing that if we change the number of spins required to facilitate a spin flip or change the topology of the lattice the properties of the model can change completely. In going from the 2SFM to the 1SFM all the interesting non-Arrhenius behavior disappears. It is hard to imagine that there is anything "magical" about requiring two neighboring regions to facilitate relaxation of another region. One might also ask whether it is possible to construct a field theory to which the 2SFM would reduce under some appropriate coarse graining procedure. If so, what other types of microscopic models lead to the same field theory? What is the relevant "length" that controls the dynamics in the 2SFM? If we identify this length, will the connection between the various models be transparent?

We are hopeful that some of these questions will be resolved in the near future. Because the 2SFM has such promising behavior, one can focus on its relaxation mechanism in detail. High-resolution computer graphics may be of great assistance in identifying this mechanism and any length that controls structural relaxation. Molecular dynamics simulations on simple liquids and projection operator techniques might also be helpful in establishing a more rigorous interpretation of a spin for a real fluid.

SUMMARY

A kinetic Ising model of cooperative relaxation in dense fluids was presented. The results of Monte Carlo simulations for linear response functions of the model were reviewed. The simulations show that the spin model possesses many of the time-dependent properties of real liquids near their glass transition. In particular, the linear response functions can be fit with the KWW expression, and the average relaxation times exhibit nonArrhenius temperature dependence that can be described by the Adam–Gibbs equation. The present paper contains the first quantitative analysis of nonequilibrium relaxation for the model. Monte Carlo simulations demonstrate that the nonlinear response of the spin system to finite temperature jumps can be accurately described by a recent phenomenological theory developed by Scherer.

ACKNOWLEDGMENTS

The author is indebted to M. Grabow for providing software that facilitated the numerical solution of equations 16–19 and to S. Brawer and G. Scherer for critiquing the manuscript. Helpful discussions with S. Brawer, S. Matsuoka, C. A. Angell, M. Goldstein, W. Van Saarloos, J. Weeks, D. Huse, C. T. Moynihan, and P. Hohenberg are gratefully acknowledged.

REFERENCES

1. FREDRICKSON, G. H. & H. C. ANDERSEN. 1984. Phys. Rev. Lett. **53:** 1244.
2. FREDRICKSON, G. H. & H. C. ANDERSEN. 1985. J. Chem. Phys. **83:** 5822.
3. FREDRICKSON, G. H. & S. A. BRAWER. 1986. J. Chem Phys. **84:** 3351.
4. BRAWER, S. A. 1985. Relaxation in Viscous Liquids and Glasses. American Ceramic Society. New York, NY.
5. WONG, J. & C. A. ANGELL. 1976. Glass: Structure by Spectroscopy. Marcel Dekker. New York, NY.
6. ANGELL, C. A. 1983. *In* Relaxations in Complex Systems. K. Ngai & G. B. Wright, Eds.: 3–11. Naval Research Laboratory. Washington, DC.
7. ADAM, G. & J. H. GIBBS. 1965. J. Chem. Phys. **43:** 139.
8. KAUZMANN, W. 1948. Chem. Rev. **43:** 219.
9. COHEN, M. H. & G. S. GREST. 1979. Phys. Rev. B **20:** 1077.
10. COHEN, M. H. & G. S. GREST. 1980. Phys. Rev. B. **21:** 4113.
11. LEUTHEUSSER, E. 1984. Phys. Rev. A **29:** 2765.
12. DAS, S. P., G. F. MAZENKO, S. RAMASWAMY & J. J. TONER. 1985. Phys. Rev. Lett. **54:** 118.
13. KIRKPATRICK, T. R. 1985. Phys. Rev. A **31:** 939.
14. ULLO, J. J. & S. YIP. 1985. Phys. Rev. Lett. **54:** 1509.
15. BENGTZELIUS, U., W. GÖETZE & A. SJÖLANDER. 1984. J. Phys. C **17:** 5915.
16. KEYES, T. 1977. *In* Statistical Mechanics Part B: Time-Dependent Processes. B. J. Berne, Ed.: 259–309. Plenum. New York, NY.
17. SIGGIA, E. 1985. Phys. Rev. A **32:** 3135.
18. DAS, S. P., G. F. MAZENKO, S. RAMASWAMY & J. J. TONER. 1985. Phys. Rev. A **32:** 3139.
19. LEE, T. D. & C. N. YANG. 1952. Phys. Rev. **87:** 410.
20. GLAUBER, R. J. 1963. J. Math. Phys. **4:** 294.
21. SHLESINGER, M. F. & E. W. MONTROLL. 1984. Proc. Natl. Acad. Sci. USA **81:** 1280.
22. MATSUOKA, S. 1985. Polym. J. (Japan) **17:** 321.
23. MÜLLER-KRUMBHAAR & K. BINDER. 1973. J. Stat. Phys. **8:** 1.
24. MOYNIHAN, C. T., P. B. MACEDO, C. J. MONTROSE, P. K. GUPTA, M. A. DEBILL, J. F. DILL, B. E. DOW, P. W. DRAKE, A. J. EASTEAL, P. B. ELTERMAN, R. P. MOELLER, H. SASABE & J. A. WILDER. 1976. Ann. N.Y. Acad. Sci. **279:** 15.
25. KOVACS, A. J. 1981. Ann. N.Y. Acad. Sci. **371:** 21.
26. NARAYANASWAMY, O. S. 1971. J. Am. Ceram. Soc. **54:** 491.
27. SCHERER, G. W. 1984. J. Am. Ceram. Soc. **67:** 504.
28. HODGE, I. M. 1985. Paper presented at symposium, Relaxation Processes in Liquids and Glasses. National Meeting of the American Chemical Society. Miami, FL.
29. SCHERER, G. W. 1986. J. Am. Ceram. Soc. **69:** 374.

DISCUSSION OF THE PAPER

I. M. HODGE (*Eastman Kodak, Rochester, NY*): The maximum in your calculated equilibrium heat capacity is not seen in Dr. Nagel's equilibrium AC heat capacity studies, and yet this appears to be essential for the correct description of configurational entropy and Adam–Gibbs dynamics. How sensitive would you expect this result to be to the two-dimensionality of your system, or to the details of the spin interactions?

FREDRICKSON: The maximum in the equilibrium heat capacity for the spin model occurs at a temperature substantially lower than the temperature range over which I performed dynamical simulations. As a result, it cannot be essential for the observed

Adam–Gibbs behavior. For the temperatures investigated, the equilibrium heat capacity is a smooth, slowly varying, and monotonically decreasing function of temperature. Because the temperature scale in the model is greatly compressed in comparison with real liquids, this behavior of the equilibrium heat capacity may be quite reasonable. Furthermore, if the model is cooled at a finite rate, it will eventually fall out of equilibrium (that is, undergo a "laboratory" glass transition) and a sharp drop in the observed heat capacity is expected.

For weak ferromagnetic interactions between spins I don't expect any qualitative changes in the behavior of the 2SFM, except in the vicinity of the critical temperature where the dynamics could be much more complicated. I have preliminary simulation data that indicate the three-spin facilitated model on a cubic lattice in three-dimensions also obeys the Adam–Gibbs equation. At this point, however, we don't fully understand the influence of dimensionality on the kinetics of the facilitated models.

J. T. BENDLER (*General Electric Research and Development Center, Schenectady, NY*): 1. Is it possible to interpret triplet and higher groupings of up spins as mobile defects?

2. How does the KWW exponent depend on the parameters of the model (such as the fundamental time scale, the concentration of up spins, and the barrier for single spin flips)? Are the results, in your opinion, independent of dimensionality?

3. Have you followed the kinetics into the glass state, or are the results only for the high-temperature behavior? (Does the KWW exponent change in the low-temperature region?)

FREDRICKSON: 1. No. We believe that the interesting non-Arrhenius behavior of the 2SFM arises from the inability of isolated clusters of up spins to diffuse in a sea of down spins. Instead of a simple defect diffusion mechanism, which leads to Arrhenius temperature dependence, it is likely that relaxation in the spin model is a result of much more cooperative motions of "strings" or "surfaces" of spin-up defects.

2. The KWW exponent does not depend on the fundamental time scale, but does depend on the dimensionless temperature as discussed in the text. The energy barrier for single spin flips, h, was absorbed in the definition of the dimensionless temperature variable, and the equilibrium concentration of up spins is a function only of the temperature $c = 1/(1 + \exp(2/T))$. The effect of dimensionality on the temperature dependence of the KWW exponent is uncertain at this point, although the qualitative behavior seen here has recently been observed for the two- and three-spin facilitated models in three dimensions.

3. It is not clear to me what you mean by the "glass state." The KWW fits in the present paper were for equilibrium time correlation functions and were obtained at temperatures where, with reasonable computational effort, I could fully equilibrate the model. One could easily form a "glass" with a fictive temperature in this same range by rapidly quenching the spin model from a high temperature. The kinetics of such a glass could presumably be described by the NAGS formalism used in the present paper and with the predetermined KWW exponents.

R. G. PALMER (*Duke University, Durham, NC*): Here again we see a model based on dynamical constraints showing glassy behavior. The PSAA "hierachically constrained dynamics" model (Palmer, R. G., D. L. Stein, E. Abrahams & P. W. Anderson. *Phys. Rev. Lett.* **53:** 958, 1984) is in effect the Bethe lattice version of this model.

FREDRICKSON: Although I am convinced that there is some connection between the present model and the PSAA model, I don't believe that the relationship is quite so simple. It is not apparent to me how one identifies the various "levels" in the PSAA model with groups of spins in the 2SFM. In the PSAA model a level can only relax if conditions on a previous level have been met. In the 2SFM on a Bethe lattice, however, a shell of spins a particular distance from the origin could be facilitated to relax both by shells of spins closer and farther away. Furthermore, the 2SFM possesses both energy and entropy barriers whereas the PSAA model incorporates only entropic constraints.

Elastic Properties of Network Glasses[a]

M. F. THORPE

Department of Physics and Astronomy
Michigan State University
East Lansing, Michigan 48824

INTRODUCTION

In the last few years a new viewpoint has emerged on the mechanical and elastic properties of network glasses. A network glass is regarded as a mechanical structure and is analyzed in a way more familiar to engineers than to physicists and chemists. We will summarize some recently established ideas on network glasses in this paper; the reader is referred to the original papers for more details.

Many network glasses are well described by the continuous random network (CRN) model first discussed by Zachariasen[1] more than fifty years ago. Network formers consist of atoms that make from two to four covalent bonds. Examples include Si and Ge, which tend to make four bonds in a tetrahedral arrangement; As, which makes three bonds; and S, Se, and O, which make two bonds. A CRN can be described by a chemical formula like $Ge_xAs_ySe_{1-x-y}$. If the total number of atoms is N and there are n_r atoms with coordination r ($r = 2$, 3, or 4), then

$$N = \sum_r n_r \tag{1}$$

and we can define the mean coordination[2] $\langle r \rangle$ by

$$\begin{aligned} \langle r \rangle &= \sum_r r n_r \Big/ \sum_r n_r \\ &= 4x + 3y + 2(1 - x - y) \\ &= 2 + y + 2x \end{aligned} \tag{2}$$

We note that $\langle r \rangle$ (where $2 < \langle r \rangle < 4$) gives a partial but very important description of the network. For a particular network with some $\langle r \rangle$, there is still a free compositional parameter. For example, As and $Ge_{0.5}Se_{0.5}$ both have $\langle r \rangle = 3$. Nevertheless, we will find that $\langle r \rangle$ is the most important single parameter in determining the mechanical and elastic properties of a CRN. For a more complete description of a CRN many other quantities are required of course. The most common is the radial distribution function (RDF) obtained as the Fourier transform from diffraction experiments. Another is the chemical clustering. Neither of these turns out to be very important in determining the mechanical and elastic properties of the network.

We regard a perfect CRN with no dangling bonds as being an ideal—analogous to

[a]This work was supported by the National Science Foundation under Grant DMR8317610 and by the Office of Naval Research under Contract N00014-80-C-0610.

the perfect crystal. In practice one can approach a perfect CRN by careful preparation techniques in which the number of dangling bonds is kept to a minimum. In some circumstances the CRN model is inappropriate and the network has effectively lower dimensionality and may be better thought of as a spaghetti-like entanglement of twisted ribbons. Because the ideas that we are discussing are quite general, they also apply to such networks.

MAXWELL'S WORK

Maxwell wrote several papers in the mid-nineteenth century on the stability of mechanical systems.[3] This work is now largely unknown to the physics and chemistry communities, but is still referred to in engineering texts.[4] The essence of Maxwell's arguments is given here. Suppose a structure is constructed from b bars and j joints as shown in FIGURE 1. The bars are of fixed length and the joints are such that there are no direct constraints on the angles. A single joint is very loose but can be made firm by completing a triangle. Consider the two-dimensional framework shown in FIGURE 1.

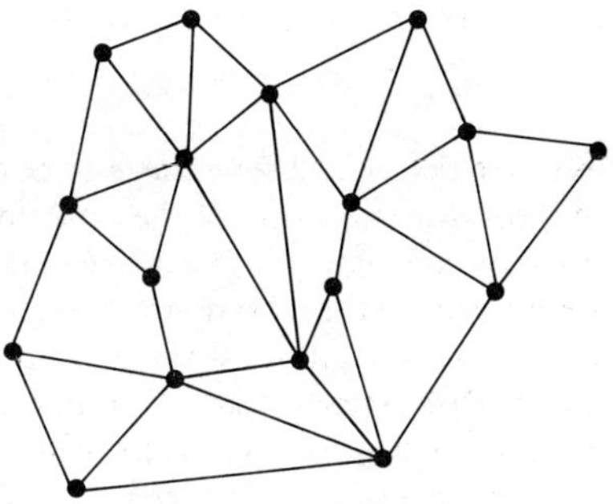

FIGURE 1. A network made up of bars (lines) and joints (dots). The astute reader will realize that this arrangement is unstable because of the presence of quadrilaterals.

The number of integral degrees of freedom is $2j - 3$ where 2 is the dimensionality and 3 comes from the two rigid translations and one rotation. The number of constraints, as would be counted by Lagrange, is b so that the network is stable if

$$b > 2j - 3 \quad (3)$$

and unstable if

$$b < 2j - 3 \quad (4)$$

These two situations are referred to as being *overconstrained* and *underconstrained*, respectively.[3,4] Similar arguments in three dimensions lead to

$$b > 3j - 6 \quad (5)$$

for an overconstrained network and

$$b < 3j - 6 \quad (6)$$

for an underconstrained network. In these expressions 3 is the dimensionality and 6 comes from the sum of the three rigid translations and three rotations. It is important to understand that the Maxwell conditions only provide an average global condition for mechanical stability and as such should be regarded in the spirit of a mean field theory—albeit a remarkably good one, as we shall see.

Another way to look at Maxwell's result is to imagine that the rigid bars are replaced by Hooke springs and to examine the eigenfrequencies of the network. This is a straightforward problem in classical mechanics.[5] The rank of the dynamical matrix that describes the system is by definition just the number of finite frequency modes. Therefore the number of zero frequency modes is the dimensionality of the dynamical matrix minus the rank. Denoting the number of zero frequency modes by F, we have

$$F = dj - R \tag{7}$$

This is an exact result. We can estimate the rank of the matrix as being equal to the number of constraints, that is

$$R = b \tag{8}$$

Each bar has a single finite frequency mode associated with it, and these are coupled together to give a total of b finite frequency modes. This is an overestimate: not all the constraints are linearly independent. From equation 7 we have

$$F = dj - b \tag{9}$$

We note that in two dimensions three of the zero frequency modes are not internal and in three dimensions six of the zero frequency modes are not internal. Thus the result given by equation 9 is closely linked to those given by equations 3–6. In a large system we can neglect the rigid translations and rotations. We shall only consider such systems for now on. Thus if $F > 0$, the system is underconstrained and is mechanically unstable under some external loads. Equation 9 eventually breaks down because, by definition, F cannot be negative. The problem arises because as the number of bars increases for a fixed number of joints, the constraints become more and more linearly dependent as is characteristic of a phase transition. A very good result is obtained in practice if we write

$$F = \max\,[dj - b, 0] \tag{10}$$

A SIMPLE MODEL

The essence of the previous discussion is that a locally floppy unit can be made rigid by introducing enough cross braces. The local floppy unit is the pin joint but can be

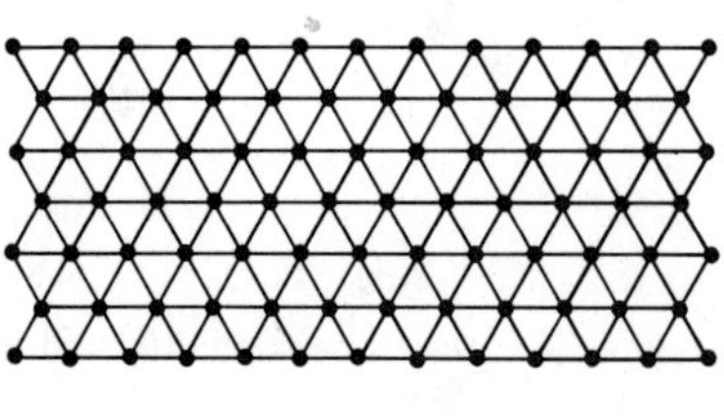

FIGURE 2. A triangular net of mass points and nearest-neighbor Hooke springs.

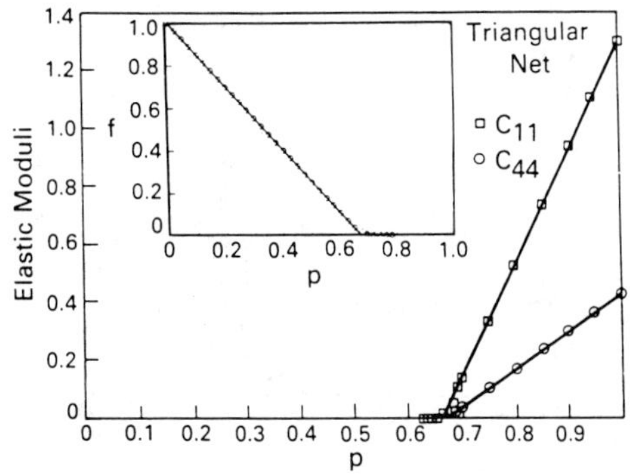

FIGURE 3. The elastic constants C_{11} and C_{44} for various concentrations p of Hooke springs on a triangular net. The insert shows the fraction of zero frequency modes. The solid lines are effective medium theory and the points are from numerical simulations. Reproduced from Feng *et al.*[7]

more complex in three dimensions, as will be discussed. As an example, consider the triangular net shown in FIGURE 2 of N mass points joined by nearest-neighbor Hooke springs that are randomly present[6] with probability p. The fraction f of zero frequencies is given by

$$F = dNf = dN - \tfrac{1}{2}Nzp \tag{11}$$

or

$$f = 1 - p/p^* \tag{12}$$

with

$$p^* = 2d/z \tag{13}$$

If $p = 0$, no springs are present and all the mass points are free particles so $f = 1$. As springs are introduced, one mode on each spring has a finite frequency, until $p = p^* = \tfrac{2}{3}$, where the number of zero frequency modes goes to zero and the structure becomes rigid. This is shown in FIGURE 3 where numerical calculations[7] are compared with the Maxwell prediction (equation 12) in the insert. The agreement is remarkably good. For completeness, equation 10 becomes for this system

$$f = \max\,[1 - p/p^*, 0] \tag{14}$$

There are small derivations from this mean-field behavior around p^*, but they are exceedingly small. The elastic constants C_{11} and C_{44} were obtained numerically by straining and relaxing the network. Note that $\rho V_L^2 = C_{11}$ and that $\rho V_T^2 = C_{44}$ where ρ is the mass density and V_L and V_T are the longitudinal and transverse sound velocities, respectively. Effective medium theory[7] predicts straight lines for C_{11} and C_{44} that also go to zero at $p^* = \tfrac{2}{3}$ in close agreement with the numerical simulations. Lest the reader think that two dimensions is special, we show similar results in FIGURE 4 for the face-centered cubic lattice in three dimensions. The agreement between theory and experiment is equally good.

This general phenomena is referred to as *rigidity percolation.* It is not enough for connections to exist across the network from one side to the other, as in *connectivity percolation*[8]; these connections must be rigid. Thus if connectivity percolation occurs

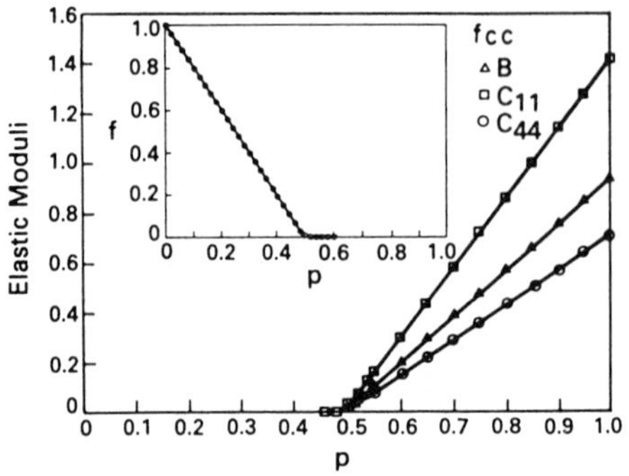

FIGURE 4. Same as FIGURE 3 except for the face-centered cubic lattice. Here B is the bulk modulus. Reproduced from Feng *et al.*[7]

at p_c, then

$$p^* > p_c$$

and

$$p^* \simeq dp_c \tag{15}$$

as discussed by Phillips and Thorpe.[9] Connectivity percolation can be thought as scalar percolation and occurs when $p_c \simeq 2/z$ (that is, equation 13 with $d = 1$); rigidity percolation can be thought of as vector percolation.

The process can be visualized as follows. For $p < p^*$, the system consists of isolated rigid regions in a floppy matrix. As p increases the rigid regions grow until they link up and percolate at $p = p^*$. For $p > p^*$, the system consists of a rigid matrix with a few floppy inclusions. This visualization of rigidity percolation was first proposed by Thorpe.[10]

In summary, the overall behavior of these systems is well described by effective medium theory, following the work of Maxwell. The key is the number of zero frequency modes. Details of the critical behavior are not well understood yet, but fortunately this is a small region of little interest in network glasses. For our discussion in the next section effective medium theory will be sufficient.

NETWORK GLASSES

A glass like $Ge_xAs_ySe_{1-x-y}$ is a three-dimensional network similar to the two-dimensional network shown in FIGURE 1. In a covalent network the bond angles as well as the bond lengths are rather well defined. Indeed, this is characteristic of covalent bonding but is insufficient to make the structure rigid. Consider amorphous Se, which is a spaghetti-like entanglement of Se chains. Because of the weak interactions between the chains, we can consider each chain in isolation. Each atom has three degrees of freedom, and we can associate one bond length constraint and one bond angle constraint with each site. Thus ⅔ of the vibrational modes have a finite frequency and ⅓ have zero frequency, and Maxwell's argument leads to

$$f = 1/3 \tag{16}$$

for Se chains. As three- and four-coordinated atoms like As and Ge are introduced, they act as cross linkages between the Se chains. After a sufficiently large number of these cross links have been introduced, the network becomes rigid. We will use Maxwell's argument to find out when this transition takes place.

As we have already discussed there is one constraint associated with each bond length. Each bond is associated with the two atoms at either end so that there are $r/2$ constraints for each r-coordinated atom where $r = 2, 3$, or 4. The angular constraints are counted as follows.[10] For a two-coordinated atom, there is a single angular constraint. As each additional bond is added, two more angles must be specified. This leads to a total of $2r - 3$ angular constraints to be associated with each r-coordinated atom. Thus

$$\begin{aligned} f &= \left(3N - \sum_r n_r \, (r/2 + 2r - 3)\right) \Big/ 3N \\ &= 2 - \frac{5}{6}\langle r \rangle \end{aligned} \tag{17}$$

where $\langle r \rangle$ was previously given in equation 2. When $\langle r \rangle = 2$, we have $f = 1/3$ as in equation 16. The transition takes place when $f = 0$ at $\langle r \rangle = 2.4$. For $\langle r \rangle < 2.4$ we have what we may call a *polymeric glass*, and for $\langle r \rangle = 2.4$ we have what we may call an *amorphous solid*.[10] The floppy and rigid regions in these networks are illustrated schematically in FIGURE 5. These ideas that network glasses can be divided into two different types was first introduced by Phillips[11,12] and further discussed by Döhler *et al.*[13] before being made more rigorous by Thorpe.[10]

A series of random networks was built by He and Thorpe[14] in order to test these ideas. The results of the simulations are shown in FIGURE 6. It can be seen that elastic constant C_{11} is determined principally by $\langle r \rangle$ and goes to zero around 2.4 (with an exponent of about 1.5—larger than the 1.0 of the central force models in FIGS. 3 & 4). Similar results are obtained for C_{44}. The fraction of zero frequency modes also closely follows the mean field prediction of equation 17.

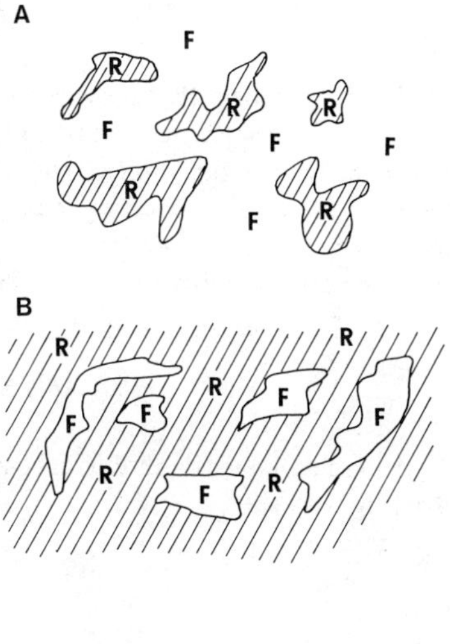

FIGURE 5. The rigid (R) and floppy (F) regions in the **(A)** polymeric glass and the **(B)** amorphous solid. Adapted from Thorpe.[10]

In computing the results of FIGURE 6, we have taken the ratio of the bond-stretching force (α) to bond-bending force (β) to be $\alpha/\beta = 5$. This is a typical value for covalent materials and does not vary much from material to material. All other forces are smaller. The most important of these other forces are the one that control the dihedral angle.

In amorphous solids where $\langle r \rangle > 2.4$, the mechanical stability and elastic constraints are determined principally by the covalent forces α and β. This is in much the same way as Coulomb forces between the ions determine the mechanical stability and elastic constants of ionic salts like NaCl. Other forces are present but are much less important.

In polymeric glasses where $\langle r \rangle < 2.4$, spontaneous collapse would take place if it were not for the stabilizing influence of the dihedral angle forces. It is these, and not

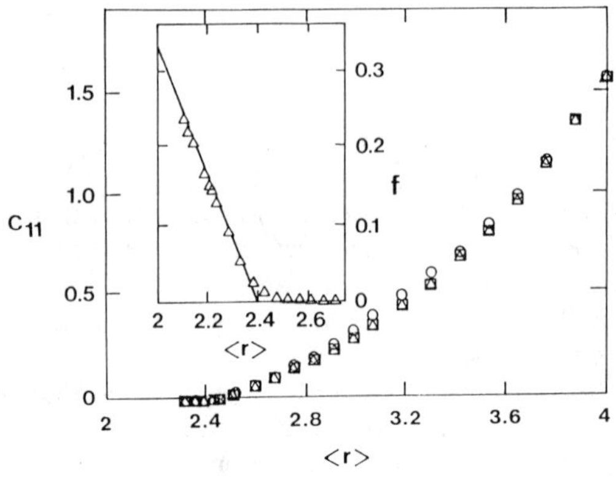

FIGURE 6. The elastic modulus C_{11} with $\alpha = 5\beta$ as a function of the mean coordination $\langle r \rangle$ in a three-dimensional model that simulates a network glass. The three symbols are for three different series of random networks. The inset shows the fraction of zero frequency modes (averaged over all three networks) compared to equation 16. Reproduced from He and Thorpe.[14]

the covalent forces α and β that determine the mechanical stability and elastic properties of the network.

CONCLUSIONS

We have shown that Maxwell's work on the stability of frames can be applied to covalent network glasses and provides a useful new conceptual framework. The key is seen to be the fraction of zero frequency vibrations. When the dihedral angle forces are included these zero frequency modes should be shifted to produce a prominent peak at low frequencies. The weight of the modes in this peak in the density of states should follow equation 17 and should be observable experimentally.

ACKNOWLEDGMENTS

I should like to thank E. Garboczi and H. He for the many happy hours I've spent working with them.

REFERENCES

1. ZACHARIASEN, W. H. 1932. J. Am. Chem. Soc. **54:** 3841.
2. PHILLIPS, J. C. 1980. Phys. Status Solidi B **101:** 473.
3. MAXWELL, J. C. 1864. Philos. Mag. **27:** 294.
4. PIPPARD, A. J. S. & J. F. BARKER. 1943. The Analysis of Engineering Structures. Arnold. London.
5. GOLDSTEIN, H. 1950. Classical Mechanics. Addison–Wesley. Cambridge.
6. FENG, S. & P. N. SEN. 1984. Phys. Rev. Lett. **52:** 216.
7. FENG, S., M. F. THORPE & E. GARBOCZI. 1985. Phys. Rev. B **31:** 276.
8. ZALLEN, R. 1983. The Physics of Amorphous Solids. John Wiley & Sons. New York, NY.
9. PHILLIPS, J. C. & M. F. THORPE. 1985. Solid State Commun. **53:** 699.
10. THORPE, M. F. 1983. J. Non-Cryst. Solids **5:** 365.
11. PHILLIPS, J. C. 1979. J. Non-Cryst. Solids **34:** 153.
12. PHILLIPS, J. C. 1981. J. Non-Cryst. Solids **43:** 37.
13. DÖHLER, G. H., R. DANDOLOFF & H. BILZ. 1980. J. Non-Cryst. Solids **42:** 87.
14. HE, H. & M. F. THORPE. 1985. Phys. Rev. Lett. **54:** 2107.

Hard Spheres and Instabilities

ROBIN J. SPEEDY

Chemistry Department
Victoria University of Wellington
Wellington, New Zealand

INTRODUCTION

Since the time of van der Waals and especially since the revival of interest in his model in the form of modern-day perturbation theories, the hard-sphere system has played a central role in modeling the structure and properties of fluids. Early studies by Bernal and Mason[1] used real ball bearings and revealed the existence of a reproducible state of random close packing with a density $z = \rho/\rho_0 = 0.858$. (Here $\rho = N/V$ and ρ_0 is the density at crystalline close packing where the spheres occupy $\sqrt{2}\pi/6$ or 74% of the space.) Bernal and Mason's observations led to the idea that liquids might tend, when cooled or compressed sufficiently, toward a locked amorphous solid state with a density and geometry similar to that of the ball bearing state. The amorphous solid state could then serve as a model for the structure and properties of glasses[2] and dense liquids.

Later studies by computer simulation established the existence of a freezing transition in the hard-sphere system.[3] Various perturbation schemes and van der Waals-like approximations can be used to make allowance for the softer continuous interaction potentials between real molecules, so that the numerical data on hard spheres provides the basis for quantitatively accurate predictions of the structure and equation of state of real molecular systems.[4]

It is hardly controversial, then, to say that an understanding of the states of matter (gas, liquid, crystal, and glass), of the transitions between them, and of their structure, thermodynamics, and dynamics, is predicated on being able to understand the corresponding properties of the hard-sphere model where the essential physics is present in its simplest form.

One could disagree with the last statement on the grounds that we do not really need to understand the hard-sphere system because numerical data on its properties are available from computer simulation and can be represented by empirical equations. But I think that if we understood hard spheres, in the sense of having a good theory, then most of the problems would be solved. They have not been solved by computer simulation.

FIGURE 1 shows the equation of state for hard spheres, as determined by computer simulation. The properties of the fluid state can be predicted accurately,[5] and the crystal phase is well represented by a free volume expansion.[6] Substantial progress has been made recently in predicting the freezing transition. The least understood part of the diagram is in the area describing the glass transition. I will focus on this area. More precisely, I will focus on the fluid state with special interest in what happens at its metastable, high-density end.

One possibility is that the fluid compresses continuously to $z_B = 0.86$ where it becomes locked into an incompressible amorphous structure at infinite pressure. But that possibility now seems to be ruled out by Woodcock's computer simulation studies, which suggest that the fluid state ends at about $z = 0.76$.[8] Hudson and Andersen[9] have shown that the glass transition in simple real liquids corresponds to this density rather than to Bernal's density.

FIGURE 2 shows the self-diffusion coefficient, D, reported by Woodcock and Angell.[7] They showed that D is a rapidly decreasing function of density, conforms to the Doolittle equation, and can be extrapolated to zero at Bernal and Mason's random close-packed density $z_B = 0.86$. Alternatively, the linear regression fit shown goes to zero at $z \simeq 0.76$. Whichever is the case, it is clear that D becomes very small at about $z = 0.76$, implying that structural relaxation times are long and that computer simulation results may be influenced by relaxation time problems near that density.

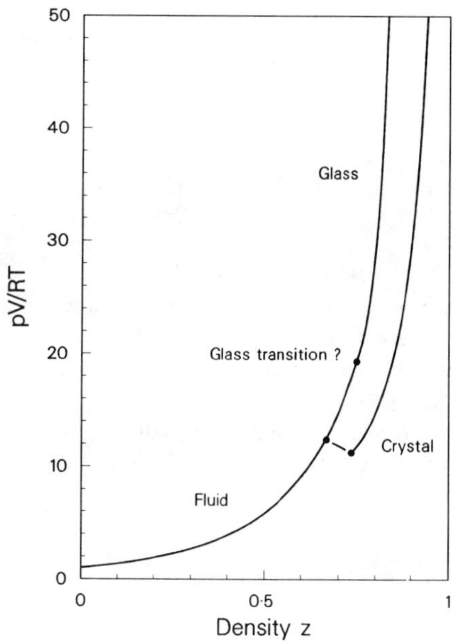

FIGURE 1. Equation of state for hard-sphere versus density ($z = \rho\sigma^3/\sqrt{2}$ is the density relative to the perfect crystal). The fluid pressures were obtained from equation 8 of reference 5; the crystal and glass pressures, from reference 6. The fluid-crystal tie line was obtained from reference 3; the glass transition, from references 7 and 8.

FIGURE 3 shows Woodcock's equation of state data for the metastable high-density amorphous state.[8] The solid line drawn through the points is a cubic spline fit used to highlight the wobbles that occur in the data. The precision of the data (based on the internal consistency of the pressure at each density) is close to the dot size where error bars are not shown. So it is clear that there are real wobbles in the results around $z = 0.75$. Remembering that D becomes effectively zero at $z = 0.76$, these wobbles can be taken as evidence that the simulation results may cease to reflect a fully equilibrated system above about $z = 0.74$.

Up to that density pV/RT is accurately predicted by an equation of state which is consistent with the known virial coefficients.[5]

A very important observation reported by Woodcock and Angell[7] and by Wood-

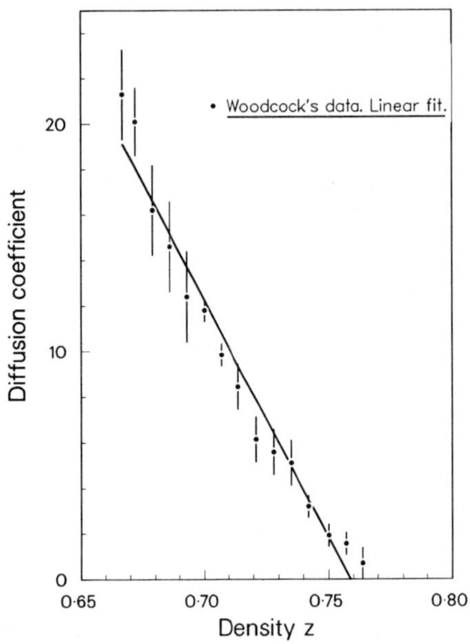

FIGURE 2. Self-diffusion coefficient in the metastable hard-sphere fluid versus density. The ordinate is 10 $D\sqrt{m/kT}/\sigma$. The data were obtained from reference 7.

cock[8] is that the metastable fluid system was stable with respect to crystallization for times in excess of a million collisions up to $z = 0.75$. At higher densities, however, it froze easily, and the results reported at higher densities were obtained from much shorter runs on rapidly compressed amorphous solid samples. Hence the larger error bars above $z = 0.76$.

If the data shown in FIGURE 3 are differentiated one can extract the configurational heat capacity

$$C_p^c = C_p - \tfrac{3}{2}R = (pV/RT)^2 \rho\, k_B T \kappa_T$$

where

$$\kappa_T = (\partial \ln \rho/\partial p)_T$$

One way to differentiate the data is to take the finite difference between adjacent points. That yields the results shown with error bars (based on the internal precision of the pV/RT values) in FIGURE 4. The other way is to represent the data by an analytical equation and to differentiate that equation to obtain smoothed values of C_p^c.

The smoothed values show clearly a discontinuity in the slope of C_p^c versus density located at $z = 0.75$. Woodcock has argued elsewhere[10] that the glass transition in hard spheres should be a third-order discontinuity with a change in the slope, but not in the magnitude, of C_p at the glass transition density. If that argument is correct then the transition would be located where the fluid and glass lines intersect.

The inclusion of C_p^c data obtained by the finite difference method emphasizes that the actual form of C_p^c in the vicinity of the transition remains uncertain. The uncertainty is doubly emphasized by noting that the scatter in the C_p^c values becomes much greater than the error bars as the apparent transition density is approached.

The next section presents an outline of a theory,[11] one that was developed before Woodcock's thermodynamic data became available, that seems to account for the qualitative phenomena reported by Woodcock and Angell.[7]

OUTLINE OF THE THEORY

The details of the theory that I wish to compare against Woodcock's data have been published elsewhere,[11] so a descriptive outline is given here.

Consider a container of hard spheres or, for ease of illustration, a two-dimensional system of hard discs. A snapshot of the system will reveal an instantaneous configuration, as shown in FIGURE 5A. On that configuration the geometrical construction shown in FIGURE 5B is made by drawing circles with a radius equal to the disc diameter about each disc center.[12] The shaded area covered by these circles is excluded to the center of another disc because the centers of any two discs must be separated by at least one diameter.

The remaining area can be called the spare volume, V_s, because it is available to the center of another disc.[12] At densities above about one-quarter of the crystalline close-packed density, the spare volume is subdivided into local regions called cavities.[13] A point that lies within a cavity is called a cavity site. The probability that a randomly chosen point lies in a cavity, $n_0 = V_s/V$, is simply related to the excess chemical potential by[12,14–16]

$$n_0 = \exp(-\beta\Delta\mu) \tag{1}$$

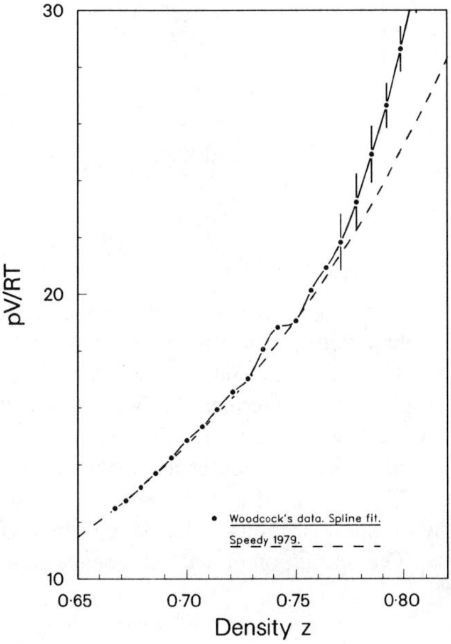

FIGURE 3. Equation of state for the metastable hard-sphere fluid and amorphous solid versus density. The experimental points were obtained from reference 8. The solid line is a cubic spline fit to the experimental points. The error bars represent only the internal precision of the calculations, given in reference 8, where this exceeds the dot size. The dashed line represents equation 8 of reference 5.

where $\beta = 1/k_BT$ and $\Delta\mu$ is the chemical potential relative to that of an ideal gas of point particles at the same volume.

Equation 1 has a simple interpretation. When a point particle is transferred from the ideal gas, where it is free to move over the entire volume V, into the hard-particle system where it can only go in the volume V_s, the change in its translational entropy is $\Delta S = k \ln (V_s/V)$ and the change in its chemical potential is just $\Delta\mu = -T\Delta S$, which gives equation 1.

If V_s or n_0 could be calculated, then all the thermodynamic properties of the hard-sphere system would be known from equation 1 and we would understand, among other things, the glass transition.

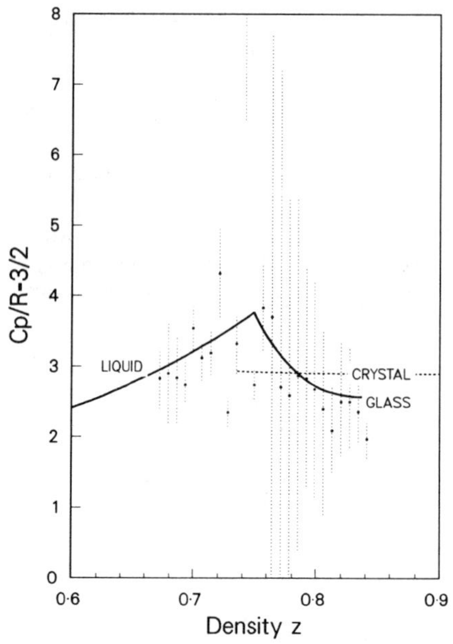

FIGURE 4. Configurational heat capacity for amorphous hard spheres at high density. Note that $C_p^c = C_p - 3R/2 = \phi^2/(\phi + z(\partial\phi/\partial z))$ where $\phi = pV/RT$. The experimental points were calculated from the pV/RT results obtained from reference 8; the finite differences between adjacent points were used to estimate $\partial\phi/\partial z$. The error bars reflect only the internal precision of the pV/RT data given in reference 8. The smooth curves were derived from analytical equations of state for the fluid (equation 8 of reference 5), crystal, and glass (reference 6).

SCALED PARTICLE THEORY

One approach to estimating n_0 is to imagine placing a (deflated) balloon at a random point in the system and blowing it up until its radius is equal to the diameter of a sphere. For the purposes of this illustration the balloon needs to have a special property: its surface can freely penetrate the physical surface of the spheres and is only repelled by their centers. The center of the inflated balloon would locate a cavity site because no sphere center lies within a diameter σ of the center of the balloon.

The amount of work, w, needed to blow the balloon up against the pressure exerted by the sphere centers colliding with its surface would be equal to $-\Delta\mu$.

The estimation of w is the essence of scaled particle theory[17,18]

$$\Delta\mu = -w = -\int_0^{\sigma} k_B T \rho(r)\, 4\pi r^2 \, dr \tag{2}$$

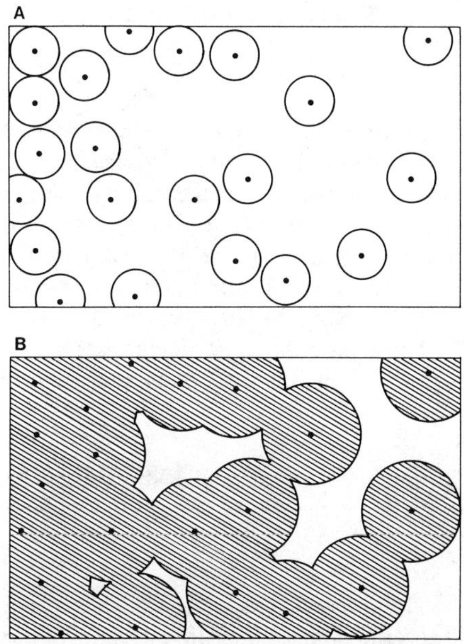

FIGURE 5. A configuration of hard disks showing the spare volume. (**A**) The location of the disk centers as dots and their surfaces as circles. (**B**) The same configuration of disk centers. The area within one diameter of a disk center is shaded. The unshaded area represents the spare volume.

In equation 2 $\rho(r)$ is the average density of sphere centers at the surface of a balloon of radius r. The factor $k_BT\rho(r)$ gives the thermal pressure acting on the surface; multiplying this by the surface area gives the force acting against its inflation; and the integration of this force over the radial distance r gives the work.

The difficulty in this approach is that the presence of the balloon influences the density of particles in its vicinity,[17,18] so the pressure acting on it is not the bulk pressure, but is determined by the conditional density $\rho(r)$, which is not known exactly for all r.

MEERON AND SEIGERT

Another approach, developed by Meeron and Seigert,[19] is to consider a pair of cavity sites, as shown in FIGURE 6. Here we start with one balloon already inflated and estimate the work required to grow a second balloon exactly one diameter away from

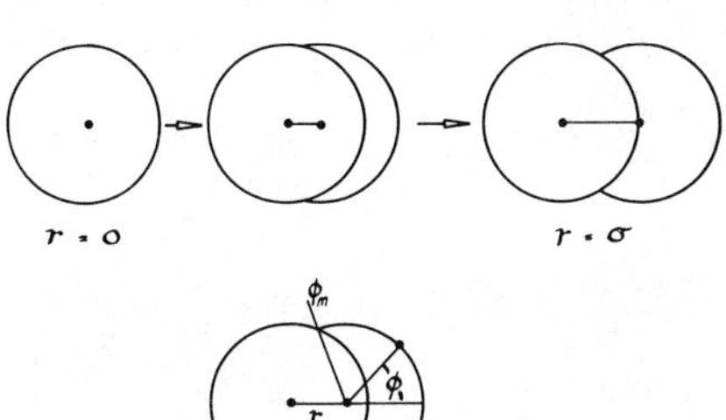

FIGURE 6. The region excluded to sphere centers by a pair of cavity sites separated by r.

the first. The work involved for this process can be written as

$$w = \int_0^\sigma \mathrm{d}r \int_0^{+\phi_m} \mathrm{d}\phi \; k_B T \rho(r, \phi) \cos\phi \; 2\pi\sigma^2 \sin\phi \tag{3}$$

In equation 3, $\rho(r, \phi)$ is the (conditional) density of sphere centers at a point ϕ on the surface of the second cavity when it is a distance r apart from the first; $k_B T\rho(r, \phi)$ gives the pressure acting normal to the surface at ϕ; the factor $2\pi\sigma^2 \sin\phi \mathrm{d}\phi$ is the area of the circle at constant ϕ; and $\cos\phi$ converts the force acting normal to that area into the force acting along the line of the two cavity centers. The integral over ϕ gives the total force acting against an increase in r, and the integral of this force over r gives the work of forming the second cavity at $r = \sigma$.

The chance that a particular pair of points separated by r are both cavity sites is $n_{00}(r)$,[11,19,20] and the work calculated above gives the ratio of the probabilities

$$n_{00}(\sigma)/n_{00}(0) = \exp(-\beta w) \tag{4}$$

The probability $n_{00}(0)$ is known. The chance of finding two cavity sites at zero separation is clearly the same as that of finding one cavity site there, so

$$n_{00}(0) = n_0 \tag{5}$$

The probability $n_{00}(\sigma)$ is also known. The pair correlation function between cavity sites is the same as the pair correlation function $g(r)$ between sphere centers because cavity sites are equivalent to spheres in the sense that the only effect they have is to exclude sphere centers from a region of radius σ. It follows that[11,19,20]

$$n_{00}(\sigma) = n_0^2 g(\sigma) \tag{6}$$

Combining the last three equations gives

$$-\ln n_0 g(\sigma) = \beta w \tag{7}$$

so that a knowledge of w provides a relation between the chemical potential (equation 1) and the pressure. The pressure and the chemical potential, however, are related independently by a Maxwell relation

$$-\partial \ln n_0/\partial\rho = \frac{q}{\rho} + \frac{\partial q}{\partial \rho} \tag{8}$$

where $q = p/\rho kT - 1 = B_2 \rho g(\sigma)$ so equations 7 and 8 can be solved simultaneously for the thermodynamic properties if w is known.

The only difficulty in this approach, as with the scaled particle theory, is that the density of particles $\rho(r, \phi)$ on the surface of the cavity pair is not known for all values of r and ϕ. This is where some assumptions or approximations must be introduced into the theory.

Meeron and Seigert[19] took the first step towards estimating $\rho(r, \phi)$. Defining the conditional correlation function $g(r, \phi)$ by

$$g(r, \phi) = \rho(r, \phi)/\rho \tag{9}$$

they noted the exact results

$$g(r, \phi) = g(\sigma) \quad (r = 0 \text{ and } r = \infty) \tag{10}$$

and investigated the consequences of assuming equation 10 to be true for all r and ϕ. This leads to an equation of state that gives the correct third virial coefficient.

One way to improve on this assumption might be to suppose that $g(r, \phi)$, although it is not equal to $g(\sigma)$ for every r and ϕ, might be expected to vary in proportion to $g(\sigma)$ as the density changes; that is

$$g(r, \phi) = K_1(r, \phi)\, g(\sigma) \tag{11}$$

The only assumption here is that $K_1(r, \phi)$ is independent of density.

The study of this assumption[11] shows that it offers little improvement, because if the definite integrals over $K_1(r, \phi)$ in equation 3 are required to yield the correct third virial coefficient, the result is the same as when $K_1(r, \phi) = 1$ for all r and ϕ. It should be noted, however, that it is possible to derive the equation of state that equation 11 implies without having to specify the form of $K_1(r, \phi)$, so there is a considerable improvement in generality despite the lack of improvement in the equation of state.

Because the above approaches do not produce an accurate equation of state, it is clear that $g(r, \phi)$ is not equal to or proportional to $g(\sigma)$, so we need another term in equation 11.

When the two cavity sites in FIGURE 6 are separated by one diameter they are equivalent to a pair of spheres in contact and $g(r, \phi)$ is then the same as the correlation function for a third sphere on the surface of two spheres. The approximation that comes to mind here for the three body correlation function is Kirkwood's superposition.[21] Superposition would imply, in the region where the third sphere can contact both of the other two, that is, where $r = \sigma$ and $\phi = \phi_m$, that

$$g(\sigma, \phi_m) = g^2(\sigma) \tag{12}$$

Equation 12 is probably a good approximation, so we make the more general assumption that, for some values of r and ϕ, $g(r, \phi)$ is likely to contain contributions that vary with density in proportion to $g^2(\sigma)$, so that

$$g(r, \phi) = K_1(r, \phi)\, g(\sigma) + K_2(r, \phi)\, g(\sigma)^2 \tag{13}$$

In other words, we suppose at every point r, ϕ, that $g(r, \phi)$ will have the same density dependence as some arbitrary linear combination of $g(\sigma)$ and $g(\sigma)^2$. The only assumption involved is that K_1 and K_2 are independent of the density of the system. No other assumptions are involved in the theory.

Equation 13 can now be substituted in equation 3 to obtain w. After some manipulations involving equations 7 and 8, one obtains the differential form of the equation of state[11]

$$dy/dx = \frac{(C_1 - 2)y^2 + C_2 y^3}{(1 - C_1)xy - 2C_2 xy^2 - 1} \tag{14}$$

where $y = q/x = g(\sigma)$, $x = B_2\rho$, and

$$C_J = \int_0^\sigma dr \int_0^{+\phi_m} d\phi K_J(r, \phi) \cos\phi\, 2\pi\sigma^2 \sin\phi \tag{15}$$

Since the $K_J(r,\phi)$ have not been specified the definite integrals in equation 15 cannot be evaluated directly. If the assumption that the K_J are independent of density is correct, however, the C_J obtained from equation 15 would necessarily be consistent with the known virial coefficients. Therefore the values of C_J were determined by requiring them to be consistent with the virial coefficients B_3 and B_4. This yields $C_1 = 1.39662$ and $C_2 = -0.02162$. Equation 14 yields the compressibility directly. Unfortunately, I have not been able to integrate it to get a closed form of the equation of state, but it can be integrated numerically to get all the thermodynamic properties.

In the next section the predictions of equation 14 will be referred to as the *GJ* equation (for $g(\sigma)^J$).[11]

PREDICTIONS OF THE THEORY

The main qualitative prediction of the theory is that the hard-sphere fluid has a well-defined limit of stability at a density $z_s \simeq 0.75$ of the crystalline close-packed

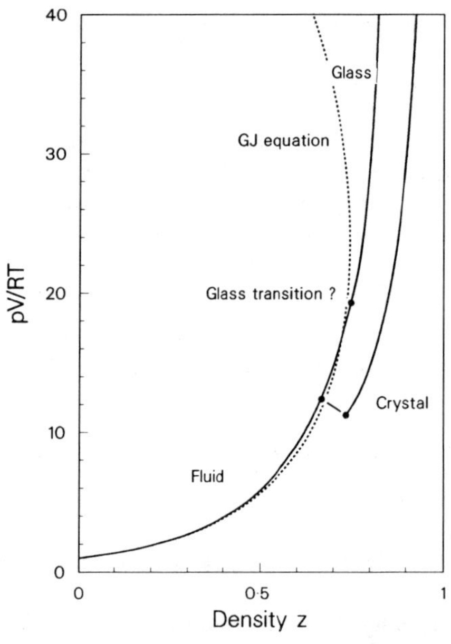

FIGURE 7. Equation of state for hard spheres. The solid lines are the same as in FIGURE 1. The dashed line is the equation of state predicted by (the numerical integration of) equation 14.

density. As this point is approached the fluid becomes incompressible, implying that its structure is locked so that the structural relaxation time should diverge at z_s. At higher densities the fluid does not exist.

EQUATION OF STATE

The above predictions can be deduced from the predicted equation of state shown in FIGURE 7. As the density of the fluid is increased, by increasing the pressure or pV/RT, a point is reached where $(\partial p/\partial \rho)_T \rightarrow \infty$ or the compressibility $\kappa_T = \rho^{-1}(\partial \rho/\partial p)_T \rightarrow 0$. At this point $z_s \simeq 0.75$ and $pV/RT \simeq 23$.

Any attempt to increase the density beyond z_s, by increasing the pressure or pV/RT, produces instead a density decrease. In other words, at higher pressures the system has a negative compressibility, a property that violates the stability condition $\kappa_T \geq 0$, which implies that the fluid ceases to exist as a stable or metastable state. At higher pressures or densities the fluid is absolutely unstable.

STABILITY

The prediction that the fluid becomes absolutely unstable at $z_s = 0.75$ can be compared against the observation made by Woodcock and Angell and by Woodcock[8] that a system of 512 hard spheres that was kinetically stable against crystallization for times in excess of a million collisions up to a density $z = 0.756$ froze after a quarter of a million collisions at $z = 0.763$. It was not possible to equilibrate fluid samples at slightly higher densities. Results reported at higher densities were from rapidly quenched samples averaged over a period of only 10 000 collisions.

Woodcock locates the homogeneous nucleation density at about $z = 0.767$,[8] which is very close to the predicted limit ($z_s \simeq 0.75$).

KAUZMANN PARADOX

Kauzmann argued that since the extrapolated entropy of liquids tends to become less than that of the crystal, something must happen to prevent a violation of the third law of thermodynamics.[22] He proposed that spontaneous crystallization is inevitable and that any discussion of the properties of the fluid beyond that point is metaphysical.

Woodcock notes that spontaneous freezing resolves the Kauzmann paradox in the hard-sphere system, but takes the view that it is still meaningful to discuss the extrapolation of the fluid equation of state to higher densities.[8]

The present theory comes down strongly on Kauzmann's side by suggesting that the equation of state is meaningless beyond the stability limit.

COMPRESSIBILITY AND HEAT CAPACITY

The dimensionless compressibility $\rho k_B T \kappa_T = \rho k_B T(\partial \ln \rho/\partial p)_T$, and the configurational heat capacity C_p^c are related by

$$C_p^c = C_p - \tfrac{3}{2}R = (pV/RT)^2 \rho k_B T \kappa_T$$

There is no indication in Woodcock's simulation results to support the prediction of the theory that κ_T and C_p^c both go to zero with an infinite slope as $z \rightarrow z_s$. If the theory is correct, however, such evidence would require very long runs over the narrow density range immediately below z_s to obtain precise data. I have noted earlier that Woodcock's pV/RT results[8] show a greater scatter than is accounted for by their internal precision in this density range, and that this is to be expected on the grounds that the structural relaxation time is long (as discussed in the next section, or as deduced from Woodcock and Angell's diffusion results[7]).

The predicted forms of κ_T and C_p are compared with values calculated (using finite differences) from Woodcock's pV/RT results[8] in FIGURES 8 & 9. The error bars drawn are those derived from the internal precision of Woodcock's results.[8] The agreement is good except for the narrow density range preceding the predicted singularity.

STRUCTURAL RELAXATION TIME

The theory, being thermodynamic, can make no direct statement about relaxation times. It predicts that the compressibility goes to zero, at a finite pressure, with an infinite slope, however, implying that the structure becomes locked. If any structural

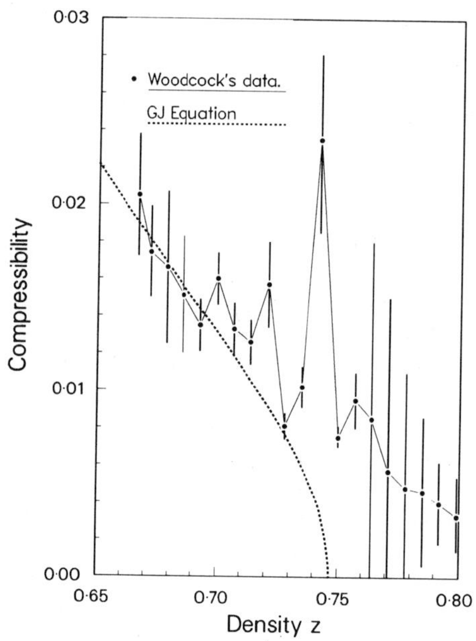

FIGURE 8. Compressibility of amorphous hard spheres at high density. The dimensionless compressibility is $\rho k_B T \kappa_T = 1/(\phi + z(\partial\phi/\partial z))$ where $\phi = pV/RT$. The experimental points are calculated from the pV/RT results obtained from reference 8; the finite differences between adjacent points were used to estimate $\partial\phi/\partial z$. The error bars reflect only the internal precision of the pV/RT given in reference 8. The dashed line shows the prediction of the GJ equation (equation 14).

relaxation could occur, the system would be able to relax toward a more dense structure in response to a pressure increase and the compressibility would then be finite. So incompressibility is necessarily associated with the absence of structural relaxation, implying that the structural relaxation time diverges at $z_s \simeq 0.75$. This conclusion is in good agreement with the linear extrapolation of the self-diffusion coefficient to zero at about $z = 0.76$ shown in FIGURE 2.

CONCLUSIONS

The assumption of equation 13, that the correlation function for the density of sphere centers at any point on the surface of a pair of sphere-sized cavities has the same

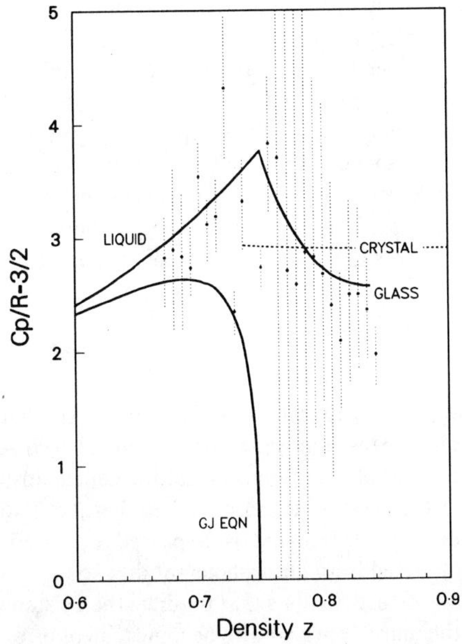

FIGURE 9. Configurational heat capacity for amorphous hard spheres at high density. The content is the same as that in FIGURE 4, only with the addition of a line that shows the prediction made by the *GJ* equation (equation 14).

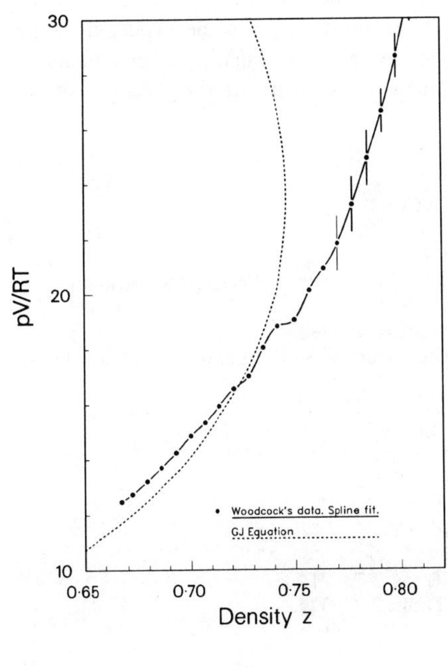

FIGURE 10. Equation of state for amorphous hard spheres at high density. The content is the same as that in FIGURE 3, only with the addition of a dashed line that shows the prediction made by the *GJ* equation (equation 14).

density as a linear combination of $g(\sigma)$ and $g(\sigma)^2$, produces an equation of state accurate over the whole fluid range. Although it is not exact (pV/RT is 4% low at the freezing density), it provides a predictive explanation for the phenomena associated with the glass transition. It is quite different in form from any other theory of the fluid in the glass transition region. It is possible that the predicted form gives a qualitatively correct account of the upper density stability limit of the hard-sphere fluid. This view can only be maintained if one is willing to accept that the pressures currently available from computer simulation are somewhat lower than those of the fully equilibrated fluid near z_s. The simulation results shown in FIGURE 10 do show some wobbles; these wobbles make the simulations suspect in the high-density region.

SUMMARY

Computer simulation studies of the hard sphere fluid and amorphous solid in the glass transition region by Woodcock and Angell[7] and by Woodcock[8] suggest that the metastable high-density fluid becomes unstable with respect to freezing at about 76% of the crystalline close-packed density. The diffusion coefficient goes to zero near this density; if freezing is bypassed by rapid quenching to higher densities, the state observed is an amorphous solid.

A theory exists that predicts the existence and location of a novel kind of instability: the hard-sphere fluid becomes incompressible at a finite pressure at 75% of the crystalline close-packed density and is unstable with respect to freezing at higher pressures.[11]

The assumptions of the theory were outlined, and its predictions were compared with Woodcock's thermodynamic data.[8] If one allows for some uncertainty in the simulation results due to the long structural relaxation times to be expected in the vicinity of the predicted instability (and implied by the vanishing diffusion coefficient), the theory accounts for the qualitative phenomena associated with the glass transition, and it does so semiquantitatively.

REFERENCES

1. BERNAL, J. D. & J. MASON. 1960. Nature **188:** 910.
2. CARGILL, G. S. 1975. *In* Solid State Physics. H. Ehrenreich, F. Sectz & D. Turnbull, Eds. Vol. **30.** Academic Press. New York, NY.
3. HOOVER, W. G. & F. H. REE. 1968. J. Chem. Phys. **49:** 3609.
4. HANSON, J. P. & I. R. MCDONALD. 1976. The Theory of Simple Liquids. Academic Press. New York, NY.
5. SPEEDY, R. J. 1979. J. Chem. Soc. Faraday Trans. 2 **75:** 1643.
6. HOSTE, R. & W. VAN DAEL. 1984. J. Chem. Soc. Faraday Trans. 2 **80:** 477.
7. WOODCOCK, L. V. & C. A. ANGELL. 1981. Phys. Rev. Lett. **47:** 1129.
8. WOODCOCK, L. V. 1981. Ann. N.Y. Acad. Sci. **371:** 274.
9. HUDSON, S. & H. C. ANDERSON. 1978. J. Chem. Phys. **69:** 2323.
10. WOODCOCK, L. V. 1976. J. Chem. Soc. Faraday Trans. 2 **72:** 731.
11. SPEEDY, R. J. 1983. Physica **121B:** 153.
12. SPEEDY, R. J. 1977. J. Chem. Soc. Faraday Trans. 2 **73:** 714.
13. SPEEDY, R. J. 1981. J. Chem. Soc. Faraday Trans. 2 **77:** 329.
14. BYCKLING, E. 1961. Physica **27:** 1030.
15. WIDOM, B. 1963. J. Chem. Phys. **39:** 2808.

16. JACKSON, J. L. & L. S. KLEIN. 1964. Phys. Fluids **7:** 228.
17. FRISCH, H. L. 1963. Adv. Chem. Phys. **6:** 229.
18. REISS, H. 1966. Adv. Chem. Phys. **9:** 1.
19. MEERON, E. & A. J. F. SEIGERT. J. Chem. Phys. **48:** 3139.
20. SPEEDY, R. J. 1980. J. Chem. Soc. Faraday Trans. 2 **76:** 693.
21. KIRKWOOD, J. G. 1935. J. Chem. Phys. **3:** 300.
22. KAUZMANN, W. 1948. Chem. Rev. **43:** 219.

DISCUSSION OF THE PAPER

F. H. STILLINGER (*AT&T Bell Laboratories, Murray Hill, NJ*): Aside from use of virial coefficient information at the last stage, nothing was present in your derivation that would restrict it to the fluid-amorphous branch. Have you tried generating solutions of the equation that would fit the crystalline branch? It would be instructive to see if such fits predict a lower density limit for stability of the crystal.

SPEEDY: In the following it is assumed that the equations of the paper can be applied directly to a nonuniform crystalline solid (with appropriate orientational and translational averaging of the correlation functions).

The assumption of equation 11, that is,

$$g(r, \phi) = K_1(r, \phi)g(\sigma)$$

where $K_1(r, \phi)$ is independent of density, leads to equation 14 with $C_2 = 0$, that is

$$\mathrm{d}y/\mathrm{d}x = (C_1 - 2)y^2/((1 - C_1)xy - 1)$$

where $y = q/x = g(\sigma)$, $x = B_2\rho$, and $q = p/\rho kT - 1$. Integration yields

$$x/x_0 = q/(q + 1)^{2-C_1}$$

where x_0 is the constant of integration. At this stage the equations for the fluid and crystal are the same. They differ only in the values of x_0 and C_1, which are determined by the boundary conditions.

For the crystal, the pressure diverges at some nonzero and noninfinite value of x, which implies that $C_1 = 1$, and that $x/x_0 \rightarrow 1$ as $q \rightarrow \infty$. Putting $z = x/x_0$, the equation of state is therefore

$$z = q/(q + 1)$$

or

$$q = z/(1 - z)$$

The result is the exact equation of state for a one-dimensional system of hard rods, first obtained by Tonks (*Phys. Rev.* **50:** 955, 1936).

In two and three dimensions it is known (Salzburg & Wood, *J. Chem. Phys.*, **37:** 798, 1962), in the high density limit ($z \rightarrow 1$), that

$$q \sim Dz/(1 - z)$$

where D is the dimensionality. So it is clear that the assumption of equation 11 does not produce a good equation of state for the crystal in two and three dimensions.

The assumption of equation 13 produces a worse result. In the limit $y = q/x \rightarrow \infty$, equation 14 becomes (when C_2 is nonzero)

$$dy/dx = -y/2x$$

This latter equation implies that the pressure can only tend to infinity from a higher (rather than from a lower) density. This result is clearly unphysical.

The conclusion to be drawn is that the methods applied to the fluid do not work when applied directly to the crystal, so a more thorough investigation that would make proper allowances for the nonuniformity of the crystal is needed.

M. ALEXANIAN (*Southern Illinois University, Carbondale, IL*): Your graph shows C_p(glass) < C_p(crystal). Would you comment on that? Also, can one define T_g as the temperature where the equality occurs?

SPEEDY: The values of C_p for the crystal and glass were calculated from empirical equations of state given in reference 6. C_p(crystal) is near to $3R$, which is the value expected at the close-packed limit. C_p(glass) should also tend to $3R$ as $z \rightarrow 0.86$ if the glass is described by a free volume expansion about that density.

Woodcock (*J. Chem. Soc. Faraday Trans 2* **72:** 731, 1976) found that C_p(liquid) and C_p(glass) were the same (empirically) at the same density where he located the glass transition. He presented an argument (that the glass transition in hard spheres should be third order—with no discontinuity in C_p on going from liquid to glass) to rationalize that observation.

In my view, the transition from hard-sphere fluid to glass should involve the sudden disappearance of density fluctuations with a consequent decrease in C_p. This would imply that the glass transition must be located in the density range where C_p(glass) < C_p(liquid).

Glass Transitions in Hard-sphere and Lennard–Jones Systems

U. BENGTZELIUS AND A. SJÖLANDER

Institute of Theoretical Physics
Chalmers University of Technology
Göteborg, Sweden

INTRODUCTION

The observation that widely different kinds of glass-forming systems seem to show some universal behavior at the glass transition point has in recent years created an increasing interest in these problems. Volume, entropy, free energy, etc. change continuously whereas a discontinuous change occurs for the compressibility, the thermal expansion coefficient, and the heat capacity. This is true for polymers, and it also seems to be true for so-called computer glasses, such as hard-sphere and Lennard–Jones fluids, which can only be brought into the glassy state through computer simulations. Because a finite cooling rate is required there is always a certain smearing of the glass transition, and this is particularly noticeable in computer experiments. Some hysteresis effects have been observed,[1] when passing through the transition point in both directions, and this strongly suggests that the vitrification is basically of a dynamical origin.

The viscosity increases many orders of magnitude when approaching the glass transition point, and characteristic non-Arrhenius temperature dependence has been found. A similar behavior is found for the self-diffusion constant, which approaches zero on freezing. This implies that structural relaxations connected with molecular diffusion are slowed down and are finally arrested when passing into the glassy state. A characteristic feature is that the slow relaxations show a nonexponential decay (in such cases the experimental data are conventionally fit to a stretched exponential). Recently a plausible explanation for such a time dependence has been given.[2] Actually, in some cases one observes two different slow relaxation processes, where one is arrested at the transition point and the other stays essentially unchanged when moving into the glass.[1]

Recently a dynamical model was proposed independently by Leutheusser[3] and by Bengtzelius *et al.*[4] that contains a mechanism for a liquid-glass transition. The basic approach is the same as one that has been used extensively and tested successfully for analyzing the microscopic motions in ordinary monatomic liquids. It is based on the Zwanzig–Mori procedure[5] for deriving formally exact transport equations combined with mode-coupling approximations, familiar from the work by Kawasaki[6] and others on critical dynamics. An important effect in liquids is the so-called cage effect, where an atom has a tendency of being trapped for some time by its surrounding atoms. It leads to a significant reduction of the self-diffusion constant. The mode-coupling approximation contains this trapping effect as a prominent feature. This becomes more and more important as we go to lower temperatures, and it leads finally to a localization of the individual atoms at a certain critical temperature or density. Since the original papers by Leutheusser[3] and by Bengtzelius *et al.*[4] appeared, Das *et al.*,[7] Kirkpatrick,[8] Götze,[9] DeRaedt and Götze,[10] Bengtzelius and Sjögren,[11] Bengtzelius,[12] and Sjögren[13] have reported on further research on this subject.

The full model proposed by Bengtzelius *et al.*[4] and later extended by Bengtzelius and Sjögren[11] has been found to predict the following: 1) discontinuous changes in the compressibility, thermal expansion coefficient, and heat capacity (the model provides an approximate relation between the Prigogine–Defay ratio and C_p/C_V for the liquid); 2) a non-Arrhenius temperature dependence for the self-diffusion constant and the viscosity; 3) two different slow relaxation processes, one of them being arrested at the glass transition point and the other one remaining on the glass side; 4) universal scaling properties for the latter relaxation process with nontrivial and nonuniversal exponents; and, on the liquid side, 5) long-term asymptotic behavior, which is characterized by nonexponential decay and which can be fitted to a stretched exponential.

The predicted positions for the glass transition points agree quite well with data obtained from computer simulations on hard-sphere and Lennard–Jones systems. The dynamical simulation data necessary for quantitative tests of the full dynamics have not yet been obtained, however. Here we report on some recent and partly unpublished results that were obtained using our model.

THEORETICAL MODEL

Our theoretical model can be phrased in various forms, but one way is to write it in a generalized hydrodynamic form with wavevector- and frequency-dependent transport coefficients. Introducing the Laplace transform of the density correlation function $F(kt) = \langle n(\mathbf{k}t)n(-\mathbf{k}0)\rangle/n$ as

$$F(k\omega) = i\int_0^\infty \mathrm{d}t\, e^{i\omega t}\, F(\mathbf{k}t) \tag{1}$$

we can write[11]

$$F(k\omega) = -S(k)\frac{\omega + M(k\omega)}{\omega^2 - k^2/m\beta S(k) + \omega M(k\omega)} \tag{2}$$

Here, $S(k)$ is the liquid static structure factor, m is the atomic mass, and $\beta = 1/k_BT$ is the inverse temperature in energy units. For $M(k\omega)$ we write

$$M(k\omega) = M_{11}(k\omega) - \frac{M_{14}^2(k\omega)}{\omega + M_{44}(k\omega)} \tag{3}$$

Similar expressions are obtained for the single-particle density correlation function, the so-called Van Hove self-correlation function, and for the transverse current correlation function. Letting $k, \omega \rightarrow 0$, we obtain the ordinary hydrodynamic form and can then identify the shear viscosities, the thermal expansion coefficient, the self-diffusion constant, the heat capacity, and the heat conductivity. The last two coefficients are identified from the asymptotic form[14,15]

$$M_{44}(k\omega) \rightarrow a\omega + bk^2 \ (k, \omega \rightarrow 0) \tag{4}$$

The Zwanzig–Mori formalism provides exact formal expressions for the various generalized transport coefficients, and these together with the mode-coupling approximations form the basis for our model. In the ordinary liquid region the $M(kt)$ functions decay rapidly because of essentially binary collisions, but as we lower the temperature, and hence reduce the diffusivity, a long time tail develops. The liquid state is characterized by having finite values for $M(k, \omega \rightarrow 0)$, implying that the local structure, hidden in $F(k, t = 0)$, disappears completely for $t \rightarrow \infty$. The transition to a glass is assumed to be connected with broken ergodicity, where vanishing diffusivity prevents the initial local structure from disappearing as $t \rightarrow \infty$. The tails characteristic of long-term asymptotic behavior in the $M(kt)$ functions arise from slow structural relaxations, and it was shown in reference 11 that the $M(kt)$ functions become, for long times, functionals of density fluctuations only. Our basic approximation is that the tail in $M(kt)$ is bilinear in $F(kt)$—no linear term exists. This has been found to be quite appropriate in the ordinary liquid region, and we argued that it should be even more appropriate for the supercooled liquid. The broken ergodicity is connected with a nonzero value for $M_{11}(k, t \rightarrow \infty)$, implying a $1/\omega$ singularity in $M_{11}(k\omega)$ for $\omega \rightarrow 0$. This leads to a reactive contribution from $\omega M_{11}(k\omega)$ also for $\omega \rightarrow 0$ and a sudden change of the sound velocity and the compressibility. Similar singular behavior in the nonhydrodynamic parts of M_{14} and M_{44} lead to discontinuous changes in the thermal expansion coefficient and the heat capacity. Long wavelength diffusive shear modes go over into propagating ones. In reference 11 formally exact expressions were given for the magnitude of these discontinuities, and they were all shown to be related in a simple way to the frozen part of a four-particle correlation function. The explicit calculations imply an approximate treatment of this correlation function and its time dependence. The changes in the heat capacity and the compressibility represent the frozen-in potential energy and pressure fluctuations, respectively. The static structure factor $S(k)$ in equation 2 is the equilibrium liquid value. Within the glass region we have to imagine that the true equilibrium still corresponds to the hypothetical liquid. Crystallization is prevented by assuming global isotropy: the liquid state is therefore the true equilibrium one for any temperature. In practice $S(k)$ has to be obtained by extrapolating continuously from the supercooled region.

Equation 2 can be transformed into

$$\frac{R(\omega)}{1 + \omega R(k\omega)} = \frac{m\beta S(k)}{k^2}\{\omega + M(k\omega)\} \tag{5}$$

where $R(k\omega) = F(k\omega)/S(k)$. The slow relaxations are here caused by the term $M_{11}(k\omega)$; the other terms may be ignored in the subsequent discussion. $M_{11}(kt)$ is approximated by[16]

$$M_{11}(k, t) = M_0(kt) + (2\pi)^{-3} \int d\mathbf{k}' V(\mathbf{k}, \mathbf{k}') R(\mathbf{k}', t) R(|\mathbf{k} - \mathbf{k}'|, t) \tag{6}$$

with

$$V(\mathbf{k}, \mathbf{k}') = (n/2m\beta)[(\hat{\mathbf{k}} \cdot \mathbf{k}')c(k') + \{\hat{\mathbf{k}} \cdot (\mathbf{k} - \mathbf{k}')\}c(|\mathbf{k} - \mathbf{k}'|)]^2 S(k')S(|\mathbf{k} - \mathbf{k}'|) \tag{7}$$

where $nc(k) = 1 - 1/S(k)$ is the direct correlation function and $\hat{\mathbf{k}} = \mathbf{k}/|\mathbf{k}|$. The term

$M_0(kt)$ includes the more rapidly decaying contributions; it is essentially irrelevant close to the glass transition point. Equation 5 is now a closed nonlinear equation for $R(kt)$. The interaction between the atoms appear in $V(\mathbf{k}, \mathbf{k}')$ only through the static structure factor, which changes with temperature and density.

Ignoring all the k dependence we recover the model described by Leutheusser.[3] We found that the integral in equation 6 is dominated by $\mathbf{k}'$ values within the first peak in the two static structure factors. Replacing these by delta functions, we recover the simplified model described in reference 4. Even though the basic underlying assumptions are quite different in the two cases, it leads to the same form for the resulting equation. It is essential to keep the full wavevector dependence, however, if one wants to draw any quantitative conclusions for comparisons with computer simulations, for instance.

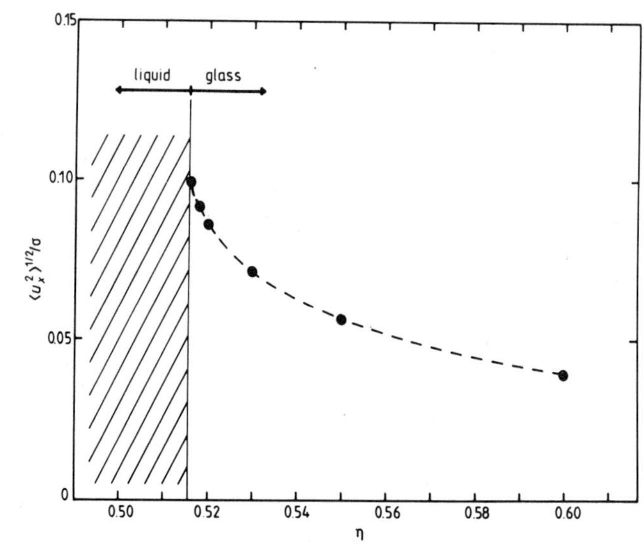

FIGURE 1. The root-mean-square displacement in units of the hard-sphere diameter versus packing fraction η for a hard-sphere system. (Taken from reference 4.)

VITRIFICATION CONDITION

The height of the first peak in $S(k)$ directly reflects the static local ordering in the fluid, which increases continuously as we lower the temperature or increase the density. The first interesting point is that equation 5 predicts a glass transition when the peak height in $S(k)$ exceeds some critical value. The particles become locked in, and the mean square displacement changes discontinuously from infinity to a small value consistent with the Lindeman criterion. FIGURE 1 illustrates this. It was shown by Götze[9] that the slope of the curve is infinite at the transition point. In an experiment

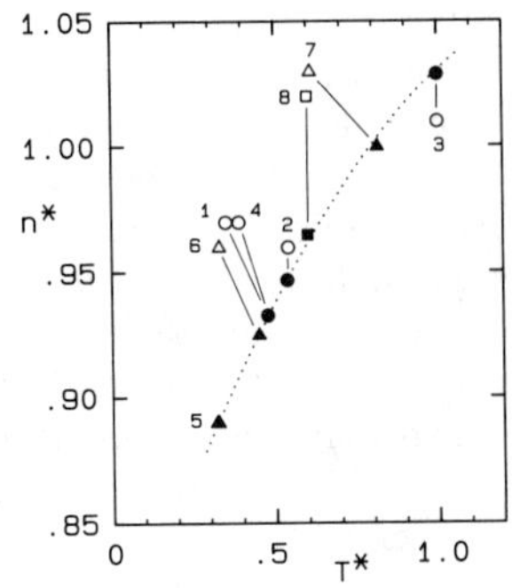

FIGURE 2. The glass transition points on a plot of T^* versus n^* for some different potentials. Open symbols mark the points determined using simulations; filled symbols mark the points determined from our calculations; solid lines connect the points that share the same potential. Density and temperature are in reduced units. (Taken from reference 12.)

this could easily be mistaken for a continuous change of $\langle u^2 \rangle$ from infinity to finite values, as in the conventional percolation problem, but this is not what our model predicts. FIGURE 2 summarizes the theoretical predictions for the glass transition points in a (T^*, n^*) plane in reduced units for some different potentials for which computer simulations have been carried out. The simulation data are denoted by open symbols. The corresponding theoretical predictions appear as filled symbols. For the hard-sphere system we obtained a critical packing fraction $n_c = 0.525$ (improved $S(k)$ relative to that in reference 4) to be compared with two simulation results: $n_c = 0.52$[17] and $n_c = 0.56$.[18]

It was mentioned before that the freezing is connected with discontinuous changes in the second-order thermodynamic derivatives. An approximate relation was obtained between the Prigogine–Defay ratio

$$\Pi = \Delta C_p \Delta \kappa_T / TV(\Delta \alpha_p)^2 \tag{8}$$

and the C_p/C_V ratio for the liquid at the transition point.[11] Equation 8 contains the changes in the heat capacity C_p, the compressibility κ_T, and the thermal expansion coefficient α_p. In FIGURE 3 comparisons are made with experimental results for three liquids for which sufficient data are available.

CRITICAL DYNAMICS

The second interesting point is that equation 5 predicts some universal slow relaxations when approaching the glass transition point from either side. There is an

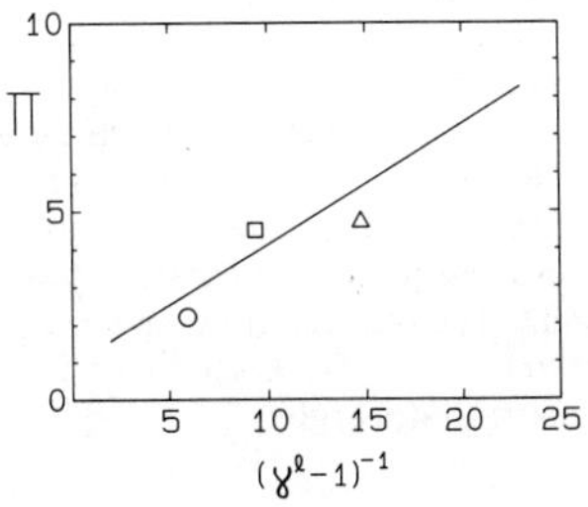

FIGURE 3. A comparison between theoretical (straight line) and experimental (symbols) results on the Prigogine–Defay ratio. Δ: B_2O_3; □: Ca $(NO_3)_2$ (40 mole percent) and KNO_3 (60 mole percent); O: polyvinyl acetate. (Taken from reference 11.)

initial microscopic time region where $R(kt)$ decreases rapidly, as in ordinary liquids and solids, and here the details of the $M_0(kt)$ function enter in a significant way. The function leads to a frequency spectrum that resembles that for the phonon excitations. Beyond this time region follows a slow decay, which on the glass side is governed by a certain universal equation. On the liquid side the slow decay splits into two time regions; the earlier region of the two has a time dependence that closely resembles the one on the glass side. The system behaves within this time region as frozen and macroscopic diffusion plays no role. Diffusion processes become dominant in the later time region. A numerical solution of equation 5, however, has shown that the decay for $R(kt)$ is not purely exponential but closely follows that predicted by a stretched exponential law.

Important results were recently reported by Götze[9] and by DeRaedt and Götze.[10] They analyzed equation 5 with an arbitrary functional relation between $M(kt)$ and $F(kt)$. They showed that $R(kt)$, when approaching the transition point from the glass side and being outside the microscopic time region, can be written as

$$R(kt) = R(k, t = \infty) + \epsilon^{1/2}a(k)\,f(\epsilon^{\alpha}t) \qquad \textbf{(9)}$$

where $f(\tau)$ satisfies the universal equation

$$-f(\zeta) + \frac{1}{\lambda}\zeta f^2(\zeta) + i\int_0^{\infty} d\tau\, e^{i\zeta\tau}\, f^2(\tau) = 0 \qquad \textbf{(10)}$$

Here, $f(\zeta)$ is the corresponding Laplace transform of $f(\tau)$ and ϵ represents the distance from the transition point, as when $\epsilon \propto (n - n_g)/n_g$, for instance. The function $a(k)$ and the parameter λ are each determined from a certain stability matrix, for which the largest eigenvalue and the corresponding eigenvector have to be calculated. In the simplified model in references 3 and 4 we have $\lambda = 1/2$, but the proper value depends on the form of $M(kt)$ in equation 5. Therefore, there is no universal value. The exponent α in equation 9 is determined from the limit of $f(\epsilon^{\alpha}t)$, for $t \rightarrow 0$, by demanding that the ϵ dependence should then disappear.

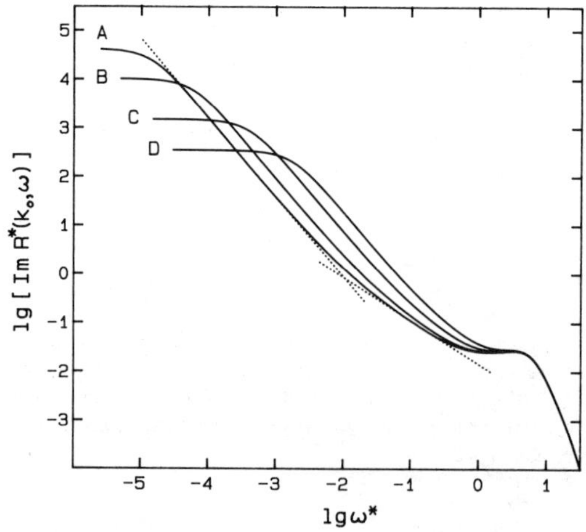

FIGURE 4. Plot of log $[ImR^*(k_0, \omega)]$ versus log ω^* (from equations 5–7) in reduced Lennard–Jones units for $k_0^* = 7.05$ and $T^* = 0.60$ on the liquid side for $n^* = 0.9599$ (A), $n^* = 0.9595$ (B), $n^* = 0.9580$ (C), and $n^* = 0.9550$ (D). The dotted lines show the two distinctive linear regions: the one on the left has a slope of -1.63; the one on the right, a slope of -0.9. The calculated glass transition point corresponds to the point where $n_g^* = 0.9602$. The reduced time unit $\tau_0 = (m\sigma^2/\epsilon)^{1/2} = 2.156 \times 10^{-12}$ sec for Lennard–Jones argon.

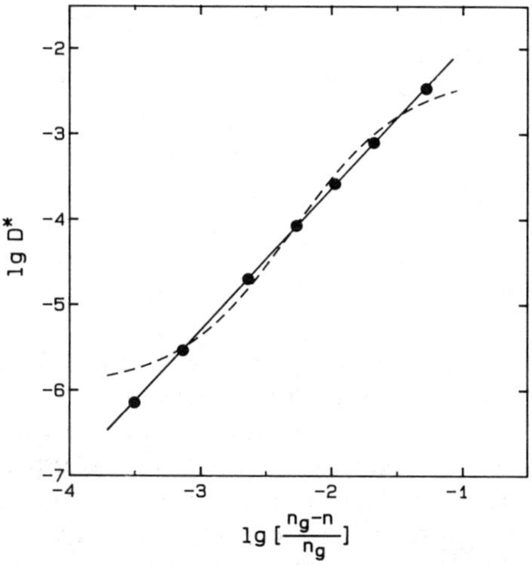

FIGURE 5. A plot of log D^* versus log $[(n_g - n)/n_g)]$ for $T^* = 0.60$ and the calculated value $n_g^* = 0.9602$ for the transition point. The circles represent calculated values for the diffusivity D^* in the density range from 0.9100 to 0.9599; the straight line shows the power law behavior with an exponent 1.66; the broken line is a fit to a Vogel–Fulcher law: $D^* = A \exp[-B/(V^* - V_K^*)]$ where $(V_K^*)^{-1} = 0.965$. In the reduced Lennard–Jones units log $D^* = -1.5$ corresponds to the diffusivity at the triple point for argon.

On the liquid side the same scaling equation holds after replacing the very first term by $1/\zeta$, provided we do not consider values of τ that are too large. For $\tau \ll 1$ this change has no essential effect, but this is not true for $\tau > 1$, and the whole equation is invalid for asymptotically large τ. Here, macroscopic diffusion processes dominate the time evolution.

We have solved equation 5 numerically for a Lennard–Jones system, using the Weeks–Chandler–Andersen procedure for calculating $S(k)$. Any error in the latter quantity will mainly influence the prediction for the position of the glass transition.

FIGURE 4 shows the imaginary part of $R(k\omega)$ in a plot of log (ImR^*) versus log ω for k at the main peak in $S(k)$. A purely Lorentzian form for small ω in our plot would give a straight line with a slope of -2. We find, as we found using the simplified model (references 3 and 4), that when the transition point is approached from the liquid side a quasi-elastic double peak structure tends to develop. The exponents, however, are different. From curve A, which corresponds to $(n_g - n)/n_g = 0.0003$, we can, for $-4.5 < \log \omega^* < -3$, extract a straight line with a slope of -1.63. This is not consistent with an exponential decay in time. Instead we find considerably better agreement with a stretched exponential, $A \exp(-at^{\beta})$, with $\beta = 0.88$. For the simplified model the numerical solution gave a slope of -1.9, whereas in this case the exact asymptotic value is -2. The discrepancy shows that one has to come very close to the transition point before the proper asymptotic value is reached.

The other component of the double peak can be seen from the approximate straight line for $-1.5 < \log \omega^* < -0.5$ that has a slope of -0.9. (This slope should be compared to the corresponding slope that was found by using the simplified model: -0.605.) This component corresponds to a power law time dependence, and it arises from those slow motions that remain on the glass side. For log $\omega^* > 0$ we are in the microscopic time region, where phonon-like excitations dominate.

We obtain results very similar to those reported recently by Ullo and Yip[19] from computer simulations, provided we assume their glass transition point to be at $n\sigma^3 \approx 1.075$. This value corresponds quite closely to the minimum in their FIGURE 4.

FIGURE 5 shows the density dependence of the macroscopic self-diffusion constant when approaching the transition point. A logarithmic plot is used, and a straight line corresponds to a power law dependence. The value we found for the exponent ($\gamma = 1.66$) should be compared to the value found using the simplified model ($\gamma = 1.76$). Note that a Vogel–Fulcher curve has been fit over our values for diffusivity.

CONCLUDING REMARKS

When considering the present model one has to keep in mind that we have made the most obvious and simple mode-coupling approximation of $M_{11}(kt)$ in equation 6. There are many improvements that could be made. Before making any, however, one would do well to compare the basic dynamical predictions and accurate simulation results. Obviously, difficult computer experiments would be required for these urgently needed comparisons. In the near future we will present a more extensive report on the outcome of our numerical calculations. Microemulsion experiments, such as those performed by Chen and Huang,[20] are important in this context, but the extent to which such experiments apply to systems as simple as the one considered in this article remains uncertain.

Even though there are striking qualitative similarities between the behavior of large classes of real glass-forming systems and the behavior of computer glasses, we do not know at present how much we can stress such similarities. For instance, experiments show that viscosity has an essentially Arrhenius temperature dependence when passing into the glass region. This indicates that the diffusion process eventually becomes an activated one. We can only speculate whether this would also be true for computer glasses. Our present model does not account for activated diffusion. Therefore, it does not shed any light on the experimental observation that certain liquids show essentially Arrhenius temperature dependence throughout and that others do not.[21] It seems likely to us, however, that an improved model (that is, one including correlations of a higher order) would, for low temperatures, introduce local potential barriers and, as a consequence of this, activated diffusion. An extension of the present model to simple multicomponent systems seems urgent, but it would lead to considerable computational difficulties.

REFERENCES

1. GOLDSTEIN, M. & R. SIMHA, EDS. 1976. Ann. N.Y. Acad. Sci. **279.**
2. PALMER, R. G., D. L. STEIN, E. ABRAHAM & P. W. ANDERSON. 1984. Phys. Rev. Lett. **53:** 958.
3. LEUTHEUSSER, E. 1984. Phys. Rev. A **29:** 2765.
4. BENGTZELIUS, U., W. GÖTZE & A. SJÖLANDER. 1984. J. Phys. C **17:** 5915.
5. FORSTER, D. 1975. Hydrodynamics, Fluctuations, Broken Symmetry, and Correlation Functions. Benjamin. New York, NY.
6. KAWASAKI, K. 1966. Phys. Rev. **150:** 291.
7. DAS, S. P., G. F. MAZENKO, S. RAMASWAMY & J. J. TONER. 1985. Phys. Rev. Lett. **54:** 118.
8. KIRKPATRICK, T. R. 1985. Phys. Rev. A **31:** 939.
9. GÖTZE, W. 1985. Z. Phys. B **60:** 195.

10. DERAEDT, H. & W. GÖTZE. 1985. J. Phys. C. In press.
11. BENGTZELIUS, U. & L. SJÖGREN. 1986. J. Chem. Phys. **84:** 1744.
12. BENGTZELIUS, U. 1986. Phys. Rev. **33:** 3433.
13. SJÖGREN, L. 1985. Preprint.
14. MAZENKO, G. F. 1974. Phys. Rev. A **9:** 360
15. FORSTER, D. 1974. Phys. Rev. A **9:** 943.
16. SJÖGREN, L. 1980. Phys. Rev. A **22:** 2866.
17. WOODCOCK, L. V. 1977. J. Chem. Soc. Faraday Trans. 2 **73:** 11.
18. WOODCOCK, L. V. & C. A. ANGELL. 1981. Phys. Rev. Lett. **16:** 1129.
19. ULLO, J. J. & S. YIP. 1985. Phys. Rev. Lett. **14:** 1509.
20. CHEN, S. H. & J. S. HUANG. 1985. Phys. Rev. Lett. **55:** 1888.
21. ANGELL, C. A., J. H. R. CLARK & L. V. WOODCOCK 1981. Adv. Chem. Phys. **48:** 397.

DISCUSSION OF THE PAPER

G. H. FREDRICKSON (*AT&T Bell Laboratories, Murray Hill, NJ*): The mode-coupling theory appears to predict singular relaxation times at a temperature where viscosities of laboratory liquids are 10^2 P. Furthermore, we normally associate the "glass transition region" with viscosities greater than 10^8 P, so how can such theories be expected to describe structural relaxation in viscous fluids or indeed the glass transition?

SJÖLANDER: We have at present quantitative predictions only for hard-sphere and Lennard–Jones systems, and we have not found any serious contradictions to available computer simulation data. The crucial tests on the dynamics, however, are still lacking.

Our predicted values for the viscosity seem to agree quite well with simulation data up to 10 P or so, and we have not found any accurate simulation data beyond that. You have to keep in mind that 10 P is roughly three orders of magnitude larger than what you find in the ordinary liquid region. Our theory predicts a rapid increase, when compressing the system further, and this implies a dramatic slowing down of the structural relaxation. When varying the temperature, however, this occurs over quite a large temperature range. We predict a weaker singularity than that obtained from the Vogel–Fulcher law.

Our approximations imply that certain thermal fluctuations are ignored, particularly those involving activated processes. This is a reason why we predict diffusivity becomes exactly zero and viscosity becomes exactly infinite at a certain point, and why we never approach Arrhenius behavior at the lowest temperature. In our present theory the diffusion mechanism is of a pushing type, where the surrounding atoms are pushed aside before the primary atom can move forward. There is no potential barrier to overcome. I believe, however, that if this process leads to a sufficiently small diffusion rate, some kind of hopping diffusion will take over and give an essentially Arrhenius temperature dependence.

An improved version of the theory should incorporate activated processes. Keeping in mind what I have said, I am not worried about the point you brought up. Of course, I cannot prove anything rigorously, and our basic predictions concerning the dynamics have to be tested against computer simulations. I hope this will be possible in the near future.

U. MOHANTY (*Boston College, Boston, MA*): 1) Using the Martin–Siggia–Rose formalism, Siggia recently showed (preprint; see also *Phys. Rev. A* **32:** 3135, 1985), starting from a hydrodynamic model, that at the glass transition one gets a nonlinear oscillator forced by a Gaussian white noise. This equation does not show any nonergodic behavior. Siggia's results are in variance with your work. Why is this the case? Which model describes the physical situation accurately?

2) In two dimensions, mode-coupling results lead to a "divergence" of certain transport coefficients. Therefore it is important to understand whether your theory has the ability to describe glassy behavior in two dimensions. Have you carried out such a calculation?

SJÖLANDER: 1) Let me first answer your question concerning Siggia, I must confess that I am very puzzled about his arguments, and I am tempted to agree with Mazenko (*Phys. Rev. A* **32:** 3139, 1985) that Siggia has missed Leutheusser's main message. In our treatment—and this contains essentially the same physical idea as that described by Leutheusser—it is the strong nonlinear coupling between density fluctuations of wavenumbers within the main peak of the static structure factor that triggers the transition. These short-wavelength fluctuations reflect the relaxation of the local structure, and they slow down dramatically at the transition point. The other fluctuations, however, are slaving (to use a term often mentioned by Haken), and they slow down as well. If one is very clever, one should, in principle, be able to renormalize away all short-wavelength fluctuations and obtain a proper equation for the long-wavelength hydrodynamic modes, as one does in connection with an ordinary second-order transition. In doing this, however, one has to be careful that the main physics concerning the short-wavelength fluctuations is not lost. I have no idea how it should be done in practice, and I am not surprised that those who considered only long-wavelength fluctuations missed the transition.

Another point is that if one does everything exactly, including also activated diffusion, one should not expect the glass transition point to become a mathematical singular point with infinite viscosity. In this sense the transition is an artifact of our approximation, but I think we have done a legitimate simplification at this stage.

2) In answering your second question I can only speculate. We have only considered the three-dimensional case so far. It appears to me, however, that the single-particle mean-square displacement becomes infinite at all temperatures and that this would imply no glass transition of the kind we have in three dimensions. The reason for this would be the same as that for a two-dimensional lattice, where the long-wavelength phonon fluctuations lead to infinite mean-square displacements. I am not aware of any results that show a glassy state can be formed by a strictly two-dimensional fluid.

S. R. NAGEL (*University of Chicago, Chicago, IL*): In your early papers you find that the relaxation times should scale with temperature as a power law with an exponent $\alpha \sim 1.8$. How universal is this exponent, and can it be as large as 12.5 to explain our ultrasonic and specific heat data on glycerol?

SJÖLANDER: When in our first paper on glass transitions we considered a simplified model, in which the true static structure factor was replaced by a δ function at its main peak, we thought of the model only as an illustrating example. I did not even suspect there was a scaling law held for the full model. Nevertheless, as already mentioned in my paper, Götze showed a scaling law that holds under very general conditions. The exponents, however, are nonuniversal. This agrees with the numerical solution of our equation. Inserting the true static structure factor has a surprisingly large effect on some of the exponents.

Considering a two-component system in which three different structure factors are involved, I would not even be able to give a rough estimate of the exponents.

There is no conceptual difficulty to extend the theory to multicomponent systems. The computational work to extract quantitative results would, however, increase considerably.

M. WEINBERG (*Jet Propulsion Laboratory, Pasadena, CA*): On what basis did you choose the form of the memory kernel in order to include the physics of glass formation? For example, how do you know that density modes of higher orders are not important when one enters the glass transition region?

SJÖLANDER: To answer your question fully would require a review of our experience using the mode-coupling approach for the dynamics of liquids, work done by many people for more than a decade. I can state here only that the form has worked remarkably well and that we now understand quite well how various physical effects enter in the theory. This gave us quite a lot of confidence in applying the same approach to supercooled liquids and to the glass transition process.

The two-mode approximation takes into account in an approximate way the motion of a single atom: how this atom pushes its surrounding atoms and how, at the same time, the atom is pushed back. As I stated in my paper, the two-mode approximation leads to a strong reduction of the diffusion constant on increasing the density or lowering the temperature, an important effect already in the ordinary liquid region. It leads to strong non-Markoffian effects. We had no guarantee that the same approximation would provide a mechanism for a glass transition, and at present we do not know whether our theory will stand the crucial tests. Certainly, in order to describe the local structure more properly and in order to bring in activated diffusion, density modes of higher orders have to be included. Such an extension may actually require a reformulation of the theory, a reformulation that would keep the physics much the same.

G. SCHERER (*DuPont, Wilmington, DE*): You showed a comparison between the location of T_g predicted theoretically and that found from molecular dynamics simulations, and the correlation was fairly good. However, the theory describes an ideal T_g, whereas the molecular dynamics results find T_g by quenching at 10^{12} K/sec, so the latter predicts T_g where the viscosity is $\sim$100 P. Therefore your theory predicts the ideal glass transition at viscosities 12 orders of magnitude lower than values observed in the laboratory.

SJÖLANDER: Our theory predicts an ideal glass transition point precisely where the viscosity becomes infinite. The transition temperature predicted by molecular dynamics is too high because the cooling rate is very rapid; the transition temperature we would estimate would be too high because we do not account for certain kinds of fluctuations. My answer to Dr. Frederickson's question applies here. I do not think a comparison between our value for the transition point and the molecular dynamics data is a sensitive test of the theory. I presented the comparisons to show that we seem to get reasonable agreement and that we get correct trends. If the values had been off by an order of magnitude, I would have been very skeptical of the whole approach. The crucial test of the theory is the dynamics close to the transition point.

C. A. ANGELL (*Purdue University, West Lafayette, IN*): In one of the figures you used as a basis for judging the relevance of the model, you compared transition temperatures predicted by the model for different potentials with transition temperatures determined by computer simulation studies on those potentials. I was wondering how you chose the transition temperatures from the computer simulations. Was it from thermodynamic criteria, namely, intersections of extensive properties like volume or enthalpy, or was it by dynamic criteria, namely, the vanishing of the diffusion coefficient? If it was the former, as I think it was, I wonder if this is the appropriate choice considering the model is predicting a kinetic transition. Would the use of the diffusivity criterion possibly remove some of the discrepancies?

SJÖLANDER: The molecular dynamics transition points were taken from the thermodynamic data. These points are somewhat ill-defined and hence introduce some uncertainties. Whether they should be compared to our kinetic transition points is questionable. This uncertainty emphasizes what I said in my answer to Dr. Scherer.

Actually, when we compared our results with the molecular dynamics data recently presented by Yip (*Phys. Rev. Lett.* **54:** 1509, 1985), we found reasonable consistency only when we placed the transition point. Our placement coincided with the minimum they found in the diffusivity, but was not consistent with their thermodynamic data.

The Kauzmann Paradox, Metastable Liquids, and Ideal Glasses

A Summary

C. A. ANGELL, D. R. MACFARLANE,[a] AND M. OGUNI[b]

Department of Chemistry
Purdue University
West Lafayette, Indiana 47907

This section could by no means be considered complete without some reference to the problem pointed out by Walter Kauzmann in his 1948 review[1] and some current comments on its possible resolution.

Kauzmann, as is well known now, drew attention to what seemed an almost general feature of the thermodynamic condition of a liquid as it cools toward its glass transition temperature, T_g. This is that the difference between the entropy of the liquid and that of the stable crystalline phase (calculated from the entropy of fusion and the excess of the liquid over the crystalline-state heat capacity) tends to vanish not far (20–50°C) below T_g, at which equilibrium (ergodicity) is lost (broken).[2] Kauzmann in his original discussion suggested that the improbable decrease of the liquid entropy below that of the corresponding ordered phase could be avoided if the nucleation probability rose to values approaching unity, that is, if the barrier to nucleation disappeared, in the region between the glass transition and the isoentropy temperature (which many in the field have since designated as the Kauzmann temperature, T_K). The alternative to Kauzmann's suggestion is that the liquid, on slow cooling, continues to lose entropy until, as it approaches the crystal value, an equilibrium transition of a higher order type occurs to yield a disordered phase of vanishingly small configurational entropy. Like a crystal polymorph this equilibrium phase could in principle conform to the third law of thermodynamics.

Which of these two possibilities is to be preferred could be judged from a careful study of the temperature dependence of crystal nucleation rates and of internal liquid structural equilibration rates. If it can be shown that the crystallization rate decreases significantly less quickly with temperature than the relaxation rate, then Kauzmann's view would be supported.

Examinations of these rates have been provided both by MacFarlane[3] and by Zelinsky *et al.*[4] We reproduce the argument of MacFarlane here to demonstrate that Kauzmann's proposition is unsupportable for at least one case, lithium disilicate, for which there are good data on the induction times for nucleation as well as data appropriate to determine structure times. We then extend the argument to suggest that this finding will prove to be general for strongly glass-forming materials, in which case we must conclude that strongly glass-forming liquids are relaxing toward a unique

[a]Present address: Department of Chemistry, Monash University, Clayton, Victoria, Australia.
[b]Present address: Department of Chemistry, Osaka University, Toyonaka, Osaka, Japan.

configurational state represented by a deep minimum in the $3N + 1$ dimensional potential energy surface for the system conceptualized by Goldstein.[5] This minimum is isolated kinetically from the one or few points of lower potential energy represented by specific crystalline structures, although on a truly infinite time scale the latter must in principle be accessible.

The argument proceeds through a discussion of classical nucleation theory, from which kinetic crystallization rates are obtained, which may then be compared with internal structure relaxation rates assessed from viscosity or bulk relaxation behavior.

Classical nucleation theory is normally formulated[6–8] for the case of a vapor condensing to form a liquid. Changes of composition and of volume are initially ignored in the standard treatments. The major assumption is that macroscopic thermodynamics can be applied to submicroscopic particles. Thus, the liquid is assumed to have the same free energy in the droplet as it has in the bulk liquid, and the droplet is presumed to be separated from the vapor phase by a sharp boundary of surface free energy, σ. This surface free energy is normally assumed to be independent of the radius of the droplet (an assumption that can be removed) and hence equal to σ_∞, the surface energy of a plane interface. Surprisingly enough, a modern density functional treatment of the problem by Oxtoby and co-workers[9] shows that the classical treatment works very well, so we continue with its use in this discussion.

If the free energy of formation of a droplet of size r is calculated from the bulk and surface free energy components, it is found that this energy passes through a maximum at some critical radius, r_c, given by

$$r_c = 2\sigma/\Delta G_v \quad (1)$$

where ΔG_v is the bulk free energy change of vaporization per unit volume. Thus there exists a thermodynamic barrier to nuclei formation; that is, the order fluctuations that produce the nucleus are not stable until $r > r_c$. Since $\Delta G_v \rightarrow 0$ as $T \rightarrow T_b$, the boiling point, we find that $r_c \rightarrow \infty$, and order fluctuation of sufficient magnitude thus becomes increasingly improbable.

The quantitative expression for the nucleation rate is obtained by assuming a Boltzmann distribution of clusters of size less than r_c and considering the flux of clusters reaching r_c to be the result of a series of monomolecular additions (condensations) to the droplet. In the case of the liquid-to-solid phase transition, this process will require a diffusional hop across the interface. Turnbull and Fisher[10] have applied absolute reaction rate theory to this problem to produce an expression for the volume nucleation rate, I

$$I = I_0 \exp(-\Delta G_a/kT) \exp(-16\sigma^3\pi^2/3(\Delta G_v)^2kT) \quad (2)$$

where I_0 is a collection of constants of magnitude on the order of 10^{36} cm^{-3} · sec^{-1} and ΔG_a is the activation energy for the jump across the interface and is presumed to be similar to that for diffusion in the liquid. It should be noted that the solid-liquid surface free energy is not experimentally measurable and thus enters this equation as an adjustable parameter. Measured nucleation rates suggest values for σ in the range 0.02–0.25 J · m^{-2}.

The Boltzmann distribution of the pre-critical clusters that is assumed in the derivation of the above equation is presumed to have already been established. In many experiments, however, the assembly is suddenly changed from a stable to a metastable condition, and the nucleation rate is then a function of time until the quasi-steady-state distribution of clusters is attained, the old distribution being much less extensive in size and frequency because of the greater energy of cluster formation at higher temperatures. This effect will be unimportant if the transient is of short duration compared with the period of observation. In condensed phases the existence of an activation barrier to the addition of atoms to the particle may mean that the distribution of nuclei is only established slowly and that a transient nucleation rate may exist for a significant amount of time. Theoretically, transient nucleation can be handled in the same way as steady state nucleation, given an assumption of the complete nonexistence of clusters at $t = 0$. The equations that the theory yields cannot be solved analytically, and several approximate solutions have been proposed. The most common of these is that of Zeldovich,[11] which gives

$$I(t) = I_s \exp(-\tau/t) \tag{3}$$

where I_s is the steady state nucleation rate and τ is a characteristic time of the transient referred to as the "induction time." This equation is valid only for small t and does not provide an accurate description of the way $I(t) \rightarrow I_s$ as t becomes larger. A more accurate solution has recently been derived by Kashchiev[12] of the form

$$I = I_s[1 + 2\Sigma(-1)^i \exp(-i^2t/\tau)] \tag{4}$$

which is integrable to yield the total number of nuclei as a function of time (an experimentally more accessible quantity)

$$N(t) = I_s\tau(t/\tau - \pi^2/6 - 2\Sigma(-1)^i \exp(-i^2t/\tau)/i^2) \tag{5}$$

For $t > 5\tau$ this reduces to the linear function

$$N(t) = I_s(t - \pi^2\tau/6) \tag{6}$$

For t as small as 0.8τ, neglecting all but the first term in the series we get

$$N(t) = I_s(t - \pi^2\tau/6 + 2\tau \exp(-t/\tau)) \tag{7}$$

which has been found by one of us[13] to provide a good approximation to the full expression. Further use of these equations will be made later.

The induction time, τ, is given theoretically by Kashchiev[12] as

$$\tau = 8kT/\pi^2 A_c Z \beta_c \tag{8}$$

where $\beta_c = -(\partial^2\Delta G/\partial n^2)n = n_c$, ΔG being the Gibbs free energy required to form a cluster of n formula units (the subscript c refers to the critical nucleus); A_c is the surface area of the critical nucleus; and Z is the number of formula units that join the critical nucleus per unit area per unit time and can be estimated from

$$Z = \nu/\lambda^2 = (kT/h\lambda^2) \exp(-\Delta G_a/kT) \tag{9}$$

where ν is the frequency of the molecular vibrations, ΔG_a is the activation energy for self-diffusion in the liquid phase, λ is a quantity of the order of the atomic dimensions (jump distance), and h is Planck's constant.

An alternative approach, used by James,[13] is to relate Z to the viscosity of the liquid through the diffusion coefficient and the Stokes–Einstein equation to yield

$$\tau = 48\sigma\eta/\pi\lambda\Delta G_v^2 \tag{10}$$

where the resultant value of τ is only approximate because of the approximations involved in the use of the Stokes–Einstein equation.

Induction times for $Li_2O\cdot 2SiO_2$ have been calculated from the nucleation rates measured by James[13] by extrapolation of the linear portion of a plot of the number of nuclei versus time back to $N(t) = 0$, the intercept giving the value of τ. These induction time measurements are probably the most extensive and accurate that have yet been made in condensed systems and seem to imply a refutation of the crystallization resolution of the Kauzmann paradox. That it is probably general for most strongly glass-forming liquids we shall show below.

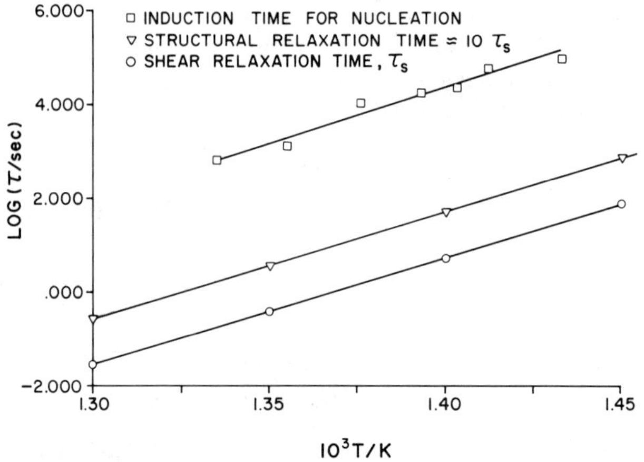

FIGURE 1. Arrhenius plots comparing crystal nucleation induction times with liquid structural equilibration times for $Li_2O\cdot 2SiO_2$. The large separation implies that the liquid always equilibrates before the nucleation rate becomes non-negligible, $I < 10^{-250}$ $cm^{-3}\cdot sec^{-1}$ (see text).

The induction time data taken from James's paper are plotted as $\log\tau$ versus $1/T$ in FIGURE 1. The Arrhenius form of the data trend is hardly surprising in view of the equations given above for the temperature dependence of τ. These show that this temperature dependence will be largely controlled by the transport term Z, which, by analogy with the other transport properties, is expected to show Arrhenius behavior, at least in the region of T_g. In fact, the viscosity of this material has been determined by Matusita and Tashiro[14] and is found to be Arrhenius in the temperature region just above T_g (451° C), obeying the equation

$$\eta = (6.52 \times 10^{-21}) \exp(5.25 \times 10^4/T) \tag{11}$$

where η is expressed in poise.

Since the structural relaxation time of a liquid is given by

$$\tau_s = \eta/G_\infty \tag{12}$$

where G_∞ is the infinite frequency shear modulus, then the structural equilibration time of the liquid after a sudden quench (found to be $\sim 10\tau_s$ for oxide glasses[15]) can be calculated from the viscosity given only a knowledge of G_∞; unfortunately, this does not appear to have been measured for this system but is found to be of the order of 1×10^{12} $N \cdot m^{-2}$ for many silicate systems near T_g and is only slightly temperature dependent in comparison with the large (experimental) temperature dependence of viscosity.[16] Thus we can obtain at least an order of magnitude estimate of the structural equilibration time in this system. These estimations are also indicated in FIGURE 1. The comparison between these and the induction times indicates that the latter are longer by three to four orders of magnitude and that the two times show no sign of converging as the lines are extended (by what appear to be reasonable Arrhenius-behavior extrapolations) into the sub-T_g region. The apparent similarity in the activation energies (calculable from the slopes of the Arrhenius plots) illustrates the similarities between the two processes involved. Both are the structural response to an initial nonequilibrium situation. In the structural relaxation case the relaxation time represents the time for establishment of the local structure of the liquid after some perturbation, whereas the induction time represents the time to establish the "equilibrium" distribution of subcritical clusters and hence is a diffusive process governed by the same activation energy as τ_s. In fact, as is usually the case, τ_s is an average structural relaxation time, there being a whole spectrum of relaxation times in the liquid governing related, but different, processes. Thus the induction process can be viewed as simply one of the elements of this spectrum of relaxation times (one which involves a larger number of atoms or molecules) and thus retains the same activation energy and is longer because of a larger pre-exponential factor.

The induction time is in fact so much longer than the structural equilibration time of the liquid that one may conclude, judging from the Zeldovich equation with $t = 10^{-3}\tau$, that is, from

$$I(t = 10\tau_s) = I_s \exp(-10^3) \approx 10^{-400}\ \text{cm}^{-3} \cdot \text{sec}^{-1} \tag{13}$$

(since I_s, from classical nucleation theory, can have a value of 10^{36} $cm^{-3} \cdot sec^{-1}$ at most, and then only under very fluid conditions: experimentally, in this system the I_s values are found to be 32 orders of magnitude below the limiting value). On this basis the nucleation rate must always be negligible at the point where the liquid has *just* reached structural equilibrium. Thus we are forced to conclude that even as T approaches T_K the liquid state will always be more accessible to the relaxing glass than is the crystalline state. This conclusion assumes, of course, that classical homogeneous nucleation theory continues to be applicable at these very high supercoolings where the critical nucleus is extremely small (<10 Å). It seems clear, however, that, at least within the bounds of classical theory and to the extent that it is applicable as $T \rightarrow T_K$, there is no escape from the impending entropy catastrophe by the route of homogenous nucleation of the crystalline state.

Any case for the existence of an ideal glass state made using data on congruently melting compounds (so as to estimate T_K, the thermodynamic limiting T_g) can be greatly strengthened by noting that the crystallization probability, but *not* T_g (nor presumably the *thermodynamic* T_g), may be greatly affected by adding a second component. This depresses the freezing temperature, hence also the nucleation rate near T_g, while leaving T_g unchanged if a like second component (for example, an isomer) is used. Although this introduces a small unknown mixing quantity into the entropy assessment, it is not the sort of temperature-dependent entropy that is involved in the Kauzmann paradox. Therefore, it can be argued that the time-scale gap between configurational relaxation and relaxation to the crystalline state (called τ_{in} versus τ_{out}

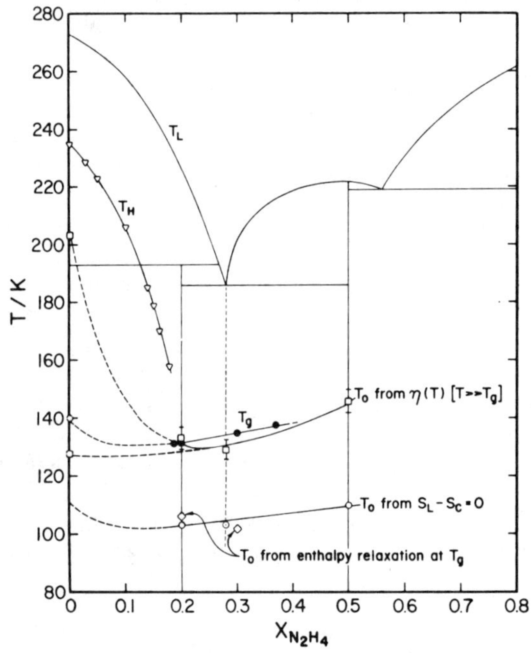

FIGURE 2. Phase diagram for the $N_2H_4 + H_2O$ system showing congruently and incongruently melting compounds: $N_2H_4 \cdot H_2O$ and $N_2H_4 \cdot 4H_2O$. Glass transition temperature (●) and "ideal" glass transition temperatures (Kauzmann temperatures) assessed from thermodynamic data for the two compounds (○) are marked. An upper bound T_K for the eutectic composition (○) is consistent with the compound values. Also shown are Vogel–Tammann–Fulcher T_0 values assessed from viscosity (high temperatures) (□) and enthalpy relaxation (low temperatures) (◇). The points for homogenous nucleation temperatures for ice crystallization (▽) were determined from DSC measurements on emulsion samples.[18] The data are discussed in reference 17.

by some authors) can be made many orders of magnitude larger than that obtained by the above argument based on $Li_2O \cdot 2SiO_2$ data.

Some previously unpublished measurements from this laboratory are relevant to this point. Heat capacity measurements and entropies of fusion, and entropies of peritectic decomposition plus dissolution (for peritectic melting) have been made[17] so that T_K could be assessed for two-compound compositions in the $N_2H_4 + H_2O$ system (see the phase diagram in FIG. 2). Measurement of the entropy of eutectic melting for the two compounds, ignoring the entropy of mixing of the two component compounds at the eutectic temperature, and liquid solution heat capacity data allow one to calculate T_K at the eutectic composition as if it were a compound.[19] This value has a small uncertainty associated with any temperature dependence of the entropy of mixing. The results of these three T_K measurements are shown in the phase diagram (FIG. 2). We believe their simple relationship argues strongly for a continuity of thermodynamic glass transitions across a binary system. These may be either of the second-order type (anticipated by Adam and Gibbs[20]) or of the continuous (infinite order though relatively sharp) type (suggested by Angell and Rao[21] and Fredrickson[22]). If a line of thermodynamic glass transitions exists, then the observation, by some long-time C_p measurement, of the onset of a thermodynamic transition free from crystallization interferences, would be best sought near the eutectic temperature where the thermodynamic drive to nucleation would be weakest.

REFERENCES

1. KAUZMANN, W. 1948. Chem. Rev. **43:** 219.
2. PALMER, R. G. & D. STEIN. 1984. *In* Relaxation in Complex Systems. K. Ngai & G. B. Wright, Eds. National Technical Information Service, U.S. Department of Commerce. Washington, DC.
3. MACFARLANE, D. R. 1982. Ph.D. diss. Purdue University. West Lafayette, IN.
4. ZELINSKI, D. J. J., H. YINNON & D. R. UHLMANN. 1983. Glastech. Ber. **56K:** 822.
5. GOLDSTEIN, M. 1969. J. Chem. Phys. **51:** 3728.
6. GIBBS, J. W. 1948. Collected Works. Vol. 1. Yale University Press. New Haven, CT.
7. FRENKEL, J. 1964. Kinetic Theory of Liquids. Oxford University Press. New York, NY.
8. LOTHE, J. & G. M. Pound. 1962. J. Chem. Phys. **36:** 280.
9. OXTOBY, D. W. 1986. Ann. N.Y. Acad. Sci. This volume.
10. TURNBULL, D. & J. C. FISHER. 1949. J. Chem. Phys. **17:** 71.
11. ZELDOVICH, J. 1952. J. Exp. Theor. Phys. (USSR) **12:** 525.
12. KASHCHIEV, D. Surf. Sci. **14:** 209.
13. JAMES, P. F. 1974. Phys. Chem. Glasses **15:** 95.
14. MATUSITA, K. & M. Tashiro. 1973. J. Non-Cryst. Solids **11:** 471.
15. MOYNIHAN, C. T., P. B. MACEDO, C. J. MONTROSE, P. K. GUPTA, M. A. DEBOLT, J. F. DILL, B. E. DOM, P. W. DRAKE, A. J. EASTEAL, P. B. ELTERMAN, R. P. MOELLER, H. SASABE & J. A. WILDER. 1976. Ann. N.Y. Acad. Sci. **279:** 15.
16. AMBRUS, J. H., C. T. MOYNIHAN & P. B. MACEDO. 1972. J. Electrochem. Soc. **119:** 192.
17. ANGELL, C. A. & M. OGUNI. J. Phys. Chem. Submitted for publication.
18. OGUNI, M. & C. A. ANGELL. 1980. J. Chem. Phys. **73:** 1948.
19. ANGELL, C. A. & D. R. MACFARLANE. Submitted for publication.
20. ADAM, G. & J. H. GIBBS. 1965. J. Chem. Phys. **43:** 139.
21. ANGELL, C. A. & K. J. RAO. 1972. J. Chem. Phys. **57:** 470.
22. FREDRICKSON, G. H. 1986. Ann. N.Y. Acad. Sci. This volume.

Thermodynamic Aspects of Glassy Crystals

Approaching the Equilibrium Glass[a]

H. SUGA

Department of Chemistry
and
Chemical Thermodynamics Laboratory
Osaka University
Osaka, Japan

INTRODUCTION

Liquids, which are the most common representative of disordered systems, can solidify in two ways: discontinuously to a crystalline solid at the freezing point or continuously to a noncrystalline solid[1] through a glass transition temperature.[2] In the latter case, prolongation of the relaxation time for molecular reorganization in the liquid hinders the system in reaching the amorphous ground state. In both cases, the entropy associated with disorder in the liquid can be removed—if not completely—to some extent. In the first process, the resulting crystal usually has zero configurational entropy and hence obeys the third law of thermodynamics; in the second process, a part of the configurational entropy remains almost constant below the glass transition temperature, thus giving rise to a residual entropy. The glass transition and the residual entropy have been observed to exist conjugationally in glassy materials and have been considered to be a reflection of the ergodic-to-nonergodic transition in the system.

This paper will demonstrate how the concept of the glassy state may be extended to any disordered system possessing some frozen-in degrees of freedom of the constituent molecules in the system. For example, this concept will be extended to those glassy crystals in which the molecules have orientational degrees of freedom but have translational invariance with respect to the center of mass. Two kinds of glassy crystals will be described: one of these kinds is realized in a metastable phase; the other kind is realized in a stable crystalline phase. Experiments on hexagonal ice, which is a classic example of the glassy crystal, will be described with its historical background. The conceptual significance of the phase transition Ih $\rightarrow$ XI observed recently in ice crystals doped with alkali hydroxides will be discussed in relation to the ordering process that might be realized in the equilibrium cooling of a system that normally has frozen-in disorder.

[a]This work was supported by contribution 101 from the Chemical Thermodynamics Laboratory, Osaka University, Osaka, Japan.

GLASSY CRYSTALS

There are several crystals[3] in which the constituent molecules have orientational disorder at high temperatures. These are called orientationally disordered crystals. The disorder generally disappears discontinuously through a solid-solid phase transition to achieve a fully ordered state. There are certain crystals for which this does not happen and for which the remaining disorder is frozen at a temperature well above a hypothetical transition temperature. This situation appears when the reorientational

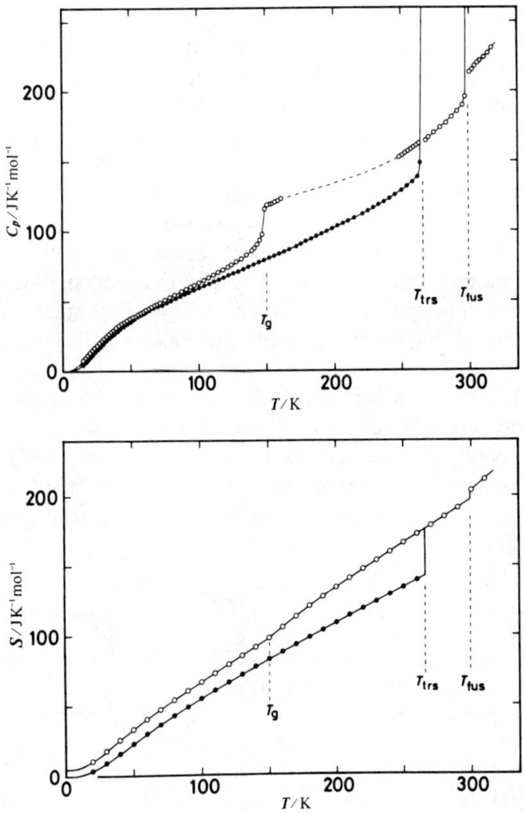

FIGURE 1. The molar heat capacity and entropy of cyclohexanol.

motion of molecules requires a high activation enthalpy compared to the relevant ordering energy. The paramount examples are carbon monoxide and hexagonal ice, for which the freezing occurs in a stable phase. The other case is again a freezing process[4] occurring in a metastable, undercooled high-temperature phase realized by bypassing the solid-solid phase transition. The process is analogous to that for the liquid case.

The first example of the second case is the cyclohexanol crystal. Calorimetric studies on the crystal are summarized in FIGURE 1. The liquid cyclohexanol[5] crystallizes into a face-centered cubic (fcc) lattice at the freezing temperature even though the crystal is composed of molecules with low symmetry. The highly disordered nature of the molecules in the cubic phase was inferred from the X-ray diffraction and NMR experiments. The cubic phase (crystal I) transforms into a low-temperature phase (crystal II) of unknown symmetry at the transition temperature. From the comparison of the calorimetric with the spectroscopic entropies determined in the ideal gas state, it is found that the low-temperature phase obeys the third law of thermodynamics.

The high-temperature phase can be undercooled easily and shows a heat capacity jump closely resembling that associated with the glass transition of a liquid. Loop calculation of the entropy along the two thermodynamic paths shows that the undercooled phase retains a finite amount of entropy at 0 K (FIG. 2). The dielectric constant[6] of cyclohexanol does not change much on freezing, indicating again the highly disordered nature of the molecular orientation. The temperature at which dielectric relaxation of the undercooled phase appears depends on the applied frequency, as in the case of undercooled polar liquids (FIG. 3). The glass transiton temperature corresponds roughly to the dielectric relaxation time of 10^3 sec, which is the typical duration of a single heat capacity measurement. All this experimental evidence clearly shows that the observed heat capacity anomaly is associated with the freezing of orientational freedom of the molecules in the crystalline lattice.

The nonequilibrium character of the glassy crystalline state arising from kinetic features was studied by an experimental technique called enthalpy relaxation.[7] Constant temperature of a calorimetric system maintained adiabatically is a condition of thermal equilibrium. This is always confirmed during the equilibration period by a heat capacity measurement of a crystal. Sometimes a spontaneous rise or fall of the calorimetric temperature is encountered (such a temperature change may be observed reproducibly in many glass-forming materials around the glass transition temperature). These temperature changes are due to the release or absorption of heat that

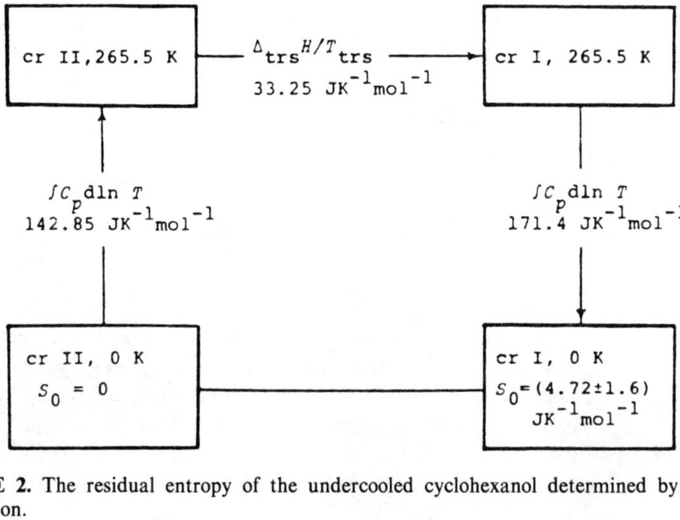

FIGURE 2. The residual entropy of the undercooled cyclohexanol determined by the loop calculation.

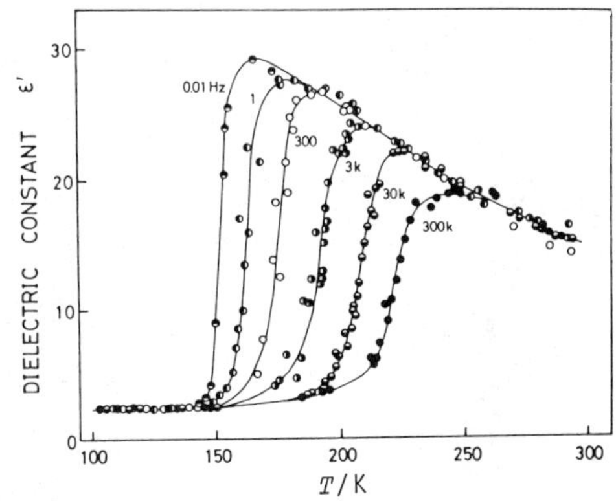

FIGURE 3. The dielectric constant of the cyclohexanol crystal.

arises during the relaxational process of the sample. At higher temperatures, the accuracy of the observation is limited because some time is required for the added energy, which causes the calorimetric temperature to rise, to become evenly distributed. At lower temperatures, the accuracy of the observation is limited by the natural temperature drift coming from incomplete adiabaticity. The temperature drift may be positive or negative depending on whether the actual enthalpy of the specimen lies above or below the equilibrium value (FIG. 4). The turning point from the exothermic to the endothermic ones will be taken operationally to be the glass transition temperature.

FIGURE 5 shows the enthalpy relaxation actually observed in a cyclohexanol crystal around 150 K, the temperature at which the heat capacity jump occurs. The dotted line ABC is an assumed equilibrium enthalpy obtained by a simple extrapolation. Relaxations were observed at 144, 141, 137, and 133 K. For example, the configurational enthalpy at 141 K changes from point I to E over 25 hr and then traces the curve EFG upon heating. The annealing process was carried out quasi-isothermally by introducing a small amount of He gas. The relaxation time τ was calculated, assuming an exponential approach of the enthalpy toward the equilibrium value during the annealing, with the following equation

$$\Delta H_c(t) = \Delta H_c(0) \exp(-t/\tau), \tag{1}$$

where ΔH_c is the excess configurational enthalpy at time t. The relaxation time increases rapidly as the temperature is lowered. Thus the undercooled cyclohexanol resembles ordinary glassy liquid in both thermodynamic and relaxation behaviors. One clear difference, of course, when compared to the conventional liquid system is the translational invariance of the molecular arrangement in the glassy state of the cyclohexanol crystal. We proposed the name "glassy crystal" for the crystalline solid having some frozen-in degrees of freedom. Thus the concept of the glass transition can be extended to any relaxation effect. (Such effects cause some processes in disordered systems to occur so slowly at low temperatures that thermodynamic equilibrium can never be reached during the length of a typical experiment.) When we published these findings in 1968, it did not always follow that people could easily imagine a thermodynamic state possessing the seemingly contradictory concepts for glasses and crystals.

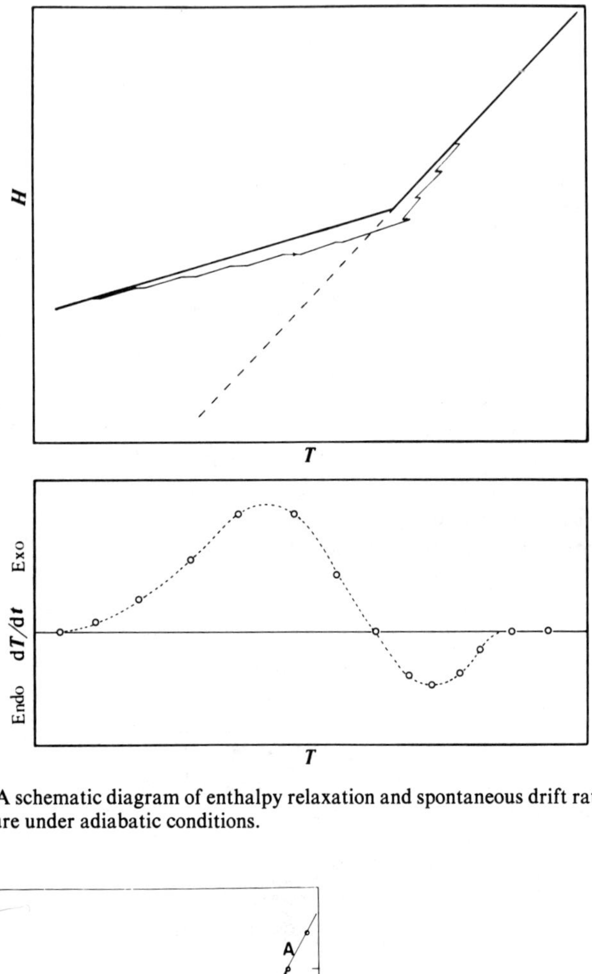

FIGURE 4. A schematic diagram of enthalpy relaxation and spontaneous drift rate of calorimetric temperature under adiabatic conditions.

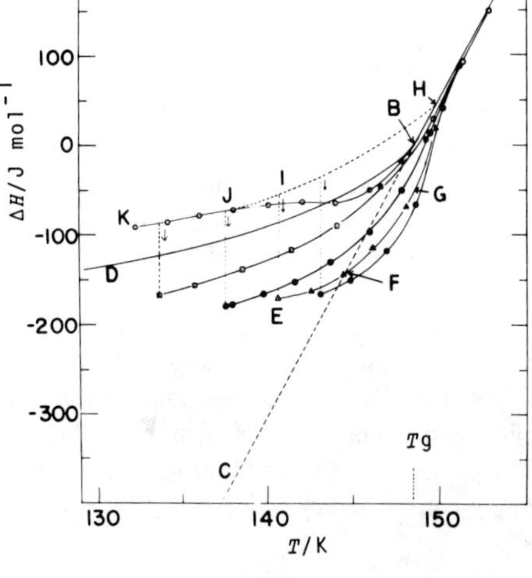

FIGURE 5. The enthalpy relaxation of the cyclohexanol crystal observed around 150 K.

The relaxational process was followed carefully as a function of time in the glassy crystalline state of 2,3-dimethylbutane.[8] Nonlinear behavior of the enthalpy relaxation was clearly observed (FIG. 6). If the linear enthalpy relaxation takes place according to equation 1, a simple differentiation shows that the plot would give a straight line. From FIGURE 6 we recognize immediately that the rate of enthalpy relaxation at the initial stage is much greater than the rate of the corresponding exponential decay having a roughly equal time constant. One possible explanation for this nonlinear behavior is to consider that the time constant is not only a function of temperature but also of the thermodynamic variables of the glassy state. This possibility was tested using the theory proposed by Adam and Gibbs.[9] Their theory may be described by the equation

$$-\dot{H}_c(t) = A \exp(B/TS_c)H_c(t) \tag{2}$$

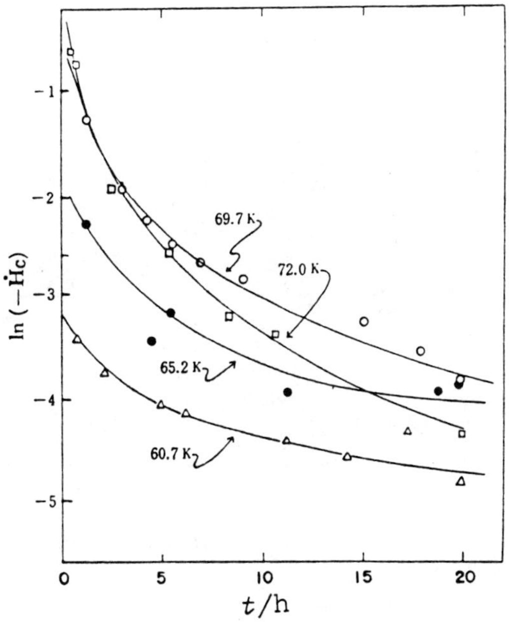

FIGURE 6. The rate of enthalpy change observed in 2,3-dimethylbutane crystal.

where A and B are constants, S_c is the configurational entropy, and

$$S_c(t) = S_c(0) - \int_0^t (\dot{H}_c/T)dt + \Delta S_{irr} \tag{3}$$

Here, $S_c(0)$ is equal to the residual entropy and ΔS_{irr} is the irreversible entropy production. As may be seen in FIGURE 7, the agreement between theory and experiment is roughly established. Another explanation is to consider a spectrum of relaxation times. For example, the nonlinear behavior could be analyzed using the Williams–Watts[10] equation.

$$\Delta H_c(t) = \Delta H_c(0) \exp[-(t/\tau)^{\beta}] \tag{4}$$

The extension of the concept of glassy state is validated by the observation of plural glass transition phenomena in a single compound with differently disordered phases. One example is shown in FIGURE 8 for ethyl alcohol,[11] for which the glassy liquid and glassy crystal states were realized by controlling the cooling rate. Since the first measurement of its heat capacity, this substance has been accepted as a typical glass-forming material. Our remeasurement shows that an extremely rapid cooling of the liquid (more than 50 K · min^{-1}) is necessary in order to realize the glassy liquid. If the cooling rate is moderate, say 2 K · min^{-1}, a metastable crystal II is formed first and then goes over to glassy crystal II through another glass transition temperature. It is interesting to note that the glass transitions of the liquid and crystal II take place at essentially the same temperature, indicating the important role of reorientational motion of molecules in determining the relaxation behavior in the vitrification process

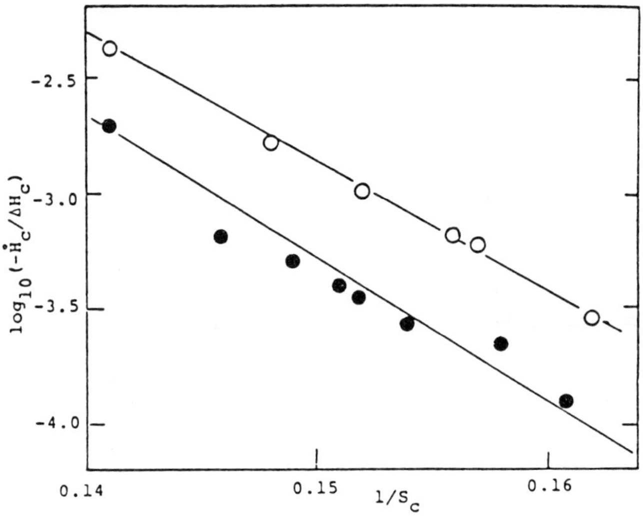

FIGURE 7. The plot of $\ln(-H_c/\Delta H_c)$ versus $1/S_c$ for the 2,3-dimethylbutane crystal. ○: 72.0 K; ●: 69.7 K.

of liquids. This was suggested already by Kauzmann from the analysis of the dielectric relaxation time of polar liquids.[2] Many examples of glassy crystals have been found in our laboratory. Seki and I reported our experimental findings in 1980.[7]

ICE PROBLEM

It is well known that many hydrate crystals obey the third law of thermodynamics, but some of them deviate from it. The residual entropy existing in sodium sulfate dodecahydrate[12] may be attributed to the randomness of proton arrangements in a four-membered hydrogen-bonded ring in the crystal. TABLE 1 gives examples of stable hydrogen-bonded crystals that exhibit glass transition phenomena. Kinetic parameters

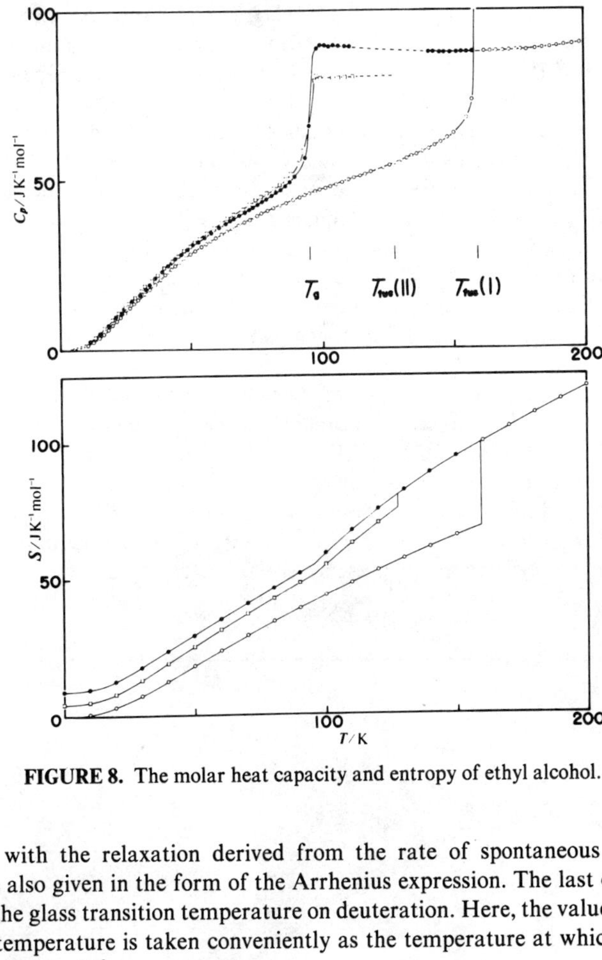

FIGURE 8. The molar heat capacity and entropy of ethyl alcohol.

associated with the relaxation derived from the rate of spontaneous temperature change are also given in the form of the Arrhenius expression. The last column is the change of the glass transition temperature on deuteration. Here, the value for the glass transition temperature is taken conveniently as the temperature at which the relaxation time becomes 10^3 sec.

TABLE 1. Kinetic Parameters Characterizing the Glass Transition of Some Hydrogen-bonded Crystals

Substance	T_g (K)	τ_0 (sec)	ΔH_a (kJ · mol^{-1})	$[T_g(D) - T_g(H)]$ (K)	Reference
$SnCl_2 \cdot 2H_2O$	159.1	8.3×10^{-15}	50 ± 2	4.6	13
$SnCL_2 \cdot 2D_2O$	163.2	3.2×10^{-14}	50 ± 2		
H_3BO_3	296.6	3.2×10^{-13}	88 ± 5	1.6	14
D_3BO_3	298.2	1.1×10^{-13}	91 ± 5		
Pinacol hexahydrate	~160				
	~140				15
Pinacol hexadeuterate	167.2	2.4×10^{-8}	34 ± 4	~8	
H_2O	113.7	1.2×10^{-7}	22 ± 4	18.9	16
D_2O	132.6	5.7×10^{-8}	26 ± 5		17

In most cases the origin of the relevant relaxation is expected to be related to the protonic disorder involved in hydrogen-bonded systems. Protons in an asymmetrical O—H · · · O hydrogen-bonded crystal are known to obey the ice rules originally proposed by Bernal and Fowler.[18] Such systems provide us with well-characterized disorder and are therefore particularly suitable for the study of ordering processes. The nature of disorder has been studied for many ices, but that for hexagonal ice Ih has been studied most extensively. Each water molecule in ice can be regarded as forming a tetrahedron with an oxygen atom at its center, with a proton at each of two of the four tetrahedral positions, and with a lone pair of electrons at each of the two remaining tetrahedral positions. The crystal arrangement of the oxygen atoms in ice Ih is the same as that of wurtzite. Thus the geometrical requirement imposed on proton donors and acceptors for the formation of hydrogen bonding is almost ideally realized in ice Ih.

The residual entropy of ice measured by Stout and Giauque[19] was explained by Pauling[20] in terms of the completely disordered pattern of the proton location, that is, in terms of the random orientation of the water molecules in the hexagonal lattice under the ice rules. The disordered nature of the proton configuration was confirmed later by a neutron diffraction experiment[21] and is known as a half-hydrogen statistical model. A large dielectric constant of ice (~100 at 270 K) requires dynamical rearrangement of the disorder. The presence of residual entropy, however, means there

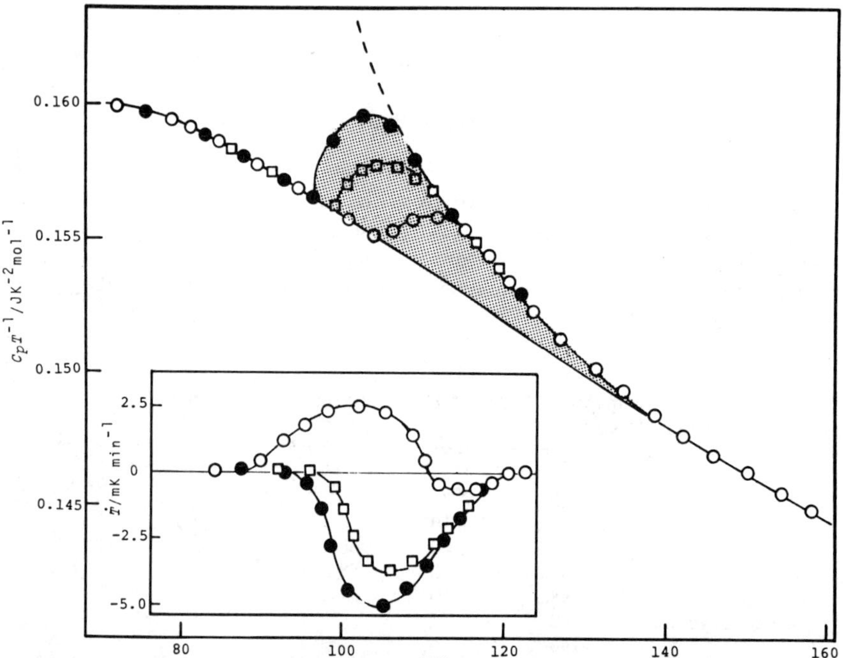

FIGURE 9. The encraty (C_p/T) and the spontaneous temperature drift of ice Ih. ○: not annealed; □: annealed at 94.4 K for 71 hr; ●: annealed at 89.4 K for 624 hr.

TABLE 2. Results of the Annealing Experiments for Ice Crystals

Ordinary Ice			Heavy Ice		
T_a (K)	t_a (hr)	τ (hr)	T_a (K)	t (hr)	τ (hr)
89.4	642	145	110.0	166	44.8
94.4	71	20.0	112.9	92	17.2
97.4	55	13.6	113.5	68	12.2
99.4	45	10.4	117.9	31	6.50
102.5	9	3.82	120.9	28	4.63
104.5	7	1.76	122.1	9	1.97
107.6	2	1.01			

NOTE: T_a: annealing temperature; t_a: annealing time; τ: relaxation time.

is a lack of thermal equilibrium below a certain temperature. In fact, a glass transition phenomenon was observed in ice crystal around 110 K.[16] The results are shown in FIGURE 9 along with the rate of the spontaneous temperature drift observed during initial 20 min of the equilibration period. The ordinate is taken as C_p/T in order to show the observed anomaly in an enlarged scale. The annealed specimens show an increased heat capacity anomaly as compared to those of rapidly cooled or less annealed specimens. The sign and magnitude of the spontaneous temperature drift rate depend on the thermal history. Such behavior is typical of glassy materials. The dotted line is the equilibrium heat capacity curve that we would expect to see if we could cool the specimen at an infinitely slow rate. Since the reorientational motion of water molecules is severely restricted by the ice rules, the motion is believed to be frozen-in[22] before the crystal reaches a hypothetical transition temperature at which a long-range order of proton configuration might be realized.

The spontaneous temperature rise during the annealing process was followed for a sufficiently long time to allow the time constant to be determined. After correcting for the temperature change that was due to the incomplete adiabatic condition, the relaxation time was determined by assuming an exponential approach toward an equilibrium value with a single relaxation time. The observation was continued for about one month at 89.4 K. TABLE 2 includes data for ordinary ice and for heavy ice, which also showed the glass transition phenomenon around 130 K.[17] The temperature at which a small heat capacity jump occurs in heavy ice corresponds roughly to a dielectric relaxation time of 10^3 sec.[23] Again, the glass transition and residual entropy were observed conjugationally. In the extended sense used in this paper, ice was not an exceptional example of a glassy material.

The experimental observation of a possible proton ordering was first reported by Dengel *et al.*,[24] who suggested that ice becomes ferroelectric at a temperature below 100 K. This report aroused heated discussion and stimulated a search for further evidence, especially from thermodynamic study of impure ices. Several scientists[25] tried this experiment, but none succeeded in finding a heat capacity anomaly that could be looked upon as being associated with a phase transition. Chamberlain and Fletcher[26] studied a thermostimulated depolarization current on HF-doped ice and suggested that the current peak at 100 K was due to a dielectric relaxation involving the reorientation of water molecules.

This conclusion is consistent with our calorimetric observation. We have all examined the same problem, but from different perspectives. Various properties as revealed by different techniques are consequences of the same reorientational motion of water molecules, whose relaxation time crosses the laboratory time scale around 110 K and reaches a geological time scale at 60 K. This latter temperature may be calculated as that for a possible order-disorder transition of protons by assuming a ferroelectrically ordered arrangement[27] at low temperatures.

The reorientation of water molecules in the ice lattice is believed to take place through the movement of a pair of orientational D and L defects.[28] The pair defects are produced when thermal agitation forces one H_2O molecule to rotate through 120° around one of its O—H · · · O axes, a motion that temporarily breaches the ice rules. The reorientation of water molecules is supposed to occur at the defect site, causing the defect site to move step by step and creating a new proton configuration. Following the suggestion of Onsager,[29] we tried to do heat capacity measurements on ice crystals doped with impurities that we hoped would create the orientational defects and hence shorten the relaxation time. Actually, the relaxation time of ice doped with 2.6×10^{-2} mole fraction HF, which will create an artificial L defect in the ice lattice, was found to become 1/30 of that for pure ice.[30] The region of anomalous heat capacity becomes more prominent and tends to occur at lower temperatures. But this dopant was not sufficient to allow a long-range order to develop in the ice crystal. After trying several dopants, we finally succeeded in observing a first-order phase transition[31] at 72 K for ice specimens doped with KOH and other alkali hydroxides. This is shown in FIGURE 10 for the specimen doped with 1.8×10^{-4} mole fraction KOH. The phase transition is characterized by the existence of a long tail for the heat capacity anomaly on the high-temperature side, that is, by the persistence of short-range order. This high-temperature tail was the one observed (partly) in the pure specimen as the relaxational heat capacity anomaly near 110 K.

The transition temperature was found[32] to be independent of the concentration and the nature of the cation of the dopant; the entropy of transition was found to be dependent on the concentration and the annealing condition. The results are summarized in TABLE 3. The largest value for the removal of entropy was 2.33 $J \cdot K^{-1} \cdot mol^{-1}$.

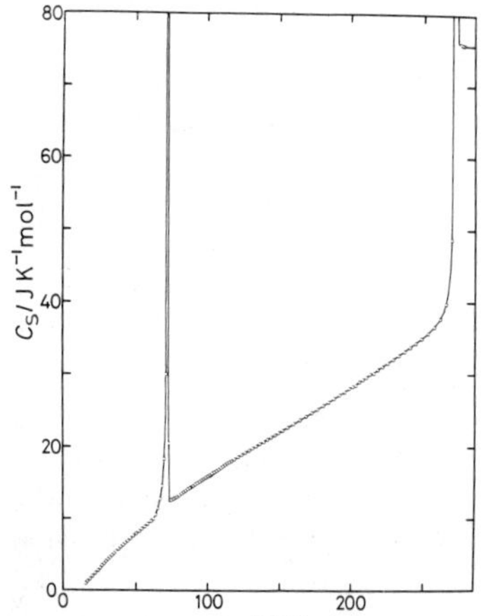

FIGURE 10. The molar heat capacity of the ice crystal doped with KOH (mole fraction: 1.8×10^{-4}).

TABLE 3. Enthalpy and Entropy of Transition of the Hexagonal Ice Crystal Doped with Various Mole Fractions of Alkali Hydroxides

t_a (hr)	ΔH_{trs} ($J \cdot mol^{-1}$)	ΔS_{trs} ($J \cdot K^{-1} \cdot mol^{-1}$)
	KOH ($f = 1.8 \times 10^{-3}$)	
170	143	1.95
256	168	2.33
72	142	1.92
72	168	2.32
	KOH ($f = 1.8 \times 10^{-4}$)	
107	92.0	1.25
252	135	1.85
	KOH ($f = 1.8 \times 10^{-5}$)	
120	36.3	0.48
106	46.7	0.64
35	20.8	0.28
	RbOH ($f = 1.8 \times 10^{-3}$)	
15	33.6	0.46
150	59.7	0.82
246	49.6	0.68
183	40.8	0.56

NOTE: t_a: time for which the samples were annealed below the transition temperature; f: mole fraction.

This value was obtained when the specimen doped with 1.8×10^{-3} mole fraction of KOH was annealed for a week at around 65 K, a temperature at which the rate of transition to the low-temperature phase was at a maximum. Thus, the phase transition removed a substantial fraction of the residual entropy. Analysis of the dependence of entropy of transition on the concentration of dopant showed that one dopant entity acted as a catalytic particle in the ice lattice by inducing the phase transition in 10^4 of the surrounding water molecules. If we visualize the spatial region of the ice crystal under the influence of a single dopant as a cube, its side contains about 20 water molecules.

The transition for heavy ice[33] doped with KOD took place at 76 K, a temperature 4 K higher than the transition temperature for specimens doped with KOH. This isotope effect is comparable in magnitude to that on ice III–IX. A neutron diffraction experiment by Leadbetter *et al.*[34] showed that the transition changed the space group of the lattice from P63/mmc to $Cmc2_1$ (orthorhombic system). This indicates a polar protonic structure along the original hexagonal axis in the ordered phase. The intensity data can be correlated to the polarization of the deuteron between the two possible sites. The occupancy fractions for the deuteron in the orthorhombic phase are 0.67 and 0.33 for the first and second sites, respectively; both fractions are 0.5 in the hexagonal phase. An anomalously broad 131 peak that newly appears in the ordered phase can be interpreted either by a partial ordering of deuteron or a coexistence of the low- and high-temperature phases. In view of the first-order nature of the phase transition, the latter model is plausible. The domain size of the ordered region estimated by line-profile of the new peak in the latter case, 40 Å, is consistent with that estimated by the calorimetric method. Presumably the ordered regions are nucleated at defect sites associated with the KOD.

It is interesting to consider why the alkali hydroxides are active and why HF is relatively inactive in restoring the mobility of protons. The hydroxyl ion replaces the water molecule in the lattice and thus introduces an ionized vertex together with an orientational L defect. Both the ionic and the L defects introduced by the hydroxyl ion can move through the lattice, unlike those introduced by HF. The HF-doped ice will produce only an L defect. The negatively charged hydroxyl ion attracts neighboring protons more easily, and thus effectively reduces the activation enthalpy for the movement of the ionic defect. It is also probable that the charged hydroxyl group disturbs the ice lattice more severely and has a greater effect on the mobility of protons. It could thus pull the lattice together, shortening hydrogen bonds and enhancing the mobility of the ionic, but not the L, defect. We can not rule out, however, an important role for cations.

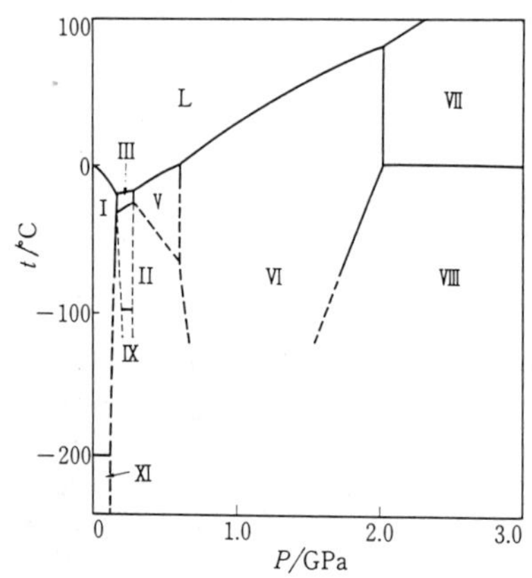

FIGURE 11. A schematic phase diagram of water. The high-pressure modification for ice X is not shown in this diagram.

We designate this ice with the ordered proton arrangement as ice XI when it is under atmospheric pressure. A tentative phase diagram for H_2O is given in FIGURE 11. The slope of the phase boundary between Ih and XI was determined experimentally by using a high-pressure calorimeter with a gaseous pressure medium.[35] The actual slope was 0.015 $K \cdot MPa^{-1}$, which is similar to the slope of the phase boundary between III and IX[36] and to that between VII and VIII.[37] The slopes are almost horizontal in the diagram, reflecting the fact that the volume change associated with the ordering of the proton arrangement is quite small compared to the corresponding entropy change in each case.

Ice Ih does not undergo the ordering transition easily. Only the addition of alkali hydroxides that cause significant deviations from the ice rules can catalyze the phase transition. If this transition turned out to be an extrinsic property of the doped ice, the entropy change associated with the dopant would be unbelievably large. The constancy of the transition temperature against changes in the kind and concentration of the dopant strongly supports the idea that the transition is an intrinsic property of ice, a property that has escaped notice for about fifty years because of the long relaxation time and that is now revealed by the catalytic action of the dopants.

The problem of the Kauzmann paradox[2] does not occur in the ice crystal because Ih is the most stable crystalline form of water under atmospheric pressure. Nevertheless, it is interesting and important to observe how the remaining entropy associated with the proton disorder is removed in an equilibrium cooling experiment. On one hand, the observation of the first-order transition verifies the wider validity of the third law of thermodynamics; on the other hand, it reveals an aspect of a possible ordering process for otherwise frozen-in disordered systems. The latter problem has long been discussed in relation to a solution of the Kauzmann paradox.

CONCLUSION

The glassy state of liquid is only a single example of a more widespread condition in which a system cannot maintain equilibrium with respect to some degrees of freedom. It is not unusual during continuous cooling for some aspects of the system to fail to maintain equilibrium in response to a change in temperature. Examples include the atomic ordering in an alloy,[38] phase separation of a solid solution,[39] conformational ordering of a molecule in a crystal,[40] and progress of a chemical reaction.[41] Spin degrees of freedom sometimes lose their equilibrium with the lattice at low temperatures. Doping techniques have been widely used, especially in magnetic systems in order to shorten the spin-lattice relaxation time. Glassy states realized in an orientationally disordered crystal diluted by an inert compound, such as a dipole glass[42] or a quadrupole glass,[43] are one of the current topics in this field.

All equilibration processes must overcome some potential barrier. Spontaneous changes are always relaxation processes. Changing the relative magnitude of the driving force and the activation enthalpy by modifying the chemical potential produces various types of freezing phenomena in systems otherwise in equilibrium or induces an ordering transition in systems otherwise in a frozen-in disordered state.

ACKNOWLEDGMENTS

I sincerely thank the late Emeritus Professor I. Nitta for his constant encouragement, Emeritus Professor S. Seki for his keen interest and stimulating discussions, and Professor J. F. Nagle for his critical reading of the manuscript. It is a pleasure to acknowledge the invaluable help I received from several colleagues. I especially thank Professor A. J. Leadbetter and his group for determining the ordered proton arrangement in ice XI by the neutron diffraction experiment.

REFERENCES

1. Zallen, R. 1983. The Physics of Amorphous Solids. John Wiley & Sons. New York, NY.
2. Kauzmann, W. 1948. Chem. Rev. **43:** 219.
3. Parsonage, N. G. & L. A. K. Staveley. 1978. Disorder in Crystals. Clarendon Press. Oxford.
4. Suga, H. & S. Seki. 1974. J. Non-Cryst. Solids **16:** 171.

5. ADACHI, K., H. SUGA & S. SEKI. 1968 Bull. Chem. Soc. Jpn. **41:** 1073.
6. ADACHI, K., H. SUGA, S. SEKI, S. KUBOTA, S. YAMAGUCHI & Y. WADA. 1972. Mol. Cryst. Liq. Cryst. **18:** 345.
7. SUGA, H. & S. SEKI. 1980. Faraday Discuss. Chem. Soc. **69:** 221.
8. ADACHI, K., H. SUGA & S. SEKI. 1971. Bull. Chem. Soc. Jpn. **44:** 78.
9. ADAM, G. & J. H. GIBBS. 1965. J. Chem. Phys. **43:** 139.
10. WILLIAMS, G. & D. C. WATTS. 1970. Trans. Faraday Soc. **66:** 80.
11. HAIDA, O., H. SUGA & S. SEKI. 1977. J. Chem. Thermodyn. **9:** 1133.
12. RUBEN, H. W., D. H. TEMPLETON, R. D. ROSENSTEIN & I. OLOVSSON. 1961. J. Am. Chem. Soc. **83:** 820.
13. MATSUO, T., M. OGUNI, H. SUGA, S. SEKI & J. F. NAGLE. 1974. Bull. Chem. Soc. Jpn. **47:** 57.
14. OGUNI, M., T. MATSUO, H. SUGA & S. SEKI. 1977. Bull. Chem. Soc. Jpn. **50:** 825.
15. OGUNI, M., T. MATSUO, H. SUGA & S. SEKI. 1980. Bull. Chem. Soc. Jpn. **53:** 1493.
16. HAIDA, O., T. MATSUO, H. SUGA & S. SEKI. 1974. J. Chem. Thermodyn. **6:** 815.
17. HAIDA, O., H. SUGA & S. SEKI. 1979. J. Glaciol. **22:** 155.
18. BERNAL, J. D. & R. H. FOWLER. 1933. J. Chem. Phys. **1:** 515.
19. STOUT, J. W. & W. F. GIAUQUE. 1936. J. Am. Chem. Soc. **58:** 1144.
20. PAULING, L. 1935. J. Am. Chem. Soc. **57:** 2680.
21. PETERSON, S. W. & H. A. LEVY. 1957. Acta Crystallogr. **10:** 70.
22. KAUZMANN, W. 1942. Rev. Mod. Phys. **14:** 12.
23. KAWADA, S. 1979. J. Phys. Soc. Jpn. **47:** 1850.
24. DENGEL, O., U. ECKNER, H. PLITZ & N. RIEHL. 1964. Phys. Lett. **9:** 291.
25. VAN DEN BEUKEL, A. 1968. Phys. Status Solidi **28:** 565.
26. CHAMBERLAIN, J. S. & N. H. FLETCHER. 1971. Phys. Kondens. Mater. **12:** 193.
27. PITZER, K. S. & J. POLISSAR. 1956. J. Phys. Chem. **60:** 1140.
28. BJERRUM, N. 1951. Science **115:** 385.
29. ONSAGER, L. 1967. *In* Ferroelectricity. E. Weller, Ed.: 16. Elsevier. Amsterdam.
30. UEDA, M., T. MATSUO & H. SUGA. 1982. J. Phys. Chem. Solids **43:** 1165.
31. TAJIMA, Y., T. MATSUO & H. SUGA. 1982. Nature **299:** 810.
32. TAJIMA, Y., T. MATSUO & H. SUGA. 1984. J. Phys. Chem. Solids **45:** 1135.
33. MATSUO, T., Y. TAJIMA & H. SUGA. 1986. J. Phys. Chem. Solids. In press.
34. LEADBETTER, A. J., R. C. WARD, J. W. CLARK, P. A. TUCKER, T. MATSUO & H. SUGA. 1985. J. Chem. Phys. **82:** 424.
35. YAMAMURO, O., M. OGUNI, T. MATSUO & H. SUGA. Submitted for publication.
36. WHALLEY, E., D. W. DAVIDSON & J. B. R. HEATH. 1966. J. Chem. Phys. **45:** 3976.
37. WHALLEY, E., J. B. R. HEATH & D. W. DAVIDSON. 1968. J. Chem. Phys. **48:** 2362.
38. SATO, H. & R. KIKUCHI. 1976. Acta Metall. **24:** 797.
39. MORIYA, K., T. MATSUO & H. SUGA. 1983. J. Phys. Chem. Solids **44:** 1121.
40. KISHIMOTO, K., H. SUGA & S. SEKI. 1978. Bull. Chem. Soc. Jpn. **51:** 1691.
41. GUPTA, P. K. & C. T. MOYNIHAN. 1976. J. Chem. Phys. **65:** 4136.
42. LOIDL, A., R. FEILE & K. KNORR. 1981. Z. Phys. B **42:** 143.
43. PRESS, W., B. JANIK & H. GRIMM. 1982. Z. Phys. B **49:** 9.

DISCUSSION OF THE PAPER

M. GRAEBNER (*AT&T Bell Laboratories, Murray Hill, NJ*): With regard to the glassy behavior of pure ice near 100 K, I would like to mention some unpublished low-temperature heat capacity data that are not typically glassy. Measurements from 0.01 to 1 K show only a phonon T^3 contribution. Any linear term is less than ~1% of the quasi-linear term in amorphous SiO_2. Thus if ice is indeed an orientational glass, it is unusual in not having a broad distribution of low-energy excitations.

SUGA: The existence of a linear term in heat capacity was clearly demonstrated in the glassy crystalline state of cyclohexanol. The low-lying energy levels may arise from the translation-rotation coupling of the molecules. In contrast, the disorder in an ice crystal is small in scale. The low-frequency vibrations of pure ice are dominated by the translational motions of oxygen atoms, as has been demonstrated by the lack of changes upon deuteration, and the disordering in the proton positions will only have a small effect in the low-frequency vibrational modes. Therefore, typical glassy behaviors may not be seen for pure ice at low temperatures.

Icosahedral Order in Glassy Metals[a]

DAVID R. NELSON

Department of Physics
Harvard University
Cambridge, Massachusetts 02118

To a first approximation, we can think of the atoms in a metal as hard spheres, like billiard balls. If we try to pack billiard balls in the plane, we find that three come together in a natural way to make an equilateral triangle. Six such triangles combine to form a hexagon, leading to a space-filling hexagonal close-packed lattice. There is no such neat progression in three dimensions, however (FIGURE 1). Four billiard balls make a tetrahedron, but there is no way to pack tetrahedra densely and fill space. Most crystals found in nature are composed of tetrahedra and octahedra, in complicated arrangements leading to lattices with a cubic symmetry. Twenty tetrahedra do combine with only slight distortions to make an icosahedron. These distortions, however, are enough to prevent a simple space-filling icosahedral crystal. Energetic arguments in favor of short-range icosahedral order in liquids were put forward by Frank in 1952.[1]

The frustration associated with tetrahedral particle packings shows up when one considers the bond lengths associated with the icosahedron in FIGURE 1: the distance between the surface atoms is about 5% larger than the distance to the center. If one tries to move the surface atoms closer together, still keeping them a constant distance from the central sphere, the cracks between surface particles in the symmetrical arrangement coalesce into a larger open space somewhere on the surface. This excess "free volume" (which is not large enough to accommodate a 13th surface atom) is an important source of degeneracy and frustration in dense-random-packing models of the glassy state. The surface atoms are "frustrated" because atoms near the gap cannot simultaneously sit at the minima of pairwise interactions with all their near neighbors. This sort of frustration is completely absent in the hexagon shown in FIGURE 1.

Understanding the statistical mechanics of simple undercooled liquids amounts to determining how the frustration is accommodated. There are many interesting solutions to the frustration problem at low temperatures. One set of solutions is embodied in the Frank–Kasper phases.[2,3] These large unit cell crystals can be regarded as regular arrays of $-72°$ wedge disclination lines in an otherwise icosahedral medium. The Frank–Kasper phases are typically binary alloys, with the larger atoms occupying disclination cores. A tangled, disordered network of disclination lines is an appealing model for the metastable states characteristic of metallic glasses.[4] The mathematical description of the statistical mechanics at finite temperatures proceeds by using an ideal, periodic icosahedral crystal, called polytope {3,3,5}, embedded in the surface of a

[a]This work was supported by the National Science Foundation, through the Harvard Materials Science Laboratory and through Grant DMR-85-14638.

four-dimensional sphere.[5,6] Sachdev has recently discussed predictions for the dynamics of undercooled liquids that result from this approach.[7]

A particularly fascinating solution to the frustration problem emerged in November 1984, when Shechtman *et al.* announced the discovery of an alloy of aluminum and manganese with long-range translational order and a perfect icosahedral point group symmetry.[8] These icosahedral "quasi crystals" are, in a sense, an intermediate stage between ordinary crystals and metallic glasses. They can be described by decorating three-dimensional aperiodic Penrose tilings with atoms.[9–12] Many alloys have now been made into icosahedral crystals, including alloys with stoichiometry of the Frank–Kasper phases. Henley and Elser have shown how the unit cell of one such Frank–Kasper phase, $Mg_{32}(Al,Zn)_{49}$, can be dissected into a finite fragment of a Penrose tiling.[13] The Penrose tiles are decorated with a particular arrangement of disclination lines.

Recently, Sachdev and Nelson have explored the relationship between short-range icosahedral order in metallic glasses and long-range order in icosahedral crystals.[14]

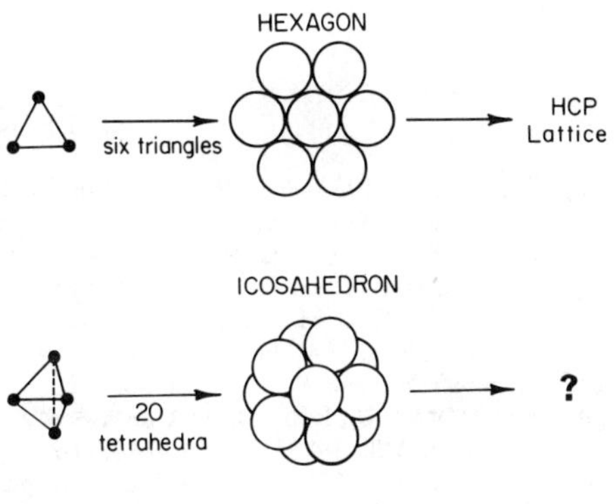

FIGURE 1. Packing of disks and spheres in two and three dimensions.

Metallic glass structure factors are assumed to describe liquids in metastable equilibrium just above the glass transition. A density functional mean-field theory is then used to search for nearby crystalline states with a lower free energy. One finds that undercooled liquids are metastable with respect to an icosahedral crystal, that is, to a crystal similar to the icosahedral phase of Al-Mn. Conventional body-centered cubic and face-centered cubic crystals have an even lower free energy. The icosahedral phase is favored by the short-range icosahedral order present in the liquid. Recent experiments showing that metallic glasses can be transformed into icosahedral quasi crystals[15,16] with rather low nucleation barriers lend support to this picture.

The ideas sketched above have been extensively reviewed elsewhere. Methods of introducing frustration into two-dimensional model systems are discussed in reference 17. Reference 18 reviews the physical ideas and mathematical machinery necessary to describe short-range icosahedral order in dense liquids. The connection between the metallic glasses and icosahedral quasi crystals is discussed in reference 19. For a review of experiments and theoretical ideas on icosahedral quasi crystals in general, see reference 20.

SUMMARY

A theoretical basis for understanding the structure and statistical mechanics of simple undercooled liquids and metallic glasses was reviewed. Short-range order in the liquid can be described by projecting particle configurations onto a periodic icosahedral crystal inscribed on the surface of a four-dimensional sphere. Fourier components of the projected particle density are used as order parameters in a uniformly frustrated Ginzburg–Landau theory of icosahedral fluctuations in flat space. The frustration forced in an excess of −72° wedge disclination lines. Although theses lines form periodic networks in the crystalline Frank–Kasper phases, they can also become entangled upon rapid cooling to form a glass.

The theory suggests that simple monatomic liquids would form glasses if cooled sufficiently rapidly, and provides a good description of the structure factor of metallic glass alloys. When the structure factors of strongly undercooled liquids are used as input into density functional theories of freezing, one finds a metastable icosahedral quasi crystal that is similar to the recently discovered icosahedral phase of Al-Mn alloys. The icosahedral quasi crystals are stabilized by the short-range icosahedral order already present in the liquid.

REFERENCES

1. Frank, F. C. 1952. Proc. Roy. Soc. London **215A:** 43.
2. Frank, F. C. & J. S. Kasper. 1958. Acta Crystallogr. **11:** 184.
3. Frank, F. C. & J. S. Kasper. 1959. Acta Crystallogr. **12:** 483.
4. Nelson, D. R. 1983. Phys. Rev. B **28:** 5515.
5. Nelson, D. R. & M. Widom. 1985. Nucl. Phys. B **240:** 113.
6. Sachdev, S. & D. R. Nelson. 1985. Phys. Rev. B **32:** 4592.
7. Sachdev, S. 1986. Phys. Rev. B **33:** 6395.
8. Shechtman, D., I. Blech, D. Gratias & J. W. Cahn. 1984. Phys. Rev. Lett. **53:** 1951.
9. MacKay, A. L. 1982. Physica **114A:** 609.
10. MacKay, A. L. 1981. Sov. Phys. Crystallogr. **26:** 517.
11. Kramer, P. & R. Neri. 1984. Acta Crystallogr. Sect. A **40:** 580.
12. Levine, D. & P. Steinhardt. 1984. Phys. Rev. Lett. **53:** 2477.
13. Henley, C. & V. Elser. 1986. Philos. Mag. B **53:** L59.
14. Sachdev, S. & D. R. Nelson. 1985. Phys. Rev. B **32:** 4592.
15. Lilienfeld, D. A., N. Nastasi, H. H. Johnson, D. G. Asat & J. W. Mayer. 1985. Phys. Rev. Lett. **55:** 1587.
16. Poon, S. J., A. J. Drehman & K. R. Lawless. 1985. Phys. Rev. Lett. **55:** 2324.
17. Nelson, D. R. 1983. *In* Topological Disorder in Condensed Matter. F. Yonezawa & T. Ninomiya, Eds. Springer Verlag. Berlin.
18. Nelson, D. R. 1985. *In* Applications of Field Theory to Statistical Mechanics. L. Garrido, Ed. Springer Verlag. Berlin.
19. Nelson, D. R. & S. Sachdev. 1986. *In* Proceedings of the Workshop on Amorphous Metals and Semiconductors. P. Haasen & R. Jaffe, Eds. Acta Metall. In press.
20. Nelson, D. R. & B. I. Halperin. 1985. Science **229:** 233.

DISCUSSION OF THE PAPER

F. H. Stillinger (*AT&T Bell Laboratories, Murray Hill, NJ*): Recent calculations by R. A. La Violette and myself have investigated coordination geometry in amorphous packings (mechanically stable arrangements) of noble gas atoms. These were generated from the stable liquid phase for both 256- and 1008-particle systems, using steepest-descent displacements on the potential energy hypersurface to provide anneal-free quenches. Results from many such packings show that icosahedral coordination is quite rare, occurring only for about 1 atom in 10^3. Disorder is associated more noticeably with coordination number fluctuations, where 11, 12, and 13 appear roughly in the proportions 1:2:1.

I have no doubt that some amorphous packings of noble gas atoms could display considerably higher concentrations of icosahedra. These must constitute, however, a distinct minority of all possible stable packings. Perhaps they are to be found preferentially in the low-potential-energy tail of the distribution, which would give them a disproportionate influence in the strongly supercooled liquid state.

NELSON: The percentage of perfect icosahedral coordinations in dense-random-packing models is also quite low. Nevertheless, about 60% of the bonds in relaxed dense random packing have five (distorted) tetrahedra wrapped around them. These pentagonal bond spindles are fragments of icosahedra. Since 12 such spindles must come together to make an unblemished icosahedron, perfect icosahedra will indeed be rather rare. The lack of perfect icosahedra is due to the tangled network of disclination lines (fourfold and sixfold bond spindles) that occupies the remaining 40% of space. The best one can do in eliminating disclinations and creating complete icosahedra probably occurs in the Frank–Kasper phases: here one finds that 90% of the bond spindles are pentagonal, and that ~60% of the atoms have an icosahedral coordination. The remaining spindles are sixfold, and are a consequence of the topological frustration of flat space.

STILLINGER: You have pointed out that geometric frustration associated with packing icosahedra can be relieved by curving three-dimensional space. It can also be abated by the presence of a free surface. Do you believe, or have evidence to indicate, that the concentration of icosahedral structures in a supercooled liquid will tend to be enhanced at its interface with the vapor?

NELSON: Since there are fewer constraints imposed on polytetrahedral particle packings at a free surface, it seems plausible that more icosahedra might occur there. A flat planar surface, on the other hand, will produce local hexagonal close-packed or face-centered cubic order, and hence many octahedra near the surface.

S. NAGEL (*University of Chicago, Chicago, IL*): You claim your topological considerations are relevant to metallic glass formation. Yet pure metals (as distinct from alloys) are very poor glass formers and most metallic glasses are alloys that still must be cooled extremely rapidly to form a glass. Does your theory explain why there are certain composition ranges (such as 20% metalloid atoms) that are relatively good glass-forming compositions? For example, would the smaller atoms lie on disclination lines that would include 20% of the atoms?

NELSON: Glass formation at cooling rates currently available in the laboratory requires that the window of temperatures between the equilibrium freezing transition and the glass transition be as small as possible. This minimizes the probability of nucleating a stable crystalline phase during the cooling process. Because the glass transition temperature is found experimentally to be only a weak function of composition, glasses should occur near deep eutectics in the temperature-versus-composition phase diagram. The theory of frustrated icosahedral order does allow one to estimate where this eutectic should occur.

A typical binary mixture equilibrium phase diagram is shown below.[b] The region rich in *small* atoms (Mg) occurs to the right; the region rich in *large* atoms (Ca) occurs to the left. As discussed in my talk, a dilute concentration of large atoms in a matrix of smaller ones produces disclination lines (sixfold bonds), which have the right sign to accommodate the topological frustration of flat space. The calcium atoms are large enough to fit inside the Z16 coordination shells discussed in my talk. There should be a maximum in the melting temperature at the concentration of larger atoms that produces a Frank–Kasper phase. A simple mean-field theory of frustration says that the average coordination number in a Frank–Kasper phase should be 13.4, which leads to the prediction of a Frank–Kasper phase at ~65% calcium. The correct number is 67%. Conversely, frustration is *increased* by a dilute concentration of small atoms in a matrix of larger ones. Disclination lines with the wrong sign (fourfold bonds) are generated. The ordering temperature must go down. A crude estimate of the concentration that produces the maximum frustration can be obtained by interchanging the large and small atoms in a Frank–Kasper phase. This argument correctly predicts the deep eutectic at ~35% calcium and hence the region of glass formation accessible in the laboratory in the above phase diagram. Although the above arguments were presented for an especially simple bimetallic glass former, I would expect similar qualitative arguments to apply to metal-metalloid systems.

NAGEL: In calculating the structure of metallic glasses you claim to get good agreement with experimental structure factors. If I understand it correctly, however, the height and width of each peak are left as free parameters. Thus the agreement between the theory and experiment is not really as good as the curves you have shown imply. What you have found is the ratio of the position of the various peaks to the first peak position. The ratios of the two maxima in the second split peak to the first peak are 2:1 and 1.7:1. These are what I would expect from simple geometrical arguments of close packing. (The number 2 is certainly not surprising, and 1.7, which is ~ $\sqrt{3}$, is also what you might expect from naive packing considerations.) Since experimental data for these values also vary by several percent (for example, Nb_xNi_{1-x}), is the agreement of your numbers with the data really convincing evidence that the curved-space picture that you present is the simplest way to think about metallic glasses?

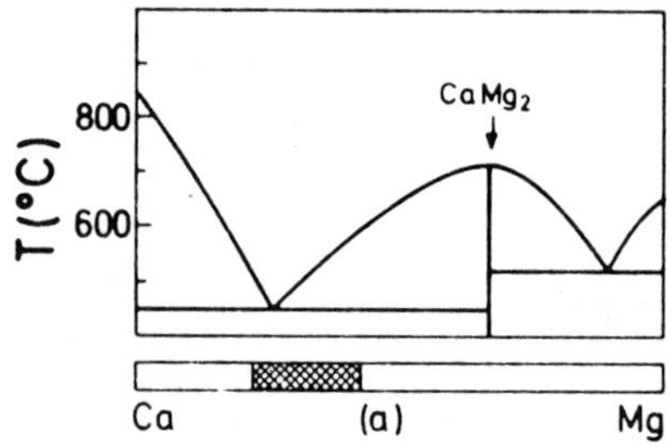

[b]This schematic of the phase diagram of Ca-Mg (see M. Hansen, *Constitution of Binary Alloys*, McGraw-Hill, New York, NY, 1958) also shows the region where glasses can be formed at currently available laboratory cooling rates. (For related figures and an extensive discussion of the criteria for glass formation, see J. Hafner, *Phys. Rev. B* **21:** 406, 1980.)

NELSON: As I stated explicitly in my talk, there are indeed two adjustable parameters in the Ginzburg–Landau theory of topological frustration for each peak in the structure factor. Similar free parameters also appear in the Ginzburg–Landau theory of superconductivity. Just as happened historically in superconductivity, there are now efforts underway to fix these parameters using more microscopic methods—for example, exact computer simulations on the unfrustrated surface of a four-dimensional sphere.

I think the prediction of the peak positions in the structure factor by the theory is nontrivial, and that you are confusing radial distribution functions with structure factors. The "simple geometrical arguments" you refer to (see, for example, J. Wong & C. A. Angell, *Glass: Structure by Spectroscopy,* Chapter 2, Figure 6, Marcel Dekker, New York, NY, 1976) apply to peaks in the *radial distribution function.* The success of such methods confirms that metallic glasses are essentially polytetrahedral particle packings. Knowledge of the positions of the first few peaks in the radial distribution function is not sufficient to calculate the maxima in the Fourier transform of the radial distribution function, however. It is this structure factor that is actually measured experimentally. Peaks in the structure factor tell us about the distances between planes of atoms, not the distances between atoms themselves. It is an interesting coincidence that the relative peak positions are approximately the same in real and reciprocal space for metallic glasses. This is clearly not true in general: compare the radial distribution function of the face-centered cubic lattice to its powder-averaged (body-centered cubic reciprocal lattice) Fourier transform!

I. M. HODGE (*Eastman Kodak, Rochester, NY*): You mentioned that two-dimensional Penrose tilings are highly degenerate. Is this also true for three-dimensional Penrose tilings in general? Do you consider three-dimensional Penrose tilings to be possible candidates for the "ideal glass" at the Kauzmann temperature T_2?

NELSON: The two-dimensional Penrose tilings are indeed highly degenerate when one locally scrambles the tiles: the "matching rules" are violated, but the Fourier transform in reciprocal space remains qualitatively the same. The same arguments apply to the "Penrose bricks" in three-dimensional tilings (V. Elser, *Phys. Rev. Lett.* **54:** 1730, 1985). My best candidate for an ideal glass would be a Frank–Kasper phase with an infinite unit cell size. Henley and Elser have recently shown (*Philos. Mag. Lett.*, in press) that a large unit cell Frank–Kasper phase ($Mg_{32}(Al,Zn)_{49}$) is closely related to a particular decoration of a three-dimensional Penrose tiling with atoms, so perhaps the two concepts are not that different.

J. C. PHILLIPS (*AT&T Bell Laboratories, Murray Hill, NJ*): You have written that in metallic glasses (such as $Pd_{0.8}Si_{0.2}$) the large-atom (Pd) framework has a nearest-neighbor Pd-Pd icosahedral coordination. Gaskell has argued that such glasses contain microclusters with the cementite (Pd_3Si) structure, where Si is ninefold coordinated and the Pd-Pd coordination shell is locally close-packed. Ordinarily it is difficult in binary alloys to distinguish icosahedral and close-packed $I(q)$ and $P(r)$ functions—there are four functions of each kind to be considered; moreover, only one truncated $I(q)$ is described in your model. Some experimental data on binary alloys, however, do show peaks in $P_{AA}(r)$ at $r = r_1$ and at $r_2 = \sqrt{2}\, r_1$, in agreement with Gaskell's model. Do you still feel that the icosahedral description applies to binary metallic glasses such as $Pd_{0.8}Si_{0.2}$?

NELSON: Gaskell's model invokes structures surrounding phosphorus and silicon atoms commonly found in the large unit cell crystalline phases of Ni_3P and Pd-Si compounds. These configurations are precisely the Z9 coordination polyhedra surrounding small atoms discussed in my talk. I would not be at all surprised if they occurred in metallic glasses. Since Z9's are nodes of three fourfold bond spindles (each fourfold bond is a distorted octahedron), there may well a shoulder at $\sqrt{2}$ times the near-neighbor Pd-Pd distance. There are also many fivefold spindles in the large unit cell crystalline phases (S. Sachdev, Ph.D. diss., Harvard University, 1985), so one might also expect these fragments of icosahedral order in the glass. To go from the crystal to the glass one must disorder the regular network of disclination lines in the crystal. Such a tangled network of disclination lines is rather different from the polycrystallites-with-grain-boundaries you often propose as models of glass. Polycrystalline models have well-known difficulties in explaining the lack of crystal nucleation while undercooling towards the glass transition. (See, for example, D. Turnbull, *J. Non-Cryst. Solids* **75:** 197, 1985).

Realistic Theories of the Origin and Molecular Structure of Glass

J. C. PHILLIPS

AT&T Bell Laboratories
Murray Hill, New Jersey 07974

INTRODUCTION

Topology

During the last seven years I have actively collaborated with experimentalists both at and outside Bell to study the molecular structure of glass and its relation to the glass-forming tendency. For many more years mathematicians have constructed models of the molecular structure of glass based primarily on geometrical considerations. When I began to study this subject it seemed to me that these geometrical models were disappointing in two basic respects. First, they did not seem to be able to relate experimental observables to the quantities they were calculating—usually the radial distribution function $P(r)$—with sufficient precision to draw any definite conclusions. In particular, these geometrical models were unable to make any statements concerning the origin of the glass-forming tendency. Second, these geometrical models usually had little or no chemical content. This meant that they were unable to suggest new kinds of experiments, or to guide experimentalists in the selection of materials that would improve the prospects for success of the new experiments developed by others.

New Wave of Realistic Submicroscopic Experiments

The new wave of submicroscopic experiments may permit the development of models that do not have the limitations of the geometrical models. The new experiments have involved a wide range of new techniques: new, that is, in the sense that for the first time definite connections are established using these techniques between structure and the glass-forming tendency. The new techniques include Raman and Mössbauer spectroscopy and low- and high-resolution transmission electron microscopy. Of course, all these techniques had been applied before to study glassy materials. The new wave, however, is distinguished by a close analysis of the experimental data that identifies its salient features. A striking feature of the analysis is detailed attention to chemical shifts that are related to the composition dependence of the glass-forming tendency in binary and pseudobinary alloys.

CONSTRAINT THEORY AND VECTOR PERCOLATION

Some time ago I discussed[1] the chalcogenide alloy network glass formers, that is, alloys of Ge or As with S, Se, and Te. These alloys have energy gaps near 2 eV that vary comparatively slowly with composition. The electronegativity differences between the cations and anions are small, and the tendency toward phase separation on a scale of less than 100 Å, which may be difficult to control or even detect in many experiments, is minimized. The central feature of the short-range order is the average coordination number *CN* of each element. For Ge, As, S, and Se one finds that $CN = 8 - N$, where N is the appropriate column of the periodic table ($N = 6$ for S and Se). For Te however, *CN* may be two or three (for example, in c-As_2Te_3). Correspondingly, S and Se are better glass formers (larger values of T_x/T_g, where T_g is the glass transition temperature and T_x is the recrystallization temperature of a melt-quenched glass alloy).

Many theories of glass formation have stressed structural complexity as the origin of the glass-forming tendency. In fact complexity is a necessary but not a sufficient condition. For example, the structure of c-As_2Te_3 is more complex than that of c-As_2(S or Se)$_3$, but the latter materials are much better glass formers.

The crucial criterion for glass formation[1] does not refer to ordinary Euclidean space but rather to configuration space and, more specifically, interatomic-force-field space. When (as in the case of many network glasses) it is possible to define bond-stretching and bond-bending constraints, glass formation is optimized at the alloy composition where N_{con} (the average number of constraints per atom) equals N_d (the number of degrees of freedom per atom, which is three in a three-dimensional space). In network glasses the bond lengths are always constrained (to within 1–2%) and some of the bond angles are constrained (to within 10°). Which of the bond angles are constrained can be determined either by studying appropriate crystal structures or by analyzing $P(r)$ as measured in diffraction studies. Two examples of the latter are shown in FIGURES 1 & 2. Incidentally, mechanical models that attempt to reproduce $P(r)$ (either in a hand-built model[2] or by computer simulation[3]) generally give poor densities and fail to distinguish between intact and broken bond-bending constraints, although the average nearest- and next-nearest-neighbor distances are fitted.

Using constraint theory one finds that the average coordination number m, which maximizes the glass-forming tendency, is 2.40 when all the bond-bending constraints are intact. This explains[1] why As_2S_3 ($m = 2.40$) is such a good glass former even

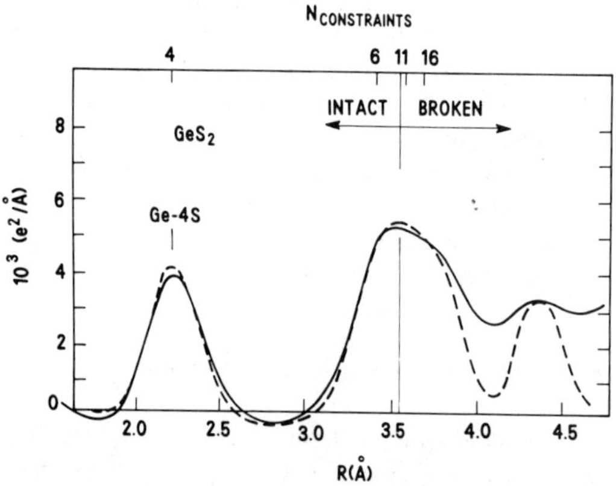

FIGURE 1. Intact (resolution limited) nearer and broken-constraint further peaks in $P(r)$ for g-GeS_2.

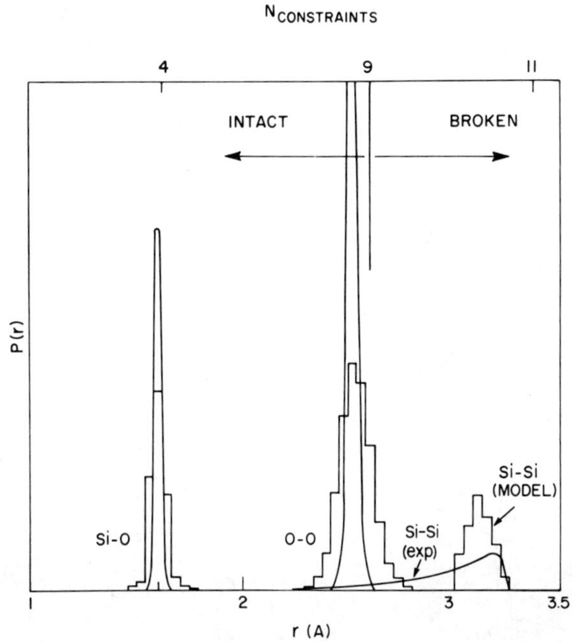

FIGURE 2. As in FIGURE 1, for g-SiO_2. The histogram represents the results of a hand-built ball-and-stick model.[2] Note that the model gives equal FWHM for the O-O and Si-Si peaks, which contradicts experimental results, and which makes the O-O peak width "zero" (resolution limited) and the Si-Si peak width at least five times greater.

though it has the composition of a naturally occurring mineralogical compound (orpiment). This explains the very narrow peak in the pseudospecific heat $\tilde{C}_p$ of As_xSe_{1-x} alloys illustrated in FIGURES 3 & 4. It explains why the maximum in the glass-forming tendency of Ge-Se alloys occurs near 20% Ge, as shown in FIGURE 5. For further discussion, see other general presentations.[4-6]

Constraint theory discusses only combinatorial aspects of network topology. It can be regarded as the first microscopic realization of Kauzmann's thermodynamic description of the glass transition as an entropy (not an enthalpy or a volume) crisis.[1] Recently, it has been shown that in mean-field theory the relation that connects force-field to configuration space, $N_{con} = N_d$, corresponds to a compositional threshold in mechanical stiffness associated with vector percolation.[7,8] It appears experimentally that similar thresholds are seen in optic mode frequencies measured by Raman scattering, although at present there is no theoretical derivation of this result.[9] An example of such a threshold is shown in FIGURE 6.

CLUSTERS IN GLASSES

The idealized continuous random network model of glasses (such as SiO_2 or the chalcogenide glasses) is generally not realized in melt-quenched or evaporated network noncrystalline solids. Instead the actual structure consists of clusters that are too small and too strained to be resolved by conventional X-ray diffraction (small- or large-angle) techniques as crystalline. This is the point of view taken by many Russian workers since Lebedev introduced the "microcrystallite" model for g-SiO_2 in 1921. (An entertaining history of the Russian view is now available.[10]) This viewpoint, however, has not gained wide acceptance in the West, where scientists have generally preferred the noncommittal "show me" attitude.

Of course, X-ray diffraction is not the only technique for studying glass structure. The availability of laser spectrometers has made Raman scattering a powerful tool for detecting microcrystallites even in fine powders, and Raman spectra are available for

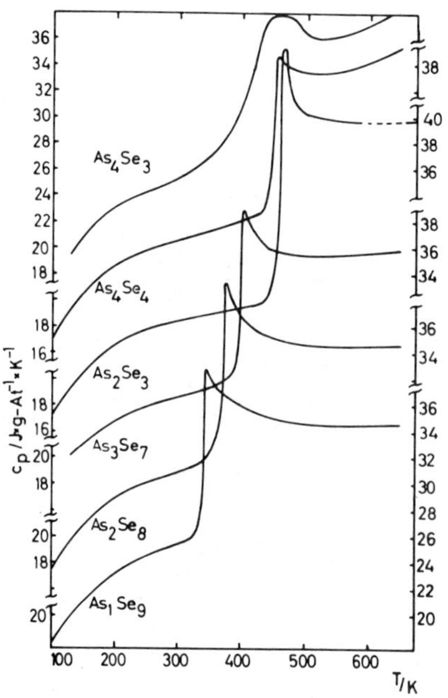

FIGURE 3. Specific heats of various As-Se alloy glasses. The width of the glass transition peak can be interpreted as a measure of microscopic chemical and mechanical (internal stress) inhomogeneity. For As_2Se_3 this width is a minimum, as shown explicitly in FIGURE 4. All alloy samples have common thermal histories.

many glasses.[11] These spectra, however, generally consist of broad bands that are not easily interpreted. This difficulty has been resolved in two ways.

First, we may choose as our network building block the smallest regular polyhedron, which is a tetrahedron, as in either GeS_2 or SiO_2. Next, we exploit the vibrational decoupling of tetrahedra, which occurs for corner-sharing tetrahedra when the corner-sharing bond angle α is close to 90°. In SiO_2 we have $\alpha \sim 150°$ (no decoupling), but in Ge(S or Se)$_2$ we have $\alpha \sim 100°$ (excellent decoupling). This decoupling produces narrow vibrational bands that correspond well to the narrow modes of vibration of tetrahedral molecules, Ge(Cl or Br)$_4$, respectively. The normally formidable problem of decoding glass Raman spectra is trivial in this case.

When this analysis is carried out,[12] and confirmed by comparing glass and crystalline Raman bands, as in FIGURE 7, a remarkable result emerges: most of the

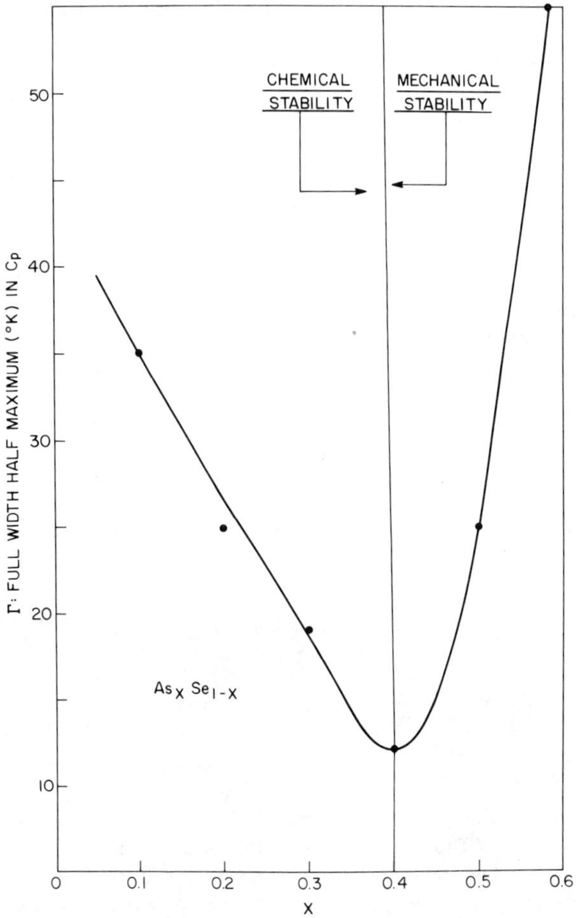

FIGURE 4. Width of peaks in the specific heat curves of As-Se alloys shown in FIGURE 3. Plotted here are the full widths (measured at half-maximum above the supercooled liquid plateau). The data show a minimum near $x = 0.4$, which is a point of both chemical and mechanical stability.

FIGURE 5. Quenching rate necessary to avoid crystallization is a measure of the difficulty of forming a glass. Minimal glass-forming quenching rates are shown here as a function of composition of Ge_ySe_{1-y} alloys. The peak indicates that fast quenching is required to avoid crystallization when the Ge-Se mixture is stoichiometrically balanced for $GeSe_2$. Horizontal lines represent the formation of glass.

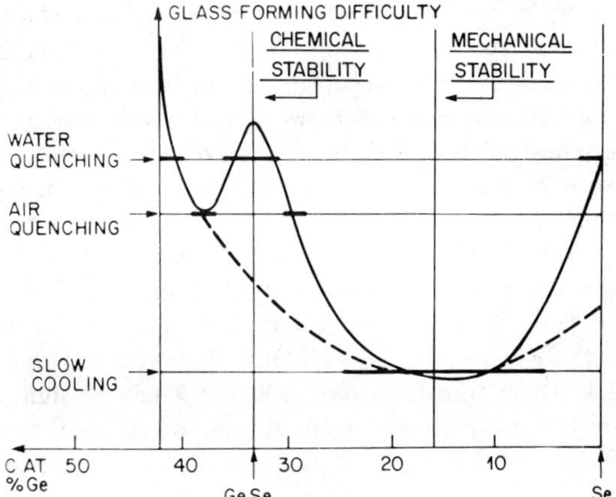

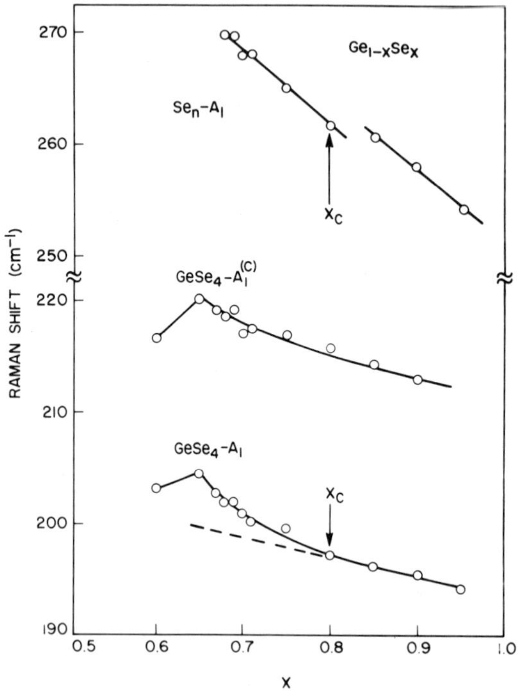

FIGURE 6. Composition dependence of certain Raman vibrational bands in Ge-Se glasses. The value $x_c = 0.2$ is critical according to constraint theory. Many more examples are given in reference 9.

glass spectrum is accounted for by the vibrational modes of a microcrystalline cluster. The internal structure of the cluster is that of the high-temperature crystalline phase, as shown in FIGURE 8. One or two extra bands, however, are present, and are associated with cluster surface and interface modes. The reconstructed cluster surfaces and interfaces contain like-atom bonds Ge-Ge and Se-Se (broken chemical order). From the composition dependence of these like-atom bond populations the cluster sizes can be estimated as 50–75 Å.

Raman spectroscopy measures bond populations. An entirely independent measurement of broken chemical order at cluster surfaces and interfaces can be obtained by Mössbauer spectroscopy, with Sn as the probe replacing Ge. This approach easily distinguishes between $M(Se_{1/2})_4$ sites and $(Se_{1/2})_3$-M-M-$(Se_{1/2})_3$ sites, for example. An excellent concordance between these two kinds of spectroscopy has by now been established for several binary and pseudobinary chalcogenide alloy glasses over very wide ranges of composition.[13–15] An example of such a concordance is shown in FIGURE 9.

One of the most dramatic results shown in FIGURE 9 is the almost complete extinction of the cluster surface Raman bond and Mössbauer site signatures in $Sn_xGe_{1-x}Se_2$ alloys near $x = 0.35$. It happens that the bond-bending forces at Sn are very weak near $T = T_g \sim 600$ K in these alloys. (Compare, for example, gray and white Sn, which transform near 300 K.) Thus the number of bond-bending constraints decreases rapidly with increasing x, and calculations show[16] that at $x = 0.40$ the ideal glass condition $N_{con} = N_d$ is satisfied. The calculation contains no adjustable parameters because it is purely topological. This confirmation of the prediction of cluster fusing at a magic composition to form a Zachariasen or ideal continuous random network is without equal in the literature on the structure of glass.

The notion that clusters of the high-temperature crystalline phase already exist in the melt-quenched glass has often been rejected by English-speaking scientists, but it has a formidable phenomenological foundation in oxide glass science, which has recently been strengthened dramatically by work on chalcogenide glasses. In the oxide glasses it has long been known, for example, that when vitreous silica is annealed and crystallized, the first devitrification product is cristobalite, the high-temperature crystalline phase stable above 1475° C. This may not seem suprising, until one realizes that this statement is valid even when the annealing temperature is 1200° C, where the stable crystalline phase is tridymite. To obtain this phase, much longer annealing times are necessary. (Equilibration can be greatly accelerated by addition of a few mole percent of Li_2O.) Thus one can reasonably infer that submicrocrystallites of cristobalite are always present in melt-quenched silica.

This experiment has been beautifully refined by Griffiths *et al.*[17] They used laser annealing gradually to crystallize the surface layer of g-$GeSe_2$. At early stages where

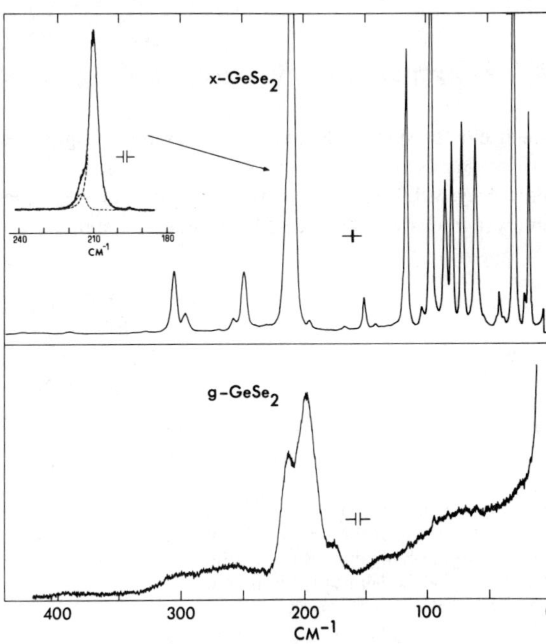

FIGURE 7. Raman spectra of c-$GeSe_2$ and g-$GeSe_2$ juxtaposed. The unit cell of the crystal contains 48 atoms, which is why the crystalline spectrum is so complex. The inset shows the fine structure of the strongest peak in the crystalline spectrum, which is actually composite. The two components correspond to corner- and edge-sharing tetrahedral A_1 (symmetric breathing) modes. This is the first Raman spectroscopic study of glass that is sufficiently accurate to make the fine distinctions necessary to resolve cluster surface and interior vibrational modes[12] and thus exploit the full capabilities of Raman spectroscopy for structural analysis of glass.

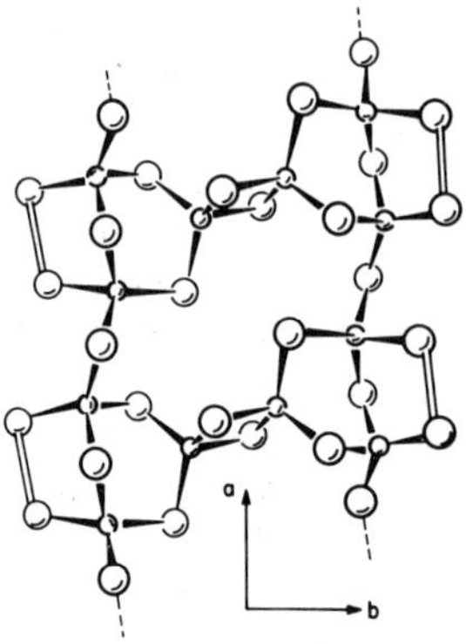

FIGURE 8. The large clusters that are the dominant structural elements of glassy $GeSe_2$ are made of stacked layer units of the kind shown here.[11] (Actually, more recent data suggest that the clusters are much wider, and consist of six chains instead of two.[13] This makes the cluster diameter about 50 Å.) Chemical order is broken at the edges where chalcogen dimers have formed. The companion Raman band, which is the 220 cm^{-1} satellite to the 200 cm^{-1} band in the Raman spectrum of g-$GeSe_2$ shown in FIGURE 7, is associated with a vibrational mode localized on these dimers.

Raman scattering gave broadened crystalline lines (FIG. 7) and the material was still X-ray amorphous the laser annealing can be stopped. If the sample is left at room temperature ($\sim T_g/2$) overnight, the submicrocrystallites revitrify. This thermal reversal at $T_g/2$ demonstrates how little difference there is between the molecular structures of the glass and crystal.

BRAGG DIFFRACTION BY AMORPHOUS SILICON

The well-defined cluster surface Raman and Mössbauer signatures identified in chalcogenide network glasses and g-SiO_2, as well as the kinetic effects generated by the presence of submicrocrystallite clusters in the annealing of melt-quenched glasses, constitute extremely persuasive evidence for submicrocrystallite structures in the

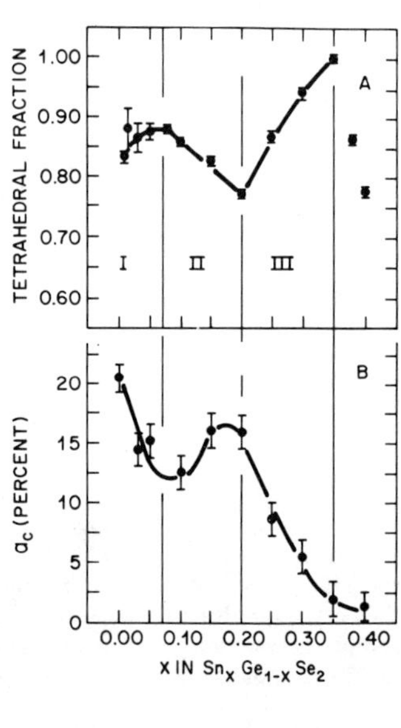

FIGURE 9. Parallel dependence on composition x in pseudobinary g-$Sn_xGe_{1-x}Se_2$ alloys of three signatures of internal surfaces: strength of the companion Raman band (FIGS. 7 & 8) and concentration of chemically disordered cation sites measured by Mössbauer spectroscopy.

vitreous state. This evidence is indirect, however, and to understand it one must make a great effort to follow the intricate patterns generated by a wide range of data obtained by many techniques. So far only a few people have been willing to make this effort; a much larger number have taken the path of least effort and adopted "random" models of glass structure such as the continuous random network (CRN) model for network glasses and the random close packing (RCP) model for metallic glasses. Polls of the community of glass scientists invariably show large majorities in favor of these "noncommittal" models.

In FIGURE 7 an effort was made to dramatize the close similarity between crystalline and glass structure by juxtaposing the two Raman spectra. An even more direct comparison can be made by physically fusing the crystalline and noncrystalline structures. For this purpose an atomically clean interface is crucial. Also we need a technique that is capable of probing the atomic structure near the interface directly.

The appropriate technique is high-resolution transmission electron microscopy (HRTEM). HRTEM images atomic columns normal to thin films. For this experiment it is natural to choose initially for the sample a c-Si-a-Si interface because Si technology is well advanced. Many previous efforts have failed to resolve submicro-

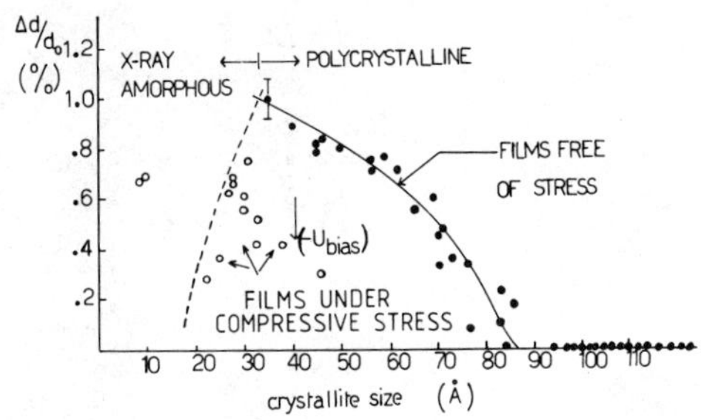

FIGURE 10. Polycrystalline Si:H films exhibit an interesting expansion (probably due to intercluster tension) for microcrystallite size <100 Å. Below 30 Å the crystallites appear amorphous. Crystallite sizes were estimated from the (111) peak width by the Scherrer formula.[18]

crystallites in a-Si thin films prepared by deposition on soluble substrates. The latter were then removed and atomic images were obtained in plan view on a-Si typically 100-Å thick. It would be difficult, however, to image submicrocrystallites in this configuration if their diameters are small compared to 100 Å. Because X-ray diffraction[18] can resolve microcrystallites with diameters as small as 30 Å (FIG. 10), submicrocrystallite diameters almost certainly are in this limit.

Let us now turn to the c-Si:a-Si (no H) configuration, which we can examine in cross section by HRTEM. The possibility now exists that the c-Si substrate will act as a template to orient microcrystalline clusters in the "amorphous" Si film deposited at room temperature. With orientational order between adjacent clusters the cross-sectional registry of atomic columns may be greatly enhanced, thus making it possible to image microcrystallites in 100-Å-thick cross sections. (The latter can be prepared with relatively little surface damage by Ar ion milling at 100 K.)

This scenario has actually been realized.[19] The correlation length λ appears to be in the range $\lambda \sim 300$ Å or more. This is much larger than the thickness (100 Å) of the cross section, and excellent lattice images are obtained (A. Ourmazd, J. C. Bean, B. A. Wilson, and J. C. Phillips, unpublished results). One of these is shown in FIGURE 11.

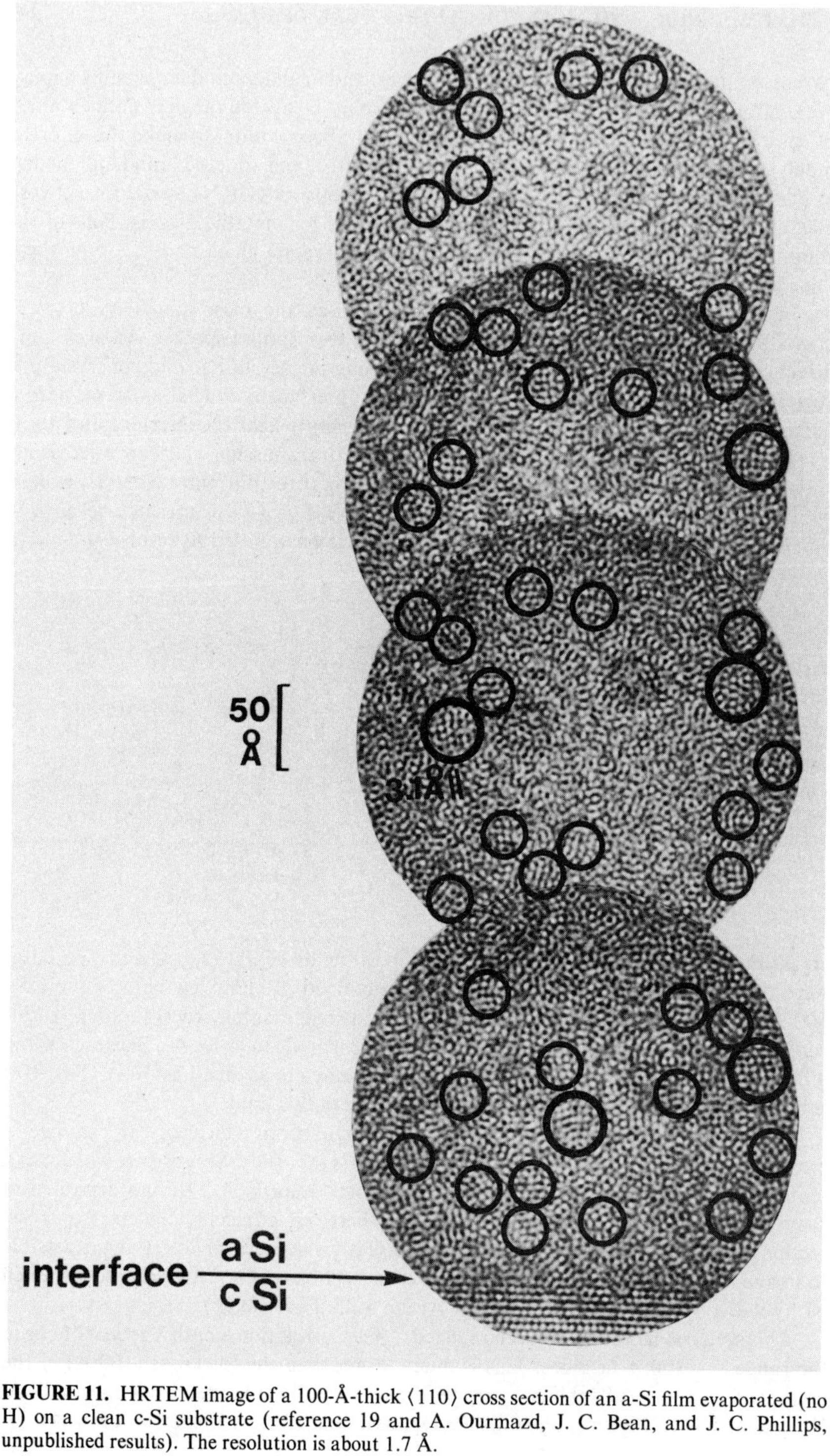

FIGURE 11. HRTEM image of a 100-Å-thick ⟨110⟩ cross section of an a-Si film evaporated (no H) on a clean c-Si substrate (reference 19 and A. Ourmazd, J. C. Bean, and J. C. Phillips, unpublished results). The resolution is about 1.7 Å.

Many submicrocrystallites, with diameters in the range 15–30 Å, are easily resolved. Close inspection of these images reveals grain boundaries (often) and even an occasional edge-partial dislocation. From these pictures there can be little doubt that submicrocrystallites dominate the structure of a-Si.

The most objective way to analyze these images is to use the photographic negatives to generate Bragg diffraction patterns. Typically regions with 50-Å diameters are used for this purpose. The results[19] are shown in FIGURE 12. Many Bragg spots are seen from the c-Si substrate—certainly out to (400), perhaps out to (600). A full complement of (111) Bragg spots is evident in this ⟨110⟩ cross section for the "amorphous" material near the interface. (This is exactly what should be seen, because the (220) and (311) rings collapse[18] to a single ring in X-ray diffraction for crystallite diameter <30 Å. The reason for this collapse is not known.) These spots gradually attenuate on a scale $\lambda \sim 500$ Å. This explains why lattice imaging has succeeded in the c-Si:a-Si configuration when it failed with a-Si deposited on blank substrates.

The simplest explanation for these observations is that submicrocrystallites (SMCs) are always formed by evaporation of amorphous Si. Their size and possibly even their crystal structure may depend on the deposition conditions. Evaporation at room temperature on a blank substrate produces SMC diameters near 15 Å. Electron spin resonance data, which measure the dangling or unreconstructed bond density as a function of annealing to crystallization, show different behaviors for a-Si on c-Si, for a-Si on SiO_2, and for a-Si:H, as shown in FIGURE 13. Surprisingly the latter two cases give evidence for a structural transition of the SMC before crystallization, whereas a-Si:c-Si shows simple crystallization. Independent data suggest that the SMC morphology may be wurtzite for the latter two cases, as shown in FIGURE 13.

The foregoing sketch is sufficient to show that there is abundant structure in "amorphous" Si. Many more experiments suggest themselves based on the amorphous/crystalline configuration, which exploit what can be called the orientational proximity effect (OPE). The OPE is perfectly general, and there seems to be no obstacle in principle that will prevent its being utilized, for example, to study cristobalite SMCs in g-SiO_2 deposited on a cristobalite substrate.

To show how powerful the OPE can be, in FIGURE 14 we show grain boundaries between SMCs and how these are affected by filtering the HRTEM image.[19] With optimal filtering the quality of the images is comparable to that obtained from specially prepared bicrystals. It becomes apparent that all the "random" models of "amorphous" materials are chiefly of interest as a description of failed experiments.

METALLIC GLASSES

Although none of my research on network glasses can as yet be applied directly to metallic glasses, there are obvious analogies on a conceptual level between the continuous random network-SMC dichotomy in network glasses and the RCP-cluster dichotomy for metallic glasses. In both cases a simplistic school has argued that in the absence of direct evidence based on diffraction, it is not only prudent but indeed advisable to adopt a noncommittal "random" model. Such a model is supposed to be a paradigm for the structure of the real metallic glass.

In the case of network glasses the simplistic approach has suffered severely in terms

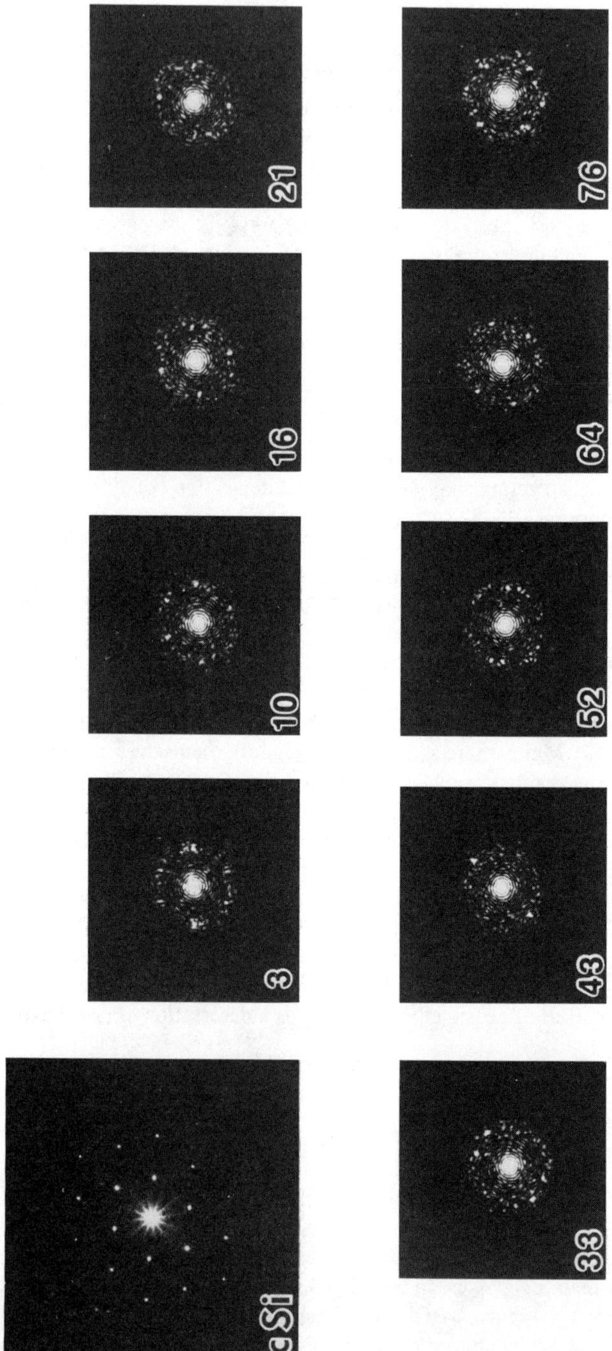

FIGURE 12. Optical diffractograms as a function of distance (in nm) from the c-Si-a-Si interface shown in FIGURE 11.

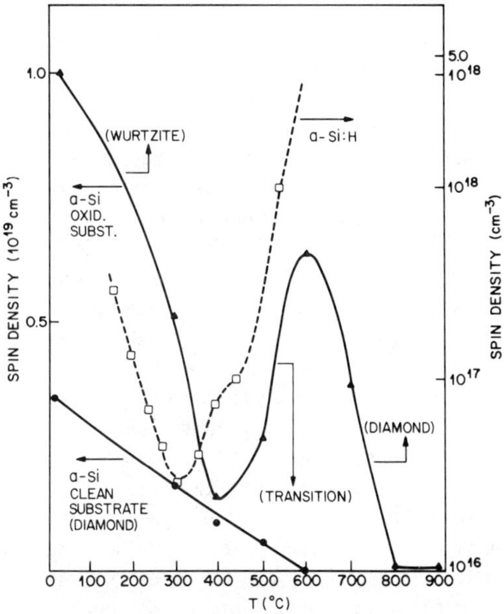

FIGURE 13. Dangling bond density (as measured by electron spin resonance) of a-Si films prepared in three ways as a function of isochronal annealing temperature.

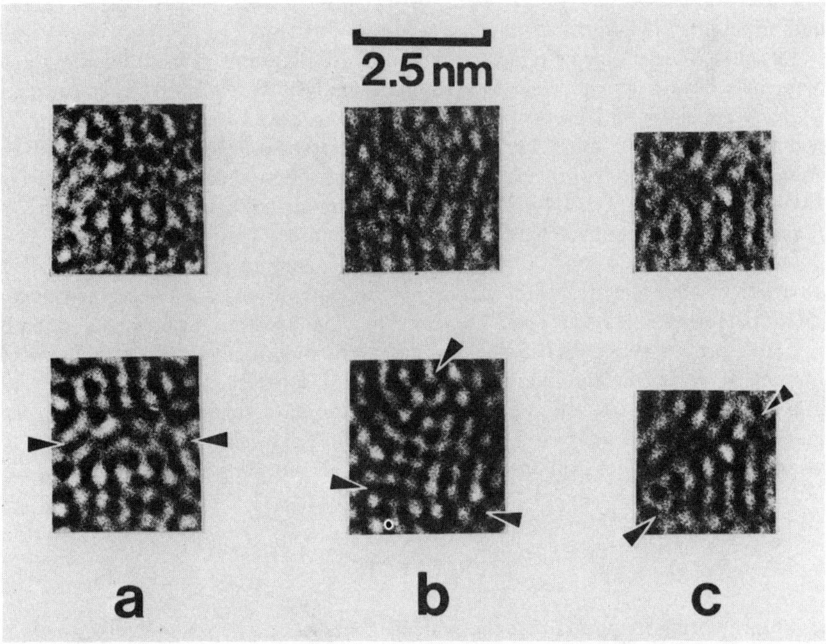

FIGURE 14. Optically reconstructed images of SMCs containing grain boundaries (arrowed). Top row: unfiltered; bottom row: filtered for reasons described elsewhere (A. Ourmazd, J. C. Bean, and J. C. Phillips, unpublished results).

of optical and more recently HRTEM diffraction experiments, as described earlier in this paper. So far comparable experiments have not been carried out for metallic glasses. Nevertheless, some points of interest should be mentioned.

The central problem in theoretical discussions of metallic glasses has been modeling the interatomic forces. For many decades mathematicians have blithely based their discussions on hard spheres, even though it is known that sphere packing is not even remotely capable of predicting the hundreds of space groups and thousands of crystal structures found in crystalline intermetallic compounds. In general, of course, metallic glasses are further from equilibrium than network glasses and experimentally more rapid quenches are required to avoid crystallization. In the better metallic glasses, however, quench times longer than 10^{-6} sec are common. With hard spheres to avoid crystallization, computer simulations generally cover times on the order of 10^{-10} sec or less.

For hard spheres constraint theory predicts a coordination number at contact of $CN = 6$, which is observed to hold in computer simulations. Although valence-force-field constraints give a good account of the crystal structures and composition dependence of glass formation for network materials, hard spheres are totally inadequate for the description of the structures of metallic compounds and alloys. The basic reason for this is that metallic radii vary with coordination number in a systematic way described, for example, by Pauling.[20] Thus structure and atomic size appear to be inextricably interwoven.

Another well-known feature of metallic glasses[21] is that they are always associated with deep eutectics, $0.2 \lesssim \Delta T/T_m \lesssim 0.4$. Hard spheres tell us nothing about equilibrium phase diagrams.

Recently these problems of structure and phase diagrams of metallic alloys and compounds have begun to yield to systematic analysis based on abstract concepts originally associated with pseudopotential theory. I have not space here to discuss these results in detail, and must be content with references.[22,23] The results may be summarized by saying that none of the various hard-sphere theories have any bearing at all on the structure of metallic glasses. These are much better understood in terms of microcrystalline cluster models.[24,25] The cluster sizes are at least 10–15 Å.

One simple observation[23] dramatizes the inadequacy of sphere-packing models. Inspection of quantum structural diagrams[22] immediately reveals a necessary condition for the formation of a eutectic. For two non-compound-forming elements A and B to exhibit even a shallow eutectic it is necessary that their crystal structures be different. If A and B atoms are modeled as hard spheres, however, their crystal structures will always be close packed. Thus in the hard-sphere approximation, no binary phase diagram will ever exhibit even a shallow eutectic, much less a deep one, and such a model cannot explain the origin of the glass-forming tendency.

PROSPECTS

At this writing it appears that the evidence for SMC clusters is growing rapidly.[26] Perhaps the most exciting avenues for further development may arise in connection with the OPE.[19]

REFERENCES

1. PHILLIPS, J. C. 1979. J. Non-Cryst. Solids **34:** 153.
2. BELL, R. J. & P. DEAN. 1972. Philos. Mag. **25:** 1381.
3. WOOTEN, F., K. WINER & D. WEAIRE. 1985. Phys. Rev. Lett. **54:** 1392.
4. PHILLIPS, J. C. 1982. Phys. Today **35** (February): 1.
5. PHILLIPS, J. C. 1982. *In* Structure of Non-Crystalline Materials. P. H. Gaskell, J. M. Parker & E. A. Davis, Eds.: 123. Taylor & Francis. London.
6. PHILLIPS, J. C. 1985. *In* Highlights of Condensed Matter Theory. F. Bassani, F. Fumi & M. P. Tosi, Eds.: 720. North-Holland. Amsterdam.
7. PHILLIPS, J. C. & M. F. THORPE. 1985. Solid State Commun. **53:** 699.
8. THORPE, M. F. 1986. Ann. N.Y. Acad. Sci. This volume.
9. PHILLIPS, J. C. 1985. Phys. Rev. B **31:** 8157.
10. PORAI-KOSHITS, E. A. 1984. *In* Phase Separation in Glass. O. V. Mazurin & E. A. Porai-Koshits, Eds.: 1. North-Holland. Amsterdam.
11. WONG, J. & C. A. ANGELL. 1976. Glass Structure by Spectroscopy. Marcel Dekker. New York, NY.
12. BRIDENBAUGH, P. M., G. P. ESPINSOA, J. E. GRIFFITHS, J. C. PHILLIPS & J. P. REMEIKA. 1979. Phys. Rev. B **20:** 4140.
13. BOOLCHAND, P., J. GROTHAUS & J. C. PHILLIPS. 1983. Solid State Commun. **45:** 183.
14. STEVENS, M., J. GROTHAUS, P. BOOLCHAND & J. G. HERNANDEZ. 1983. Solid State Commun. **47:** 199.
15. RUFFOLO, D. & P. BOOLCHAND. 1985. Phys. Rev. Lett. **55:** 242.
16. PHILLIPS, J. C. 1983. Solid State Commun. **47:** 203.
17. GRIFFITHS, J. E. 1982. Phys. Rev. B **25**: 1272.
18. VEPREK, S., Z. IQBAL & F. A. SAROTT. 1982. Philos. Mag. B **25**: 137.
19. OURMAZD, A., J. C. BEAN & J. C. PHILLIPS. 1985. Phys. Rev. Lett. **55:** 1599.
20. PAULING, L. 1960. The Nature of the Chemical Bond: 413. Cornell University Press. Ithaca, NY.
21. DONALD, I. W. & H. A. DAVIES. 1978. J. Non-Cryst. Solids **30:** 77.
22. VILLARS, P. 1985. J. Less-Common Met. **109:** 93.
23. PHILLIPS, J. C. 1986. J. Mater. Res. **1:** 1.
24. GASKELL, P. H. 1979. J. Non-Cryst. Solids **32:** 207.
25. COCCO, G. 1984. J. Non-Cryst. Solids **61** & **62:** 577.
26. GASKELL, P. H. 1985. Nature **317:** 285.

DISCUSSION OF THE PAPER

R. CAPELLETTI (*Ohio University, Athens, OH*): One obvious question to ask of a "Newtonian" theory is how it accounts for dynamics. In particular, can you sketch how the constraint theory, which does so well in predicting glass-forming tendency, will deal with the diverging viscosity?

PHILLIPS: Constraint theory is a mean-field theory that assumes mechanical equilibrium. As a glass-forming liquid is supercooled, the viscosity diverges according to the Vogel–Fulcher law. The physical picture of this divergence, as suggested for instance by Kauzmann, is that as the temperature is lowered clusters of larger and larger size form in the liquid. The steric hindrance of rotation by these clusters explains the divergence of the viscosity qualitatively.

I have tried to extend the Kauzmann picture of this process to include configuration entropy and specific heat, which may be as large as the Dulong–Petit vibrational contributions (*J. Non-Cryst. Solids* **44:** 17, 1981). These clusters are analogous to fluctuations in equilibrium statistical mechanics and so, strictly speaking, they lie outside the scope of constraint theory. The latter, however, does explain how such large clusters can form and reconstruct their surfaces in such a way as to grow so slowly that the supercooled liquid does not crystallize.

Panel Discussion

C. A. ANGELL (*Purdue University, West Lafayette, IN*): This is our final session, and it is intended as a commentary on and a review of some of the important points this conference has helped to bring to the forefront, as well as an opportunity to draw attention to points of importance to the phenomenology of relaxation and configurational arrest leading to the glassy state that may have been underemphasized or neglected. I am going to start the session by inviting the panelists to comment on the conference papers and related issues, after which I hope residual problems or omissions can be exposed and succinctly elucidated by our speakers and others. Perhaps I can begin with our old and valued colleague and relaxation phenomenon expert, Cornelius Moynihan.

C. T. MOYNIHAN (*Rensselaer Polytechnic Institute, Troy, NY*): There are two things that struck me particularly. The first was Dr. Whalley's talk, where he suggested the existence of a phase transition between two amorphous phases. Dr. Nagel did point out that the liquid-vapor phase transition is an amorphous phase transition, with which we are all familiar. Nonetheless, Whalley's suggestion seems remarkable, and I was a little bit surprised that there was not more comment on this. I still think Whalley may merely be seeing some sort of stabilization process rather than a real phase transition. It would be very convincing, if this happened to be a transition that has to be nucleated, if one could produce a partially transformed material. On the other hand, this could be a transition that does not require nucleation, such as a martensitic transition. One usually thinks, however, that a well-defined crystal structure is needed for a martensitic transition. I hope investigations will continue on Whalley's phenomenon. It may be unique to a few systems, but I thought it was very remarkable.

The other thing that struck me was said both in the paper by Dr. Frederickson and in some mention that was made of curve fitting of relaxational data using the Adam–Gibbs theory. This was that one could predict the nonlinear character of the relaxation from data taken strictly in the linear regime or on the equilibrium liquid. This sounds very nice and elegant, but in a sense it is a little bit disturbing if you look at it from the irreversible thermodynamic standpoint. The region where you get linear relaxation is right in the vicinity of the minimum of the free energy surface. In linear irreversible thermodynamics one predicts nonequilbrium free energies via a Taylor series expansion about the minimum in which only the lowest order nonzero terms are retained. The nonlinearity presumably comes about because further from equilibrium you need to retain higher order terms and these reflect some sort of asymmetry in the bottom of the free energy surface. Consequently, one in general should not be able to predict the nonlinear behavior from the linear behavior. This prediction of nonlinear behavior from parameters characterizing the linear regime seems to be coming out of several treatments, but I wonder whether it is really expected.

ANGELL: Before seeking any response to those comments we would like to hear from Professor Anderson, who needs no introduction.

P. W. ANDERSON (*Princeton University, Princeton, NJ*): Several difficulties and minor disasters prevented me from getting to as much of this conference as I would have liked.

There was one thing I wanted to say that really has to do with a lot of the papers but is a bit extraneous to the material of the conference. I am now calling it the Bardeen identity; I first came upon it 32 years ago in trying to fit the motion of ferroelectric domain walls. There is this very strange identity. The function exp $\{-1/x\}$ happens to be equal to $(2/e)^2x - 0.13$ for about one decade. It is a remarkable fit over the region around $x = \frac{1}{2}$. It has something to do with there being a point of inflection in the middle and two zeros of the third derivative and a cluster of zeros of the fourth derivative that appear in the same region. All this is just a matter of accident. I call it the Bardeen identity because he is trying to use a tunneling theory that gives you things like exp $(-1/E)$ for certain data, and because other people are trying to use a coercive force theory that gives $E - C$, where E is the electric field.

We had exactly the same thing in 1952 except it wasn't a tunneling theory, it was a nucleation theory, and in that case it happened to be $e^{-1/x}$ that was right.

Now let's call $T - T_0 = x$, and write the Bardeen identity $e^{-1/(T-T_0)} = T-T_0^1$, which implies $e^{-12/(T-T_0)} = (T-T_0^1)^{12}$. Thus we may see that Dr. Nagel cannot tell whether he's got a Vogel–Fulcher or a scaling law.

Actually, even Kohlrausch runs into the same kind of problem: you cannot quite tell whether you have exp $(-t^\beta)$ or a constant minus $1/t^\beta$. And that is why there has been difficulty in fitting some of the relaxation data on the spin glasses. Unfortunately, there is no *a priori* reason to believe either theory is right; you just have to look at the parameters. I happen to believe that Dr. Nagel is right and that scaling is nonsense in the case of glasses. It is very suspicious, in fact, that that number 12 is required because you have to take the 12th root in order to get back to the Bardeen identity.

On the other hand, Dr. Palmer says, and I believe, that spin glasses fit better to the power law, but you have to be very clever to make these fits really work, and I think you have to have physical reasons for believing one or the other is really right.

ANGELL: What's your physical reason for believing that Dr. Nagel is right?

ANDERSON: In which case? In the glass? My physical reason has to do with the fact that in the glass I really suspect that Arrhenius behavior at high temperatures is what you should be going towards; moreover, I do not see why the scaling theory should fit so far outside of its critical region.

So, just thinking about a few papers, I began to realize that there was a coincidence, but it is a well-known coincidence.

There was a whole series of papers on hierarchical models for relaxation, and I think the more you think about it the more you realize—perhaps I am being a little prejudiced here—that some kind of hierarchy has to be responsible for the glass, for the actual details of going through the glass transition and these anomalous relaxations and so forth.

As I said in my comments, you can find exp $\{-(t)^\beta\}$ with β equals a half, as a result of diffusion of the activation energy: β equals a quarter, as a result of reptation, and β equals a third, as a result of percolation in three-dimensional space. There has to be some theory, however, some kind of hierarchial theory that gives you Kohlrausch behavior for the relaxation data and that gives you something like a Vogel–Fulcher law.

I was impressed by the remarks about the isotope effects in these lithium glasses: is it really all that uncommon?

For the last talk, Dr. Phillips's talk, I have a problem I could have pointed out right after the talk. Isn't it true that silicon is in the ill-condensed matter area and that glass is in limbo? I do not see why something that is true of silicon is true of something that's in limbo. Frankly, I have always believed amorphous silicon was microcrystalline: it hates to be glassy. I do not think it is ever anything like a glass.

I believe I was long before Phillips in saying, at least in public talks, that silicon is not a glass and that glass is glass: transparent, brittle, comes from a melt, and is a supercooled liquid. And I believe it probably has something to do with constraint theory, as he says also.

Okay, I think that is enough.

ANGELL: In the general discussion Dr. Phillips can have the first shot. Now to Professor Nelson.

D. R. NELSON (*Harvard University, Cambridge, MA*): Well, as I am a mere Ptolemaic physicist (according to Dr. Phillips!) I hesitate to make any comments, especially in view of my youth. I will instead pose questions. These questions are directed at all the chemists in the audience. I do not know much about organic glass formers, but I think they are very interesting.

The question has to do with typical organic glass formers: perhaps somebody can explain them to me or comment. Now, there can be bad or good organic glass formers just as in the other kinds of glass we have heard about. You take something like benzene, a nice sort of simple-looking thing: it is a bad glass former in general. On the other hand, if you take benzene and merely hang a methyl group on the end, you have toluene, which is a very good glass former. I gather from Dr. Angell that you can make benzene glassy but only by trickery. The same thing happens if you look at xylene. Ordinarily, para xylene is a very bad glass former, but if you simply take one of the methyl groups and cant it over to one side to make meta xylene, it becomes a glass former. I have the following questions.

If I make conventional toluene glass and I somehow by trickery (using microemulsions) make a glass out of benzene as well, are the structures going to be similar? Is the dangling methyl group terribly important in the actual structure? Or does it merely slow down the dynamics and allow things to get entangled, sort of like a hook on the end of the molecule? It is a little bit like the issue that Dr. Gaskell raised about phosphorus atoms in nickel. That is, are they there to prevent premature crystallization or are they really a fundamental piece of the structure? Would you expect the order of the centers of mass of benzene and toluene glass to be the same? In addition, would these organic glasses typically have order in their centers of masses similar to a Lennard–Jones glass in a computer simulation or would they be something fundamentally different?

ANGELL: I'd like to respond to at least a part of this collection of questions, because I have a couple of firm ideas about organic glasses. The first idea, which I believe we have quantified, is that toluene is a good glass former, not primarily because of any entanglements in the liquid state (though there is no question that molecules are jammed at the glass transition) but rather because symmetry considerations prevent them from forming stable lattices, leading them to melt at temperatures that are low relative to their boiling points, that is, to the intermolecular forces that determine their viscosities, and hence their crystallization kinetics. For instance, the lower melting point, which makes meta xylene more readily vitrifiable than the ortho or para isomers, can be traced to a lattice energy for the meta isomer, which is just 1 kcal/mole smaller than for either of the others (C. A. Angell *et al.*, *J. Chim. Phys.* **82:** 267, 1985; J. Wong and C. A. Angell, *Glass: Structure by Spectroscopy,* Chapter 1, Marcel Dekker, New York, NY, 1976).

The second idea, which follows directly from the first, is that there may be more organic glasses than we know about because many compounds that are too heavy to be purified by distillation do not enter the chemical catalogues because they resist crystallization and hence are not easily obtained as pure chemicals by recrystallization.

NELSON: So it would follow then that the kind of glass that you would get from benzene would be very much the same as toluene glass, since all you're doing is just suppressing crystallization?

ANGELL: Yes. I might add that argon is an extreme case and that it fuses at 0.96 T_b (the boiling point), where the viscosity is ~50 times less than that of toluene at its melting point. The viscosity changes so relatively slowly with decreasing temperature in this case that the crystallization rate rises to a value some 10^{20} greater than that of toluene before the maximum rate (minimum "escape" time, τ) is passed (see the references mentioned above). This is why it requires the quenching rate of computer simulation "experiments," which are ~10^8 faster than ordinary laboratory quenching rates, to vitrify it.

MOYNIHAN: Possible examples, though perhaps extreme ones, are the alkanes. The straight-chain hydrocarbons are among the most crystallizable of liquids. It is very difficult to supercool these things more than a couple of degrees. You just take a few carbons off the end and stick them on the middle to form a branch, and you get a branched alkane—very difficult to crystallize. Again, I think the reason is probably just what Dr. Angell said: it is probably difficult to get the crystal nucleus organized to start with; you have some sort of entropy barrier there.

ANDERSON: I think years ago, when I began to think about glasses, it was this fact that the organic glasses are just these medium-sized molecules of peculiar shapes that turned me away from any kind of general geometrical theory of the things, into the direction of a kinetic and what has now become a hierarchial theory, the theory that had to be basically dynamic and kinetic.

NELSON: Certainly, one must think about dynamics as well as structure. I hope with the high-resolution synchroton experiments we can get a better idea of how complicated organic molecules are arranged in space. Perhaps the centers of mass will not be so different from the atomic positions in a metallic glass.

ANDERSON: The really nice glass is glycerol: we should all face it.

NELSON: Yes, right. That is all I have to say.

ANGELL: Our cochairman, Dr. Goldstein, has had longer contact with the theme of this conference than anyone else in the auditorium. His wind-up comments will be of value.

M. GOLDSTEIN (*Yeshiva University, New York, NY*): I would like to comment on something I heard very little about at this conference: the so-called β-relaxation. We look for general features of glasses, recognizing that so many different kinds of substances form glasses. I am not sure there is any one general feature that is universally present in every single glass, but the occurrence of the secondary relaxation a short distance below the occurrence of the glass transition itself is as universal as any other feature. If, for example, you are using a measuring frequency of 1 kHz you will see the secondary relaxation perhaps 10° or 20° below the main transition. This seems to occur, not in every single glass former, but probably in about three-quarters of all glass formers, and this is regardless of whether they are simple molecular liquids, fused salts, or whatever. It is therefore a nearly general feature, and it would be nice to have some greater degree of attention drawn to it.

It did cross my mind when I heard Dr. Thorpe talk about structures in which there is a certain amount of floppiness left over once the rigidity has formed; it sounded to me more like a β-relaxation kind of thing than the two-level systems that are associated with the low-temperature anomalies. The other time that it appeared in this conference was in Dr. Loidl's very detailed molecular discussion of the dipolar and quadrupolar relaxations in the mixed cyanide, bromide crystals, which have many of the characteristics of the main relaxation–β-relaxation pair. But I think it is a subject that deserves notice.

Another comment suggested by something in Dr. Frederickson's talk: many thermodynamic models proposed for liquids or disordered systems give an entropy-temperature relation where there is at some temperature a point of inflection (corresponding to a maximum in specific heat). Obviously a linear extrapolation of the entropy from this point will have a T-axis intercept (T_0) at some positive T. It would be useful to have a criterion to tell us whether the model does or does not show a "Kauzmann paradox."

I would like to suggest the dimensionless ratio S/C_p at T_g (the point of inflection or the temperature identified on other grounds as the glass temperature). A simple geometric argument shows this to be equal to the ratio $(T_g–T_0)/T_g$. For glass-forming liquids, measured values of $\Delta S/\Delta C_p$ (these are excess quantities: $\Delta S = S$(liquid) − S(crystal) at T_g; $\Delta C_p = C_p$(liquid) − C_p(glass)) are typically about 0.2.

ANGELL: Thank you. Now to a general discussion.

J. C. PHILLIPS (*AT & T Bell Laboratories, Murray Hill, NJ*): I just wanted to comment that we are well aware that amorphous silicon is not a glass because when it is annealed it does not melt into a supercooled liquid but instead it crystallizes explosively. In high-resolution electron microscopy, however, lattice images are formed by columns of atoms in oriented submicrocrystallites. These experiments, though straightforward in principle, are complex in practice, and many disappointing results have been obtained over the last twenty years (for example, by Howie's group at Cambridge). In these circumstances you exercise the option available to those working in a field that is still in its infancy and you choose a sample that is the best available because of modern technology. Clearly amorphous silicon and amorphous germanium are the ideal starting points. After you have perfected the geometry and learned to deal with the many practical problems that come up in any new experimental arrangement, you then pass on to other systems. Of course, the ideal system that heads our list is SiO_2, where we would like to study the interfacial region between crystobalite and vitreous silica.

Although I am on the subject of vitreous silica, I would like to say that I have always thought that the organic glasses (and I do not want to offend any of the organic chemists in the audience) are essentially trivial. I mean you stir up an organic mess and the more side groups you tack on in asymmetric positions the more you would expect to inhibit crystallization.

If you asked me to name the prototypical glass, I, as an old-fashioned person, would simply name window glass, whose basic component is SiO_2.

ANDERSON: One comment I want to make is that many people think of vitreous silica as the prototypical glass, but it is not a very good glass. Its glass-formation point is too high, and there is a lot of evidence that there is vacancy diffusion and Arrhenius-type behavior confusing the glass transition in this case. Vitreous silica containing some of this contemptible junk-like sodium and potassium and other glass formers is a much better canonical glass than pure vitreous silica, a much better example of the glass.

To defend the organic chemists I think they can make glycerol extremely pure and made it extremely pure long before the physicists made silicon pure. What is more, Giauque measured glycerol first. I believe he made on it the first specific heat measurements on any glass.

PHILLIPS: Modern technology is all very well, but one should explain the results that were found in the 1920's before one deals with modern technology.

ANGELL: I think this raises a question that this conference should address, especially in view of some of the contributions on systems that seem in many ways to demand to be classified as "glass" and yet in other ways do not. We should ask what are the criteria on which we should insist before recognizing some structural "freezing" phenomenon as a "vitrification" leading to a "vitreous" or "glassy" state. For instance, everyone accepts that SiO_2 and its weak relative BeF_2 are glasses, yet they are not typical glasses because they show only weak or unnoticeable heat capacity anomalies as relaxation ceases. This means one cannot identify the Simon–Kauzmann vanishing excess entropy problem in these cases. And yet I would think that this is one of the more demanding problems in the whole of supercooled liquid phenomenology. If we admit BeF_2, which, when pure, shows no heat capacity anomaly or secondary relaxation, why shouldn't we also admit the solid solution KBr-KCN, which shows no C_p anomaly but does show both primary and secondary relaxations with the normal relation between them? Then there are Professor Suga's observations on the "glassy crystals." Many of us would agree that many of the "glassy crystals" systems are extremely close in phenomenology to typical liquid-formed glasses, but then you have cases like the jamming of relaxation in ice I, where it is hardly reasonable to include it as an example of vitrification. I wonder if we could make some attempt to formulate a definition of glass formation that we can all be happy with? Let me direct the question to Dr. Moynihan.

MOYNIHAN: I have no really good definition of a glass in the broadest sense. Moreover, a good definition may not be feasible in the context of current thought. I believe that the most important feature of a glass is that it is a system that is kinetically stalled in a thermodynamically unstable state—unstable as opposed to metastable, as opposed to stable. One might of course be able to achieve this situation in the gas phase by mixing gases together that really want to react, except that the kinetics are a little slow. Hence we probably should confine the term "glass" to condensed matter at least. Beyond that one has all sorts of examples of systems that on rapid quenching, pressurization, or other perturbation do not have time to come to thermodynamic equilibrium internally and are frozen into an unstable state. Are these glasses or aren't they?

My own personal feeling is that maybe we ought to confine the term "glass" to the traditional amorphous structures or things that are pretty much like them. This means that plastic crystals might fall into this category, or not too far beyond it. The important thing is that people use words clearly and precisely, so that they can communicate with one another accurately. If somebody wants to call a thermodynamically unstable system a glass, but the system is not unstable in the way that fused silica or polystyrene are, it is important that this be communicated clearly. This way we may hope to minimize confusion.

J. P. SETHNA (*Cornell University, Ithaca, NY*): I think you should probably delay making final definitions until after you have understood what the important common features are that glasses share. In the meantime, you should describe the materials in ways that avoid confusing the issues. I certainly would not want to call KBr-KCN a glass, even though it shares the glassy low-temperature properties, because it obviously does not share many other important traits, especially those involving the glass transition. Calling it glassy seems unwarranted.

D. FISHER (*AT&T Bell Laboratories, Murray Hill, NJ*): Since others have brought up the question of KBr-KCN, it seems to me that it is much closer to spin glass systems than glasses. This is because there is a natural separation of time scales given to one by God in that the diffusion time for rearranging the bromine and cyanide ions is much longer than any other time scale. So we assume this motion is completely frozen-out, making KBr-KCN a system with quenched disorder much like spin glasses. In this way, it is rather a different kettle of fish from true glasses that do not have the natural enormous separation of time scales of the quenched systems.

A. LOIDL (*Institut für Physik, Universität Mainz, Mainz, Federal Republic of Germany*): That is only partially true. The difference between the spin glass systems and KBr-KCN, for instance, is that you have a strong rotation-translation coupling and that the freezing or the slowing down of the "spins" destroys the long-range translational order. At low temperature you have no long-range orientational order and no long-range translational order. I feel in this respect it is very different from the spin glass systems. And that is my point: that it is in between spin glasses and canonical glasses. Maybe it is a new type of a glassy system.

ANGELL: It definitely has quenched-in disorder?

LOIDL: That is correct, yes.

ANGELL: Well, shall we turn to another issue?

S. R. NAGEL (*James Franck Institute, University of Chicago, Chicago, IL*): We have heard several different theoretical models for what glasses are like, what the glass transition is, and why we get the Kohlrausch–Williams–Watts distribution. As an experimentalist, I am left with a very puzzled feeling. I do not know with what experiments the theorists are proposing to test these models. What experimental tests distinguish between them? So, for example, in the hierarchical models, how do I tell that theory apart from the theory of Grest and Cohen or Ngai's theory or Schlesinger's picture? Are there any experiments that can be done to elucidate the difference between them? Also, for the various theories of the glass transition (the hydrodynamic theories, the free volume theories, the facilitated spin theories), what experiments will give us a clue as to which, if any, is correct? As Dr. Anderson was saying, you cannot tell very much just by looking at the scaling law behavior. In particular, you cannot tell the difference between scaling law and Vogel–Fulcher behavior. Are there other things that are going to be experimentally testable in these theories?

GOLDSTEIN: I wonder if I could make a remark on that. One of the places where there is some experimental information but not enough has to do with the question of whether there really is a viscosity divergence below the glass transition. Most of our empirical equations for fitting viscosity indicate it, but several experiments done about a decade ago in several different laboratories have shown that if you make a very careful analysis on at least some liquids as they approach the glass transition there is an indication that the behavior becomes Arrhenius. We do not know whether this is true for a large class of liquids. I do not think there are any examples of a return to Arrhenius behavior with polymer systems, but perhaps this question has not been carefully examined for them. But I think it is a very important gap in our experimental knowledge of the transition.

R. G. PALMER (*Duke University, Durham, NC*): I do not think there is an immediate answer to Dr. Nagel's question. I think that a historical view is perhaps appropriate here, in that a few years ago the state of the art for the theorist was that there were several very common ubiquitous phenomena—the Vogel-Fulcher law, the Kohlrausch law, the Kauzmann paradox—that demanded an explanation. No viable explanations were derived from these phenomena, however.

In the last few years we have seen many toy models that have come from basic assumptions in configuration space, or defect diffusion, or whatever. Many of them now seem to display Kohlrausch-like behavior. I do not think there is a way of distinguishing between them at this stage on an experimental basis. I think we are lucky to have any of them. I think that, looking at them collectively, perhaps the one thing you notice is that almost all of these models are saying that what you basically need are dynamical constraints. That is what I would be pushing. In one way or another you can see that in essentially all of them. But we are a long way from making a good microscopic theory that is capable of many experimental tests.

FISHER: I have a fundamental question concerning the various theoretical approaches that we have seen attacked from different ends. What are the length scales involved in the glass transition, whether or not one believes there is a real transition? I think that the point that Dr. Phillips was making is that something rather microscopic is going on with very short scales, scales involving the constraints of the numbers of bonds and so on. On the other hand, the hydrodynamic theories have assumed that freezing occurs on rather long scales. Furthermore, some of the Ginsburg–Landau-type theories that Dr. Nelson alluded to also assume that some rather long scales are involved. One of my worries about the hierarchial models is that I am not sure how they tie into real space or into objects outside of phase space at all.

There seems to be some fundamental question here about whether the length scales that are important in understanding the glass transition are a couple of atomic spacings or many tens or hundreds of times longer, and about how large a blob of something one needs before it has a dramatic change in its dynamical properties as it is cooled down. It seems to me that the answer one would get from the different approaches would be rather different.

ANGELL: Well, the question of whether the long-range correlations come into the static picture or only into the dynamics is a critical issue. I know of one suggestion for an experiment that might make some distinction, and about which Dr. Frederickson might say a word. I was thinking of that prediction of excess light scattering during relaxation that I have heard associated with your model.

G. H. FREDRICKSON (*AT&T Bell Laboratories, Murray Hill, NJ*): Actually, that prediction is not mine, but is something that Steve Brawer has been working on. The basic idea is that there is a relevant length in a nonequilibrium fluid near a glass transition that is much larger than any static correlation length at thermodynamic equilibrium. In the spin model this length might be identified with the size of a "patch" of spins in the vicinity of a local perturbation that must undergo some sort of cooperative rearrangement to relax the perturbation. Steve's prediction is that if this dynamic length is associated with density-density fluctuations in a nonequilibrium liquid, it should be seen in quasi-elastic light-scattering experiments on samples that have been subjected to finite temperature jumps in the vicinity of the glass transition. It is not clear to me, however, that this length will necessarily manifest itself in the pair correlation function of the density.

ANGELL: Let me clarify a little. As I understand your model, you account for the dynamics without requiring any long-range static correlations, whereas, if I understand what Dr. Stillinger was telling us, his model builds up static correlations over considerable distances. Now this is a clear kind of distinction between the two, and there should be an experimental means of distinguishing which is correct. It is important because we have to account not just for the dynamics but also the thermodynamic aspects of the glass transition phenomenon.

FREDRICKSON: The crucial point here is how one interprets these two models. In mine a spin can be thought of as a region of fluid whose size is the same order of magnitude as the equilibrium correlation length in the fluid at the appropriate temperature. The legitimacy of neglecting ferromagnetic interactions between spins in the model is a consequence of this identification. Dr. Stillinger's "square" model can also be given an Ising-like interpretation, but with long-range interactions between spins. As long as one stays away from the condensation temperature in the model, however, the equilibrium correlation length is very likely not too large. An important question is whether the square model possesses a "dynamic length" that is longer than this equilibrium length at relevant temperatures. If this is the case, then it might possess kinetic behavior similar to that of the facilitated model.

ANDERSON: I'd like to go back to Dr. Palmer's remarks. Ten years ago there were no theories. Even now I think we are being a little too polite to the theories. I think that at some point one should observe that most of them have one or two glaring defects, so glaring that they are just wrong, just obviously wrong. On the other hand we have hierarchical theories containing entities that are not physical, and that you have not yet identified the physics of, but at least they are not wrong. And I myself prefer the theories that are at least not wrong.

FISHER: But they are not even wrong, though.

ANDERSON: Well, I do not think that is true. I think once you have the concept of constraint in, you have something which is at least relevant. Most of the theories that existed ten years ago, five years ago, and even some that have been invented in the last two or three years are obviously wrong. One does not necessarily need an experiment to decide among theories. One may look at the theory and say that even if the experiment confirmed that theory you still would not believe it. This seems to me to be applicable both to the hydrodynamic theory, and to the older free volume theory.

PHILLIPS: I would just like to say that the cluster sizes that I am talking about are quite large: 50 or 75 Å. The glass transition really has to do with hindered rotation, that is, with the freezing-in of the rotational motion of these quite large clusters—where the astronomically large viscosity comes from.

The other thing is that if that kind of graininess exists in your glass, you will have static structural disorder on that scale and on all longer length scales. So if you do an experiment that probes things at still larger lengths and jiggles the thing in one way or another, you can indeed get a response because the system is frozen into a nonequilibrium state. You will get some kind of response at those still longer lengths. In some respect, however, the length that is of interest from a structural point of view is not a long length: this really is different from equilibrium thermodynamics. The length that is of interest from a structural point of view in glass is the shortest length, so to speak, at which the nonequilibrium effects appear in the structure because that is the largest energy and because that is the thing that takes you furthest from equilibrium.

I. M. HODGE (*Eastman Kodak, Rochester, NY*): We have heard a lot about nonexponential decay functions and so on, but I would like to point out that this is not a particularly unusual feature or a feature restricted to the glass transition. We get these in ordinary "fluid" fluids. I would like to point out that a much more unusual feature of the glass transition and of relaxation in glasses is the nonlinearity, and I would like to see perhaps some more theoretical attention given to this issue.

ANGELL: Can I ask for a response to that question from perhaps the most knowledgeable in this area on the panel, Connie Moynihan?

MOYNIHAN: I am rapidly becoming not the most knowledgeable in this area. The nonlinearity is an important aspect of the glass transition, but it is not unexpected. One always has a linear region for transport laws. If the driving force for the process—the electric field or temperature jump—gets too big, the linear laws do not work anymore. The big problem with the glass transition is that we are exciting structural changes by jumping temperatures or pressures, and it turns out that one encounters the nonlinear region very quickly—temperature jumps of a couple of degrees and pressure jumps of about a kilobar. In dielectric measurements one can go all the way up to electric fields of 10^4 or 10^5 V/cm before one sees any non-ohmic or nonlinear behavior. Similarly, one has to go to quite high shear stresses with ordinary fluids before one sees any non-Newtonian behavior. The nonlinear behavior associated with the glass transition has to be addressed because it is an essential feature of it. I do not think, however, the nonlinearity is unusual or unexpected.

ANGELL: We have now used up all our time, and we can draw this to a conclusion. I would like to thank very heartily everybody who contributed to these proceedings, either by giving talks or asking questions or simply by coming along. I hope we have all benefited from it.

Marty, do you have any concluding remarks to make?

GOLDSTEIN: Just to join Austen in thanking all of you.

The Glass Transition in the Hard-sphere Fluid

MOORAD ALEXANIAN[a] AND SUBIR K. BOSE

Department of Physics
Southern Illinois University at Carbondale
Carbondale, Illinois 62901

The glass-forming tendency in the hard-sphere model may be studied with the aid of molecular dynamics (MD) computer experiments.[1,2] The underlying thermodynamic description of the glass transition follows[3] from the temperature dependence of the excess entropy $S_c(P_0,T)$ for fixed pressure $P = P_0$. A minimum of S_c at $T = T_g$ corresponds to the glass transition temperature. This characterization of T_g gives the correct kinetic aspect of the transition. The configurational entropy S_c is determined just as in a previous work.[4] A particular MD run covering both the metastable (supercooled fluid) phase and the nonequilibrium (amorphous solid) phase, however, is fitted by the single formula[3]

$$PV/NkT = A/\alpha^{\mu} + B + C\alpha$$

for $0 \leqq \alpha \leqq V_f/V_a - 1$ with $A > 0$ and $0 < \mu \leqq 1$, where the relative free volume $\alpha \equiv V/V_a - 1$, the freezing density[5] $N\sigma^3/V_f = 0.949$ for spheres with diameter σ, and N/V_a is the high-density limit of the amorphous solid phase. (For the solid phase,[6] $0 \leqq \alpha \equiv V/V_0 - 1 \leqq V_m/V_0 - 1$, $\mu = 1$, $A = 3.00$, $B = 2.56$, and $C = 0.56$, where the crystalline close-packed volume $V_0 = N\sigma^3/(2)^{1/2}$ and the melting volume $V_m = 1.35 \cdot V_0$.) The MD data for the 32-particle system with $V_a = 1.1506$ (see Table 9 in reference 1) gives, for a linear least-squares fit of $PV/NkT - 3/\alpha^{\mu}$ with $\mu = 0.963$, $B = 0.877$ and $C = 5.84$. In a similar fashion, the MD data for the 500-particle system with $V_a = 1.162 \cdot V_0$ (see Table 1 in reference 2 for $\rho\sigma^3 \geqq 0.9$) gives $\mu = 0.980$, $A = 3.00$, $B = 0.8707$, and $C = 4.914$.

The configurational entropy is evaluated readily[4] with the aid of the equations of state in the supercooled fluid-amorphous solid and crystalline solid branches. The glass transition is found at $\rho_g\sigma^3 = 1.02$ for $N = 32$ and at $\rho_g\sigma^3 = 0.98$ for $N = 500$. The result for the excess entropy S_c is shown in FIGURE 1. The heat capacity, $C_p/Nk = (3/2) - (P^2/NkT)(\partial V/\partial P)_T$, follows from the equation of state on differentiation and is shown in FIGURE 2. Our results show a similar behavior of T_g on the heating or the cooling rate q, namely $dT_g/d|q| > 0$, as found for real glass-forming materials. The average cooling rate[4] $q = 1.3 \times 10^{11}$ K/sec for the $N = 32$ data for argon's physical parameters and gives rise from FIGURE 1 to $T_g = 0.09 \cdot PV_0/Nk = 62.5$ K because[5] $PV_0/NkT_m = 8.27$ and $T_m = 84$ K. The average cooling rate $q = 1.6 \times 10^{12}$ K/sec for the $N = 500$ data[2] in argon units and leads to $T_g = 0.105 \cdot PV_0/Nk = 72.9$ K. It is interesting that the slower quench rate in the case $N = 32$ gives a lower residual entropy for the

[a]Present address: Department of Physics, University of North Carolina at Wilmington, Wilmington, North Carolina 28403.

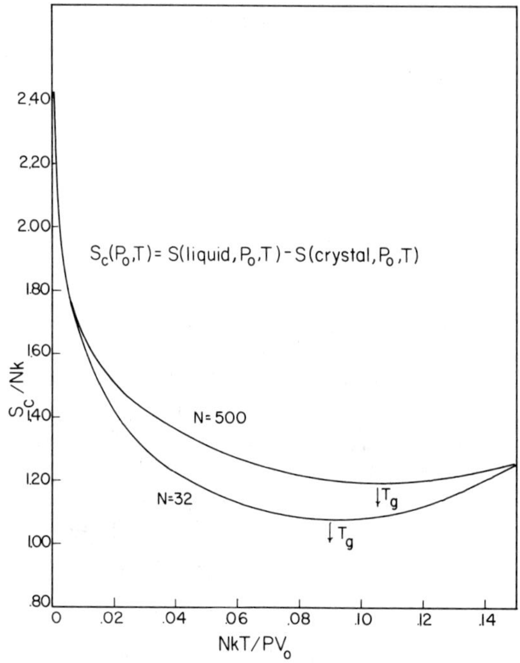

FIGURE 1. Configurational entropy S_c versus NkT/PV_0 (T at constant P) for an N-particle system of hard spheres. The first order phase transition temperature $T_f = 0.121 \cdot PV_0/Nk$ with $V_0 = N\sigma^3/(2)^{1/2}$, and T_g refers to the glass transition temperature. For both curves, $S_c \rightarrow -3 \cdot \ln (NkT/PV_0)$ as $NkT/PV_0 \rightarrow 0$.

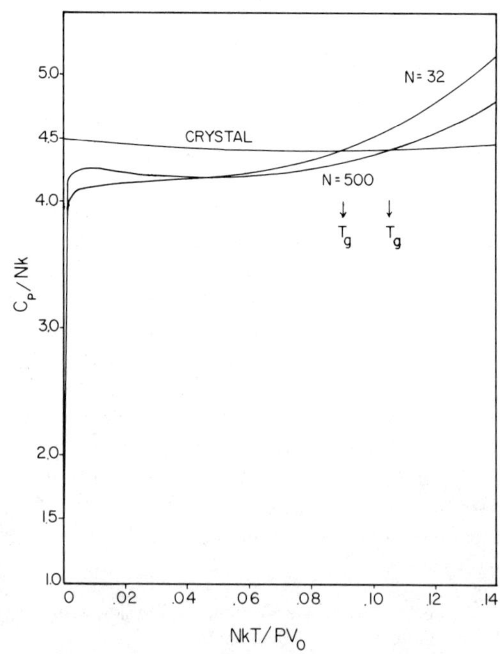

FIGURE 2. Heat capacities at constant temperature for the crystal and glassy states for an N-particle system of hard spheres. The curves are obtained by numerical differentiation of the MD data. For the glassy state with $N = 32$, $C_p/Nk \rightarrow 3/2 + 3.25 (NkT/PV_0)^{0.0384}$ as $NkT/PV_0 \rightarrow 0$. For the glassy state with $N = 500$, $C_p/Nk \rightarrow 3/2 + 3.13 \cdot (NkT/PV_0)^{0.0204}$ as $NkT/PV_0 \rightarrow 0$.

amorphous solid—in addition to a glass transition at a somewhat lower temperature—than the faster quench rate in the case $N = 500$. This is expected because slower quench rates should lead to an enhanced relaxation, a glass transition at a higher density, and a lower residual entropy for the amorphous solid. Therefore, the differences between $N = 32$ and $N = 500$ are due primarily to differing cooling rates rather than finite size effects. Note that the Kauzmann[7] point of zero excess entropy does not arise because $S_c > 0$ for $0 \leqq T \leqq T_f$, where T_f is the freezing temperature. The latter is a direct consequence of the hard-core nature of the potential, which leads to $S_c/Nk \rightarrow +\infty$ as $T/P \rightarrow 0$.

REFERENCES

1. ALDER, B. J. & T. E. WAINWRIGHT. 1960. J. Chem. Phys. **33:** 1439.
2. WOODCOCK, L. V. 1976. J. Chem. Soc. Faraday Trans. 2 **72:** 1667.
3. ALEXANIAN, M. 1985. Phys. Rev. A **31:** 4018.
4. GORDON, J. M., J. H. GIBBS & P. D. FLEMING. 1976. J. Chem. Phys. **65:** 2771.
5. HOOVER, W. G. & F. H. REE. 1968. J. Chem. Phys. **49:** 3609.
6. ALDER, B. J., W. G. HOOVER & D. A. YOUNG. 1968. J. Chem. Phys. **49:** 3688.
7. KAUZMANN, W. 1948. Chem. Rev. **43:** 219.

Barrier Distributions and Defect Migration in Glasses

JOHN T. BENDLER

Polymer Physics and Engineering Branch
General Electric Research and Development Center
Schenectady, New York 12301

MICHAEL F. SHLESINGER

Physics Division
Office of Naval Research
Arlington, Virginia 22217

One approach to modeling dispersive transport in amorphous solids is to employ continuous-time random walks (CTRWs) to simulate spatial and energetic disorder through a pausing-time density $\psi(t)$ for walker motions.[1] Scher and Montroll first applied such methods to describe dispersive charge motion in amorphous semiconductors.[2] Formal justification and clarification followed, and it is now clear that CTRWs and generalized master equations are equivalent descriptions. In practice, the advantage (if any) of a CTRW model arises from the possibility of using physical insight to suggest useful new approximations.

Shlesinger and Montroll[3] found a CTRW version of Glarum's[4] defect diffusion model that gave the fractional exponential form for a dipole orientation survival probability $\phi(t)$

$$\phi(t) = e^{-(t/\tau)^{\alpha}} \tag{1}$$

Bendler and Shlesinger derived a momentless defect $\psi(t)$ by requiring that the defects had to jump over a distribution of free energy barriers in the course of their progress toward the dipole.[5] Their expression (ignoring unimportant constants) for the dipole survival in three dimensions is

$$\phi(t) = \exp\left[-c(\lambda_0 t)^{\alpha}\right] \tag{2}$$

where c is the density of defects and λ_0 is the rate of defect hopping over the smallest barrier. In the case of bond conformer defects in polymers, the minimum barrier is the dilute solution one due solely to intramolecular interactions (~5–15 kcal or so). Comparing equations 1 and 2, the defect diffusion expression for τ is

$$\tau_{DD} = 1/(c^{1/\alpha}\lambda_0) \tag{3}$$

For a barrier of 9 kcal at room temperature, $\lambda_0 \sim 10^{14} \times 10^{-8} \sim 1 \times 10^{6}$. In the barrier dispersion model,[5] it is assumed that configurational fluctuations generate a Poisson distribution of barriers above $E_{\min}$. Thus the distribution of $\delta E \equiv E - E_{\min}$ is

$$\rho(\delta E) = qe^{-q\delta E} \tag{4}$$

The parameter q determines whether transport is normal or dispersive because the exponent $\alpha = kTq$.[5] If the mean barrier fluctuation $\langle \delta E \rangle$ is less than kT, there is a mean hopping time for the defects. When the average $\langle \delta E \rangle$ exceeds kT, no mean time for motion exists and the event times form a fractal set. Notice that α does not depend on the minimum barrier, E_{min}. Also, because $kT \sim 0.6$ kcal at room temperature, rather modest fluctuations produce a dramatic change in the dynamics.

For the above example with an exponent $\alpha = .5$ (that is, $q = 2kT$) and a defect concentration of .1%, equation 3 gives $\tau_{DD} = 1$ sec. Thus megahertz defects at this concentration produce dipole relaxation of about one per second. Because there is no defect barrier renormalization in the model, Williams–Landel–Ferry behavior presumably arises from a drastic decrease in defect density near T_g.

REFERENCES

1. MONTROLL, E. W. & G. H. WEISS. 1965. J. Math. Phys. **6:** 167.
2. SCHER, H. & E. W. MONTROLL. 1975. Phys. Rev. B **12:** 2455.
3. SHLESINGER, M. & E. W. MONTROLL. 1984. Proc. Natl. Acad. Sci. USA **81:** 1280.
4. GLARUM, S. 1960. J. Chem. Phys. **33:** 1371.
5. BENDLER, J. T. & M. F. SHLESINGER. 1985. Macromolecules **18:** 591.

Behavior of Supercooled and Liquid Phase $LiCl \cdot nH_2O$

Thermal and Structural Aspects

A. ELARBY-AOUIZERAT,[a] J. DUPUY,[a] J. M. LÉTOFFÉ,[b] P. CLAUDY,[b] AND P. CHIEUX[c]

[a]*Université Claude-Bernard Villeurbanne, France*

[b]*Institut National des Sciences Appliquées de Lyon Villeurbanne, France*

[c]*Institut Laue-Langevin Grenoble, France*

The phase diagram for the $LiCl \cdot nH_2O$ system has already been established,[1–4] and has been so refined that it may be used to choose heating and cooling rates such that one may obtain equilibrium and nonequilibrium phases.[5] The said refinement permitted the confirmation of the existence of two eutectics between the four compounds shown in the phase diagram: the first lies between H_2O and $LiCl \cdot 3H_2O$; the second, between $LiCl \cdot 5H_2O$ and $LiCl \cdot 2H_2O$. The temperatures for these eutectics are below the glass transition temperature, T_g, over the entire concentration range. Clearly, this system presents a typical example of a glass-to-liquid transition caused by heating a sample that is at T_g. If one begins to heat a sample that is in the liquid state and at T_g, one may subsequently detect the crossing of an extrapolated peritectic line by using differential scanning calorimetry (DSC) and neutron diffraction spectroscopy (NDS). FIGURES 1 & 2 show results obtained with these techniques.

Near T_g, DSC results show that the C_p of the glass depends on the annealing time, τ_N, at a given temperature, T_N; furthermore, differences between determinations for T_g are within the margins of experimental error. It follows that the enthalpy of the glass does not vary, despite earlier reports to the contrary.[6]

During the glass transition, a variation in the first maximum may be detected by using either DSC or NDS. DSC detects a thermal effect that accompanies the

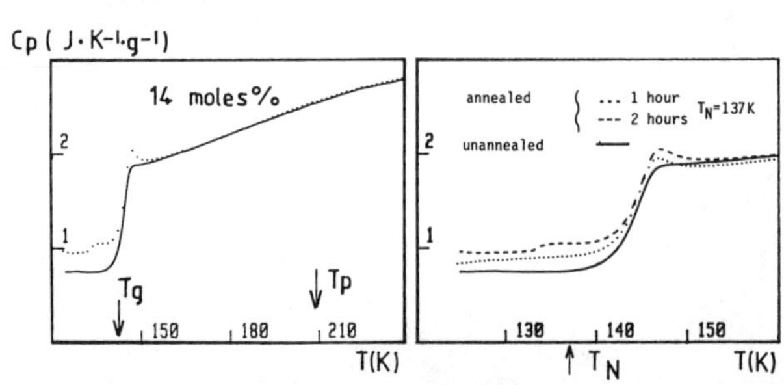

FIGURE 1. *Left panel:* NDS results (near the first eutectic); *right panel:* DSC results (at 0.033 K · sec^{-1}).

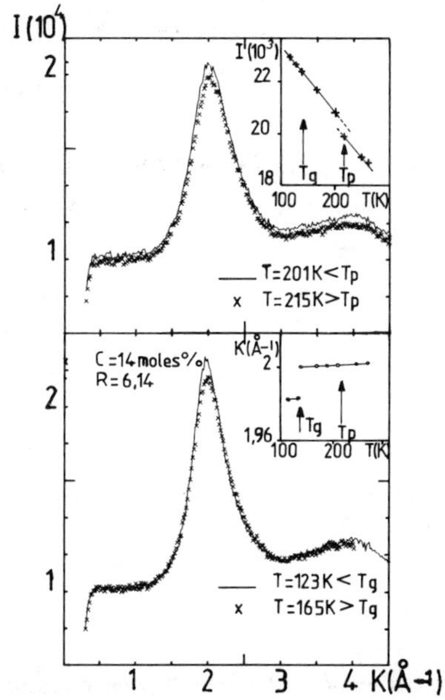

FIGURE 2. *Top panel:* DSC results (at 0.033 $K \cdot sec^{-1}$); *bottom panel:* NDS results (near the first eutectic).

transition; NDS, a change in microscopic order that accompanies the transition. The magnitude of the variation detected using NDS is larger than that detected using DSC. The glass transition is characterized by only the K_{max} variation, without any loss of the diffraction signal.

Near T_p, DSC results show a C_p change (a third-order transition) whereas NDS results show an overall decrease. Both are characteristic of the disappearance of a compound in the system, and both are consistent with the definition of a peritectic line in the solid state.

Near the glass transition, some compaction of the microstructure of glass may be expected, and some compaction is, in fact, observed in transitions between crystals and liquids. At the crossing of a peritectic line, however, the liquid may show modified dynamic behavior, and such a modification could be induced by some change in an interaction of the system (J. K. Kruger, private communication).

REFERENCES

1. Elarby-Aouizerat, A., J. F. Jal, J. Dupuy, P. Chieux, A. Wright & R. Parreins. 1982. J. Phys. (Paris) **43**(12): C9–C205.
2. Elarby-Aouizerat, A., J. F. Jal, J. Dupuy, P. Chieux, A. Wright & R. Parreins. 1982. J. Phys. (Paris) Lett. **43:** L355.
3. Dupuy, J., A. Elarby-Aouizerat, J. F. Jal, P. Chieux, P. Claudy & J. M. Létoffé. 1984. Riv. Staz. Sper. Vetr. **5:** 63.
4. Dupuy, J., A. Elarby-Aouizerat, J. F. Jal, P. Chieux, P. Claudy & J. M. Létoffé. 1985. Paper presented at the Third International Conference on the Structure of Non-Crystalline Materials, July, at Grenoble.
5. Elarby-Aouizerat, A., J. M. Létoffé, J. F. Jal, P. Claudy, P. Chieux & J. Dupuy. Manuscript in preparation.
6. Zarzycki, J. 1983. Les Verres et l'Etat Vitreux. Masson. Paris.

Vogel–Fulcher Time Scale for Glass Freezing

A New Derivation on the Principle of Optimal Network Kinetics

ULF LARSEN[a]

H. C. Ørsted Institute
University of Copenhagen
DK 2100 Copenhagen, Denmark

The empirical law of Vogel and Fulcher describes a temperature-dependent kinetics time scale

$$\tau^* \sim \tau_0 \exp\left(\frac{T_1}{T - T_0}\right) \tag{1}$$

for $T \rightarrow T_0^+$. The law has been found to apply to the freezing of many glass systems (TABLE 1), and may represent a universality class in the contemporary sense. Universal behavior, as expected, submits to secondary effects close to the singular point, that is, below the glass transition temperature, which is somewhat larger than T_0.

Our new interpretation of this disorderly freezing phenomenon involves a theorem on certain stochastic process.[1-3] This theorem made an exact derivation of equation 1 from simple and universal concepts possible.

The singularity at T_0 seems typical of systems with an underlying randomly hierarchical structure. A new kind of transition, which would be one from null-recurrent kinetics to transient kinetics, may be the significance of this singularity.

A system with a random structure may be understood to have an infinity of degenerate ground states and energy barriers separating these states. Detailed balance restricts the process to optimal network kinetics, with a certain energy level probability density $\phi(E)$.[2,3] This principle is illustrated in FIGURE 1.

In the glass-forming universality class we presume that $\phi(E)$ is the convolution of hierarchical densities: $\omega(\epsilon) = \epsilon_0^{-1} \exp(-\epsilon/\epsilon_0)$, as in FIGURE 1, pertaining to a variable number of active units (such as monomers, clusters, or domain walls) with a Poisson distribution. The average number in such a flock is n. The active unit energy scale is $\epsilon_0 = k_B T_0$.

The exact time scale

$$\tau^* = \tau_0 \Phi\left(\frac{1}{k_B T}\right) \Phi\left(-\frac{1}{k_B T}\right) \tag{2}$$

includes an elementary time scale, τ_0, and, for simplicity, is based on a Monte

[a]Address for correspondence: H. C. Ørsted Institute, University of Copenhagen, Universitetsparken 5, DK 2100 Copenhagen, Denmark.

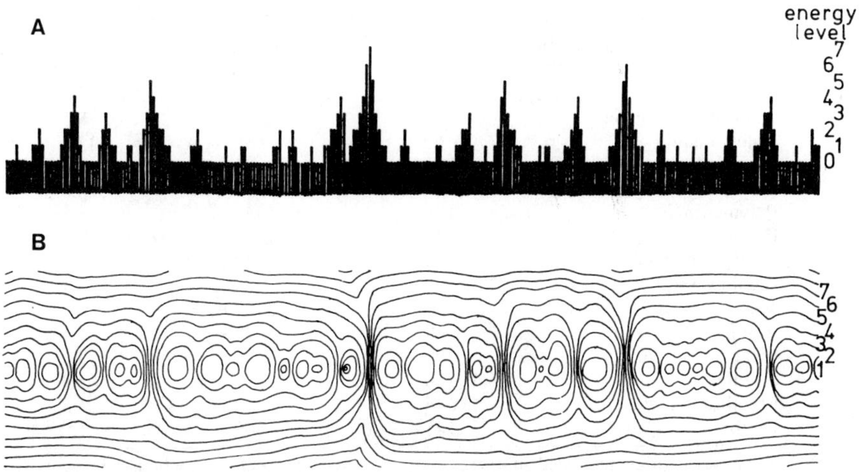

FIGURE 1. A random terraceous landscape, viewed (**A**) in profile and (**B**) from above. The landscape illustrates how the optimal network is formed and how network formation may be caused by the energy bias of detailed balance and the scarcity of energy resources (delays are relatively long if energy requirements are large, and are relatively short if only $k_B T$ is required). The energy bias tends to restrict the stochastic process to a locally one-dimensional network of optimal reaction coordinates, which in the present example is a straight line path. The distribution of energy levels along these paths is random (the distribution in this example, for each unit, is the exponential typical of hierarchical terraceous landscapes) but has short-range correlation. According to reference 3 only the level density matters in the long run. More passages along the optimal network act as barriers (that is, cause significant delays) if energy resources are scarce (that is, if the temperature is low). There are barriers of all sizes, but the larger ones are correspondingly less frequent. The hierarchical compartmentalization of the configuration space is thought to be characteristic of all systems in the present universality class. T_0 is a measure of the energy scale of this structurally determined optimal network, not a barrier height, and n is the average number of active units that are moving on it.

Carlo-type process.[2,3] The diffusion constant is $D^* = 1/\tau^*$, and Φ is the two-sided Laplace transform of ϕ. By the compounding procedure we get $\Phi(s) = \int_{-\infty}^{\infty} \mathrm{d}E\, \phi(E) \exp(-sE) = \exp(-ns\,\epsilon_0/(1 + s\epsilon_0))$.[4] Hence the time scale for glass freezing is

$$\tau^* = \tau_0 \exp\left(\frac{2n}{(T/T_0)^2 - 1}\right) \tag{3}$$

In the $T \rightarrow T_0^+$ limit we get the Vogel–Fulcher law (equation 1) with $T_1 = nT_0$. Neither T_0 nor T_1 is an energy barrier in the conventional sense.

As indicated in TABLE 1, one always finds $n \gtrsim 1$, so there is at least one active unit performing in any system category. In a less globally connected system, that is, one without long-range forces, there are flocks of many units; in a system with long-range interactions, it appears that the entire system, or at least a large part of it, is the active unit.

The performance of the active units may be compared to migrations of flocks of animals in a mountainous landscape (see the hierarchically terraceous landscape in

TABLE 1. Examples of Systems in which the Vogel–Fulcher Law Has Been Detected at Temperatures above the Glass Transition Temperature

Category	T_0 (K)	T_1 (K)	n
Amorphous spin glass	3.8	74	19.5
Alcohol liquid	60	806	13.4
Basaltic lava	885	9201	10.4
Insulating spin glass	.43	4.4	10.0
Polymer melt	178	1443	8.1
Resin	230	1786	7.8
Glassy spin glass	3.09	23.9	7.7
Liquid crystal	168	1070	6.4
Metallic glass	513	3180	6.2
Andesitic lava	1249	5078	4.1
Electrolytic solution	190	675	3.6
Salt melt	320	621	1.9
Metallic spin glass	45	68	1.5
Rubber	161	165	1.0

NOTE: The categories are listed in descending order with respect to n, which represents the average number of active units. Relatively large values of n are seen for systems with short-range interactions (such as magnetic direct exchange); small values of n, values as small as 1, are seen for systems with long-range interactions (such as electrostatic interaction, dipolar magnetic interaction, RKKY indirect exchange interaction, and network cross-linking). The values of T_0 and T_1 are typical and are quoted from the literature.[4]

FIG. 1). The analogy seems true to the extent that it includes the effects of scarcity of energy resources (low temperature) and detailed balance.

When $T > T_0$ energy resources are adequate to keep the active units itinerant, despite hurdles along the optimal network. At $T \leq T_0$ barriers are too frequent and too high for the resources to sustain global kinetics, except in a transient manner, which could involve history dependence. Diffusion ceases, thus disclosing anomalous time dependences of temporarily confined motion in local regions of a configuration space that is compartmentalized on any energy or time scale.

REFERENCES

1. LARSEN, U. 1984. Bounds on the longest time scale in master equations. J. Phys. A: Math. Gen. **17:** 267–276.
2. LARSEN, U. 1984. New exact approach to models of diffusion in one-dimensional random systems and optimal network kinetics in spin glass models. Phys. Lett. A **105:** 307–310.
3. LARSEN, U. 1985. Time scale of thermally activated diffusion in random systems: A new law of thermal activation. Phys. Rev. B **32:** 1772–1777.
4. LARSEN, U. 1986. Glass freezing as kinetics time scale singularity of new significance. Phys. Lett. In press.

Anelastic and Plastic Response of Polymers

D. G. LEGRAND, W. V. OLSZEWSKI, AND J. T. BENDLER

Polymer Physics and Engineering Branch
General Electric Company
Schenectady, New York 12301

INTRODUCTION

The mechanical and optical properties of glassy polymers vary with time, temperature, and history, as well as with static strain or stress. These relationships are the outward manifestations of internal relaxations. Clearly, it is important to understand the mechanisms for internal relaxations; furthermore, it is useful to understand the role of chain structure and processing. The current work addresses relaxation and recovery of polymeric glasses both experimentally and theoretically. Emphasis is placed on a series of novel recovery experiments to evaluate the response of different amorphous polymers, and an analysis is carried out using a simple nonlinear superposition principle. The immediate goal is to determine if the kinetics of recovery is described by the fractional exponential (Kohlrausch–Williams–Watts) function; eventually a molecular-level description may be found in terms of chain-conformer response functions. The prospect for such an interpretation is favorable because similar theoretical parameters (exponents, relaxation times, and prefactors) are found in each experiment (that is, in experiments on stress relaxation, strain, and birefringence recovery) for all polymers studied. A quantity that is found to be of primary interest is the relaxed (asymptotic) modulus below the glass transition temperature.

As noted in the early works of Alfrey[1] and Stein,[2] the deformation of glassy polymers involves localized distortions of bond angles and distances that subsequently relax by both side- and main-chain conformational changes. A variety of optical techniques, such as birefringence, infrared dichroism, and fluorescence depolarization, have been used to detect such mechanisms.[1,2] Birefringence is one of the most useful because of its simplicity and sensitivity. Questions remain, however, regarding the sign and magnitude of birefringence as a function of strain or stress.[3–5]

The basic assumption of the present work is that the kinetics of stress relaxation and the kinetics of recovery are the same; that is, they are both assumed to be governed by the same distribution of relaxation times or molecular correlation function or both. This is rigorously true for a glass that obeys Boltzmann superposition and for the nonlinear superposition model presented below. Recovery kinetics proved to be convenient for data analysis and model fitting and are hence emphasized throughout. (Schapery described circumstances in which kinetic models could only be evaluated by studying recovery data.[6])

EXPERIMENTAL

The materials studied included cast polymethyl methyacrylate and extruded sheets of bisphenol A polycarbonate, polyetherimide, and polybutylene terephthalate. The

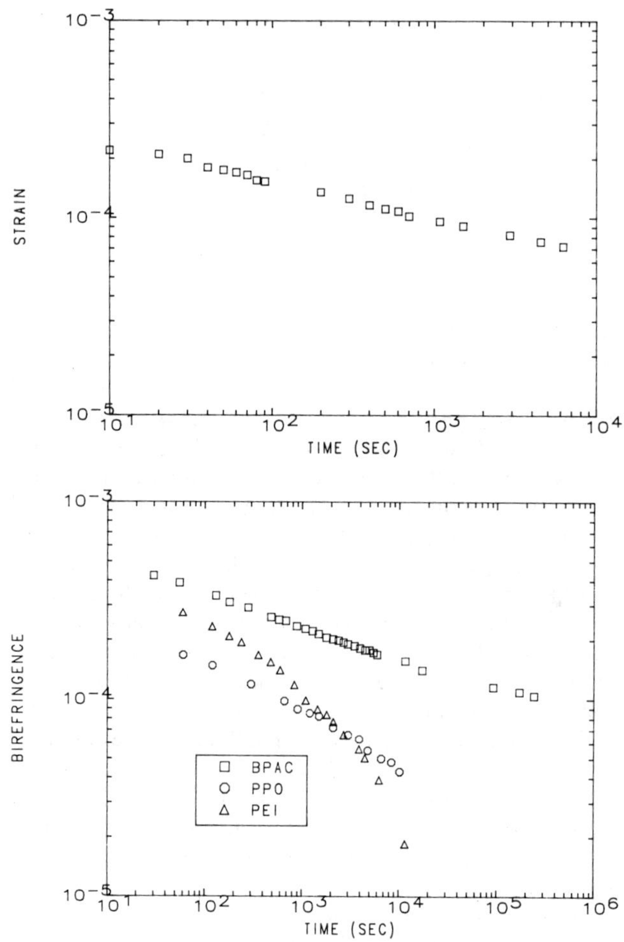

FIGURE 1. *Top panel:* strain recovery of bisphenol A polycarbonate. *Bottom panel:* recovery of polymers.

experiment consisted of imposing a deformation on the sample during a time t_i followed by a hold at fixed length for a time t_h while the force was measured. The system was then returned to its starting position (zero force), and the length and birefringence were measured as a function of time.[7]

THEORETICAL

To model memory effects and nonlinearity, a simple modification of superposition (see Leaderman's superposition[8]) was adopted

$$\sigma(t) = E_u \epsilon(t) - [E_u - E_r] \int_0^t g(t - \tau)\epsilon(\tau)d\tau \tag{1}$$

where the unrelaxed E_u and relaxed E_r moduli depended on the initial $\epsilon(0)$ and final $\epsilon(\infty)$ strains, respectively. The advantage of equation 1 was that it allowed the initial value problem to be handled by Laplace transforms.[9] If $\phi(t)$ was the bond correlation function, the Laplace transform of the strain was[10]

$$\tilde{\epsilon}(s) = \left[\frac{s^{-1}(1 - e^{-st_h}) + \rho\,\tilde{\phi}(s,t_h)}{1 + \rho s\,\tilde{\phi}(s,\infty)}\right]\epsilon_{\mathrm{hold}} \tag{2}$$

where the equation for the relaxation strength ρ was

$$\rho \equiv \frac{[E_u - E_r]}{E_r}$$

Step-function Laplace transforms were defined in equation 2.[10] Use of the Kohlrausch–Williams–Watts function for $\phi(t)$ gave the "early-time" (large s) form for recovery

$$\epsilon(t) = \epsilon_{\mathrm{hold}}\,[\rho/(\rho + 1)]\,[1 - e^{-(t_h/\tau)^{\alpha}}]\,e^{-[t/(\rho+1)\tau]^{\alpha}} \tag{3}$$

RESULTS

Typical strain and birefringence recovery data are shown in FIGURE 1. The fractional exponents, α, and relaxation times, τ (evaluated as described elsewhere[7]), were used to analyze the stress relaxation data for the values of the unrelaxed and relaxed moduli. These are presented in TABLE 1.

DISCUSSION

The values of α in TABLE 1 do not vary much between polymers, nor do they appear to depend on strain, hold time, or thermal history. In contrast, the values of the unrelaxed and relaxed moduli depend on the material, strain, and thermal history. The relaxed moduli are all significantly larger than those of the material in the rubbery

TABLE 1. Parameters for the Various Polymers

Material	History	Strain	Hold Time (sec)	α	τ (sec)	$E_u - E_r$ (GP)	E_r (GP)
BPAC	Press polished	.01	5000	.15	400	2.08	2.30
		.02	5000	.15	400	2.64	1.96
		.03	5000	.15	400	3.65	1.64
	Quenched	.01	5000	.15	400	7.04	1.98
		.02	5000	.15	400	8.64	1.67
		.03	5000	.15	400	14.41	1.24
PEI2	Extruded	.02	5000	.18	100	3.84	2.41
PPO	Injection molded	.02	5000	.15	400	5.94	1.80
PMMA	Cast	.02	5000	.14	135		

NOTE: GP: gigapascal; BPAC: bisphenol A polycarbonate; PEI2: polyetherimide; PPO: (2,6 dimethyl)polyphenylene oxide; PMMA: polymethylmethacrylate.

state, a finding consistent with theoretical results found for a simple two-state conformer model.[7,9] "Long-time" experiments currently in progress also support this prediction, and raise the question of whether the *amount* rather than the *rate* of relaxation varies with strain, sample preparation, and molecular weight in glasses.

REFERENCES

1. ALFREY, T., JR. 1948. Mechanical Properties of High Polymers. Interscience. New York, NY.
2. STEIN, R. S. 1956. *In* Die Physik der Hochpolymeren. H. A. Stuart, Ed. Vol. **4.** Springer. Berlin.
3. LEGRAND, D. G. 1970. Macromolecules **3:** 764.
4. LEGRAND, D. G. 1973. *In* Structure and Properties of Polymer Films. R. W. Lenz & R. S. Stein, Eds. Vol. **1.** Plenum. New York, NY.
5. HONG, S., S. Y. CHUNG, R. F. FEDORS & J. MOACANIN. 1983. J. Polym. Sci. Poly. Phys. Ed. **21:** 1647.
6. SCHAPERY, R. A. 1969. Poly. Eng. Sci. **9:** 295.
7. BENDLER, J. T., W. V. OLSZEWSKI & D. G. LEGRAND. 1985. Am. Chem. Soc. Prepr. **26**(2): 90.
8. LEADERMAN, H. 1943. Elastic and Creep Properties of Filamentous Materials and Other High Polymers. Textile Foundation. Washington, DC.
9. BENDLER, J. T. 1984. J. Stat. Phys. **36:** 625.
10. BENDLER, J. T. 1985. AIP Conf. Proc. **137:** 227.

Structural Relaxation Behavior of Iron-based Metallic Glasses upon Annealing

I. MAJEWSKA-GLABUS

Laboratory of Metallurgy
Delft University of Technology
Delft, the Netherlands

The atomic structure of metallic glass may be frozen into a highly disordered state during quenching, and may then be driven toward metastable equilibrium by low-temperature annealing. This latter phenomena—structural relaxation—marks both reversible and irreversible effects on many physical properties and may be analyzed in terms of a model[1] that discriminates between two subprocesses: 1) chemical short-range ordering, in which compositional changes are described with the use of α (a Warren–Cowley type of chemical short-range ordering parameter), and 2) topological short-range ordering, which is characterized by changes in local density conditions and the loss of excess free volume during annealing.

This model suggests theoretical predictions for structural relaxations that have both a topological component and a chemical component. These predictions are more consistent with experimental results when spectra of activation energies for the reversible, chemical component are considered.[2]

As part of a systematic study of structural relaxation effects in iron-based metallic alloys, differential scanning calorimetry and electrical resistometry were applied to investigate structural relaxation kinetics in amorphous $Fe_{40}Ni_{40}B_{20}$ and $Fe_{40}Ni_{40}P_{20}$. The isochronal measurements of apparent specific heat were performed for as quenched samples and for samples that were isothermally annealed after the quench. Each of the latter samples was annealed for the same amount of time and at 55, 60, 65, 70, 75, and 80% of its glass transition temperature, T_g. The enthalpy relaxation effect for preannealed materials was found to consist of a broad irreversible exothermal peak. Such peaks were found for FeNiP and FeNiB: the former peak had a width of 250 K; the latter peak, a width of 230 K. For each preannealed sample, before the exothermal effect, an endothermal peak developed; the position of each such peak shifted toward higher temperatures with increasing annealing temperatures. Evidence of differences in relaxation behavior between BeNiP and FeNiB was clearly established: the first endothermal peak for FeNiB appeared after annealing at 75% of T_g; the first endothermal peak for FeNiP appeared after annealing at only 60% of T_g. The size of the effect was about four times larger for FeNiP. The differences in shape and magnitude of the Δc_p effects for FeNiB and FeNiP indicate a smaller contribution from topological structural relaxation and a much smaller contribution from chemical structural relaxation in the case of FeNiB. All these results justify the conclusion that FeNiP relaxes faster than FeNiB.

Electrical resistance measurements were performed for FeNiP using an AC Kelvin double bridge. The isothermal variation of the resistance was measured *in situ*, with an overall accuracy of 50 ppm, while the sample was cycled stepwise between 446 K and 543 K. A slow irreversible increase of the resistance (free volume disappearance) was observed. These data were successfully analyzed with a recently developed numerical program, which, by applying the unconstrained optimization method, evaluates six material constants and parameters appearing in the model equations.

REFERENCES

1. VAN DEN BEUKEL, A. & S. RADELAAR. 1983. Acta Metall. **31:** 419.
2. MAJEWSKA-GLABUS, I., B. J. THIJSSE & S. RADELAAR. 1984. J. Non-Cryst. Solids **62 & 62:** 553.

Nonexponential Relaxation of Viscous Supercooled Liquids[a]

M. CRISTINA MARCHETTI[b]

Department of Physics
City College
City University of New York
New York, New York 10031

The slowing down of structural rearrangements on molecular length scales has recently been incorporated into a microscopic dynamical theory of dense liquids by the use of a generalized mode-coupling formalism (see Kirkpatrick[1] and the references therein). The generalized mode-coupling theory that takes into account molecular scale effects, in contrast to the standard hydrodynamic mode-coupling theory, is employed here to describe the nonexponential relaxation of shear and compressional disturbances in viscous supercooled molecular liquids. The basic idea is to recognize that in a dense fluid, density fluctuations decay very slowly at wavenumbers q_0 near the first maximum of the static structure factor $S(q)$. On such length scales the decay occurs via self-diffusion. The renormalization of the transport coefficients due to nonlinearities in the slowly decaying dynamical variables is then evaluated in a standard way.

Recently several authors[2–5] found that a self-consistent description of the nonlinear coupling of density fluctuations in a dense simple fluid predicts a transition to a glass-like phase where dynamical processes are frozen. The viscosity diverges with a weak power law singularity at a temperature higher than observed in real glass-forming fluids. These dynamical theories of the glass transition are therefore inadequate to describe the behavior of real glass-forming liquids in the highly viscous regime. The present theory differs from the latter in that it is self-consistent, but uses the experimental values of the static transport coefficients as an input.

The propagation speed and attenuation coefficient of shear and sound waves in a liquid are related to the real and imaginary parts of the frequency-dependent shear and longitudinal viscosity, $\mu_S(\omega)$ and $\mu_L(\omega)$, respectively. The mode-coupling contribution to $\mu_{S,L}(\omega)$ is calculated in a standard way.[6] For intermediate wavevectors the relevant nonlinearities are those involving density fluctuations, leading to

$$k^2 \mu^{mc}_{S,L}(\omega) = \lim_{\epsilon\to 0^+} k^2 \frac{k_B T}{2} \int \frac{dq}{(2\pi)^3} \int_{\gamma-i\infty}^{\gamma+i\infty} \frac{dz_1}{2\pi i} V_{S,L}(q)\Phi(q,z_1)\Phi(q,\epsilon - i\omega - z_1) \qquad (1)$$

where $V_{S,L}(q)$ is the two-mode-coupling vertex, given elsewhere[7] in terms of the static

[a]Some of the work for this paper was done at Rockefeller University. The work done at Rockefeller University was supported by U. S. Department of Energy Grant DE-AC02-81ER10807; the work done at City College was supported by U. S. Department of Energy Grant DE-FG02-84ER45058 and by U. S. Army Research Office Grant DAAG29-85-K-0112.

[b]Present address: Department of Physics, University of Illinois at Chicago, Box 4348, Chicago, Illinois 60680.

correlation functions of the fluid, and $\Phi(q,z)$ is the normalized dynamic structure factor

$$\Phi(q,z) \approx \frac{1}{z + q^2 k_B T / m S(q) \gamma_L(q,z)} \tag{2}$$

Here $\gamma_L(q,z)$ is a generalized dissipative kernel. For long wavelengths it reduces to $\gamma_L(q,z) \approx q^2 \mu_L(z)/\rho$, with ρ the density; for $q \approx q_0$ it becomes independent of q, since on molecular length scales momentum is no longer a conserved quantity, and $\gamma_L \approx k_B T/mD$.

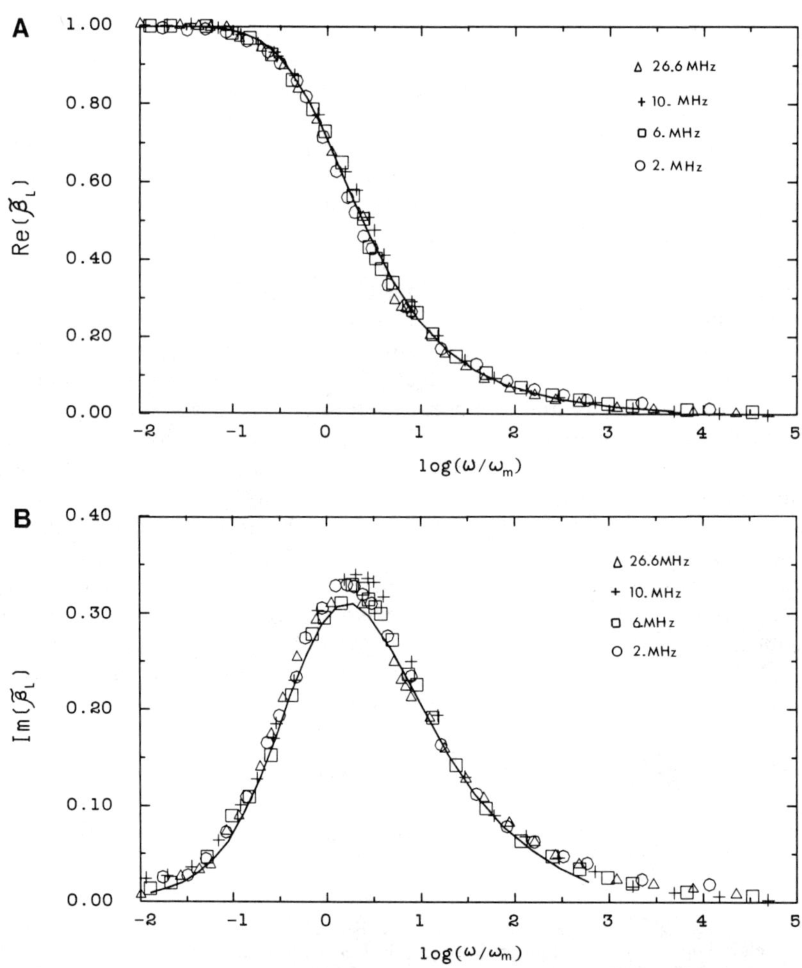

FIGURE 1. The calculated real (**A**) and imaginary (**B**) parts of the normalized compliance, $\tilde{\beta}_L(\omega)$ (solid curve), as functions of log (ω/ω_m), with ω_m^{-1} a characteristic relaxation time (see Jeong *et al.*[8] for the definitions). The data are from Jeong *et al.*[8]

Equations 1 and 2 have been solved iteratively by using the experimental value of the static shear viscosity as an input on the right-hand side of equation 1. The static structure factor has been approximated by choosing one appropriate for a fluid of hard spheres. Further details have been given elsewhere, together with a more extensive discussion of the results.[7] The theory provides a good fit of shear and ultrasound data in a variety of liquids over the entire relaxation region, as shown in FIGURES 1A & 1B. In contrast with some of the previous mode-coupling theories of the glass transition, the full wavevector dependence of the mode-coupling vertices is retained here. This dependence is crucial to providing the continuum of relaxation times needed to obtain nonexponential relaxation.

REFERENCES

1. KIRKPATRICK, T. R. 1984. Phys. Rev. Lett. **53:** 1735–1738.
2. LEUTHEUSSER, E. 1984. Phys. Rev. A **29:** 2765–2773.
3. BENGTZELIUS, U., W. GOETZE & A. SJOLANDER. 1984. J. Phys. C **17:** 5915–5934.
4. DAS, G. F., G. F. MAZENKO, S. RAMASWAMY & J. TONER. 1985. Phys. Rev. Lett. **54:** 118–121.
5. KIRKPATRICK, T. R. 1985. Phys. Rev. A **31:** 939–944.
6. POMEAU, Y. & P. RESIBOIS. 1975. Phys. Rep. **19:** 63–139.
7. MARCHETTI, M. C. 1986. Phys. Rev. A **33:** 3363.
8. JEONG, Y. H., S. R. NAGEL & S. BATTACHARYA. 1986. Phys. Rev. A **34:** 602.

Density Wave Theory of the Liquid-Solid Interface[a]

U. MOHANTY

Department of Chemistry
Boston College
Chestnut Hill, Massachusetts 02167

G. JONES

Department of Physics
University of Notre Dame
Notre Dame, Indiana 46556

A molecular theory of the solid-liquid interface is presented that makes use of structural information about the liquid phase and that avoids using the traditional square-gradient approximation. The theory is an extension of the Jones–Mohanty[1] density wave theory of the liquid-solid transition to the interface problem. The thermodynamic potential difference for a nonuniform solid-liquid system is reduced analytically to a tractable form in terms of the "order parameters" of the theory. Minimization of the thermodynamic potential with respect to the order parameters yields the interfacial density profile.

A formula is obtained for the surface free energy of a diffuse nonplanar crystal-liquid interface; the result resembles a generalized form of the discrete Gaussian model and can be reduced to the SOS model used by Chu and Weeks[2] in their study of the roughening transition. We find that the surface energy acquires *periodic* components due to *broken* translation symmetry of the crystal lattice. The surface is thus characterized by a rational set of Miller indices. An explicit expression is derived for the amplitudes of the periodic potential that pin the surface, and a relation is found between the surface tension and the orientation-dependent surface stiffness. The formulas for the pinning amplitudes provide ways to include fluctuations, in the roughening transition, that are not present in the usual spin-lattice models.

The theory is also used to derive an expression for the rate at which a critical nucleus of the crystalline phase grows from its melt. We derive the Kotchine boundary conditions[3,4] that describe the equation of motion for the growing crystalline drop under the assumption that the mechanism for the dissipation of latent heat occurs by thermal diffusion away from the interface and into the supercooled liquid. We find that the "dynamical prefactor" k, which governs the initial growth rate for a critical droplet, is controlled by the thermal conductivity of the supercooled liquid rather than by the activation energy for diffusion as predicted by the classical nucleation theory of

[a]This work was supported by Petroleum Research Funds and the National Science Foundation.

Turnbull and Fisher.[5]

$$k = \frac{2\lambda T\sigma}{l^2(\rho(R))^2} \frac{1}{R^{*3}} \tag{1}$$

l is the latent heat per molecule, λ is the thermal conductivity, σ is the surface tension, $\rho(R)$ is the density of the solid evaluated at the radius of the drop, and R^* is the radius of the critical droplet. Note that parameters like l, σ, R^*, and the densities of the solid and liquid phases near freezing (which come in the evaluation of $\rho(R)$) are given explicitly by the molecular theory of the liquid-solid interface as developed in this work and by the freezing theory of Jones and Mohanty.[1]

REFERENCES

1. JONES, G. & U. MOHANTY. 1985. Mol. Phys. **54:** 1241.
2. CHU, S. T. & J. D. WEEKS. 1976. Phys. Rev. B **14:** 4978.
3. TURSKI, L. A. & J. S. LANGER. 1980. Phys. Rev. A **22:** 2189.
4. WILMANSKI, K. 1975. Arch. Appl. Mech. **27:** 517.
5. TURNBULL, D. & J. C. FISHER. 1949. J. Chem. Phys. **17:** 71.

Energetics and Variations in Structure in Framework Aluminosilicate Glasses, Melts, and Crystals

ALEXANDRA NAVROTSKY

Department of Geological and Geophysical Sciences
Princeton University
Princeton, New Jersey 08544

Aluminosilicates in the series $M^{n+}_{1/n}$ AlO_2-SiO_2 (M: alkali or alkaline earth) form a series of fully polymerized structures with Al substituting for tetrahedral Si and M filling interstitial cavities. Many properties vary systematically with Al/(Al + Si) and with the nature of M. The non-framework cation, M, interacts with the oxygens bonded to two tetrahedrally coordinated cations (T: Si or Al). The resulting perturbation of the framework is reflected in spectra and in thermodynamic and physical properties. With increasing ionic field strength (z/r) of M, the enthalpy of mixing becomes less exothermic, the enthalpy of vitrification increases, the increase in C_p at T_g becomes more pronounced, viscosity decreases, and crystallization occurs more readily.[1]

Information on bond length and angle distributions is obtained from X-ray radial distributions and analysis of ^{29}Si magic angle spinning NMR data (TABLE 1). With increasing Al/(Al + Si) the average T-O bond length increases significantly because Al-O is longer than Si-O, and the Si-O bond length may increase slightly. The average T-O-T angle narrows with increasing Al/(Al + Si) and with greater perturbing power of the cation M. Although vibrational spectra cannot be interpreted uniquely, and the distance range over which they give information (local coordination versus longer range ring statistics) is still under discussion,[8,9] the overall pattern of spectral shifts supports increasing perturbation and irregularity of the tetrahedral framework with increasing Al/(Al + Si) and with increasing z/r of M.

Ab initio STO-3G molecular orbital calculations on aluminosilicate clusters[10] support these bond length and bond angle trends. A tetrahedral bond length perturbation parameter (ΔTO) consistent with these calculations and with observed bond lengths in crystalline framework aluminosilicates[1,10] can be defined and used to correlate both enthalpy of mixing and enthalpy of vitrification data (FIG. 1).

Recent ^{29}Si NMR spectra of silicate melts at high temperature show one narrow line suggestive of very fast interconversion of silicate species.[11] The ^{27}Al NMR spectra of a sodium aluminosilicate glass quenched from high pressure suggest that the change in Al coordination from fourfold to sixfold occurs above 6 GPa,[12] that is, at pressures significantly higher than those for corresponding transitions in crystals. Very substantial densification of the melt and glass, presumably via changing ring size and bond angle distributions, occurs in the 1–2 GPa range.

TABLE 1. Bond Distances and Angles in Aluminosilicate Glasses

Composition	T (K)	Si-O or T-O Bond Length (Å)	T-O-T Bond Angle (degrees)	Method[a]	Reference
SiO_2	298	1.59 (Si-O)	151	RDF	2
	298	1.62 (Si-O)	146	RDF	3
	298	1.60 (Si-O)	156	NMR	4
	2053	1.62 (Si-O)	140	RDF	5
$0.1CsAlO_2$-$0.9SiO_2$	298	1.60 (Si-O)	152	NMR	4
$0.1RbAlO_2$-$0.9SiO_2$	298	1.60 (Si-O)	150	NMR	4
$0.25LiAlO_2$-$0.75SiO_2$	298	1.61 (Si-O)	143	NMR	4
$0.25NaAlO_2$-$0.75SiO_2$	298	1.63 (T-O)	146	RDF	6
	298	1.61 (Si-O)	148	NMR	4
$0.25KAlO_2$-$0.75SiO_2$	298	1.63 (Si-O)	146	RDF	6
	298	1.61 (Si-O)	144	NMR	4
$0.25Pb_{0.5}AlO_2$-$0.75SiO_2$	298	1.61 (Si-O)	145	NMR	4
$0.25Ba_{0.5}AlO_2$-$0.75SiO_2$	298	1.61 (Si-O)	142	NMR	4
$0.25Sr_{0.5}AlO_2$-$0.75SiO_2$	298	1.61 (Si-O)	142	NMR	4
$0.50Ca_{0.5}AlO_2$-$0.50SiO_2$	298	1.66 (T-O)	143	RDF	6
	298	1.62 (Si-O)	136	NMR	4
$0.50NaAlO_2$-$0.50SiO_2$	298	1.67 (T-O)	143	RDF	7

[a]RDF: radial distribution function, which gives the average T-O (T = Al or Si) bond distance; NMR: ^{29}Si magic angle spinning NMR, which gives the average Si-O bond distance.

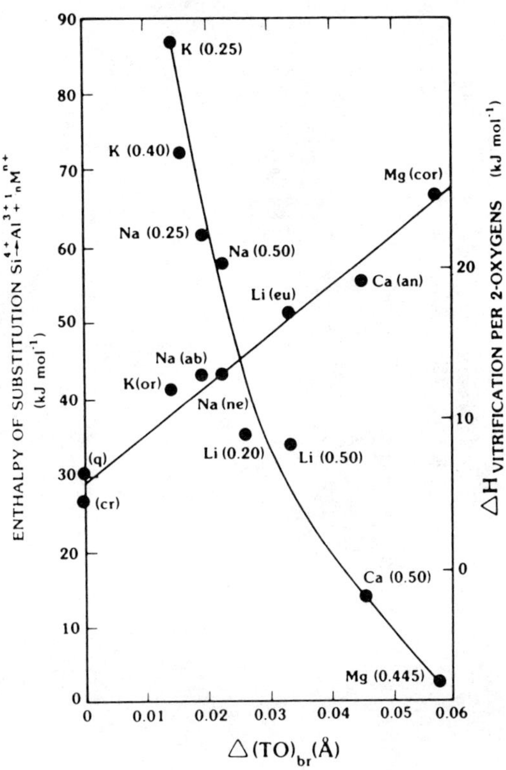

FIGURE 1. Enthalpy of the substitution reaction (left-hand scale and points labeled with mineral names) and of vitrification (right-hand scale and points labeled with chemical symbols) plotted versus the bond length perturbation parameter. Adapted from Navrotsky *et al.*[1]

REFERENCES

1. NAVROTSKY, A. *et al.* 1984. Phys. Chem. Min. **11:** 284.
2. KONNERT, J. *et al.* 1973. Science **179:** 177.
3. MOZZI, R. L. & B. E. WARREN. 1969. J. Appl. Crystallogr. **2:** 169.
4. OESTRIKE, R. 1985. Ph.D. diss. University of Illinois. Urbana, IL.
5. WASEDA, Y. & J. M. TOGURI. 1978. Metall. Trans. **9B:** 595.
6. TAYLOR, M. & G. E. BROWN. 1979. Geochim. Cosmochim. Acta **43:** 61.
7. TAYLOR, M. & G. E. BROWN. 1979. Geochim. Cosmochim. Acta **43:** 1467.
8. MCMILLAN, P. *et al.* 1982. Geochim. Cosmochim. Acta **46:** 2021.
9. SEIFERT, F. *et al.* 1982. Am. Mineral. **67:** 606.
10. GEISINGER, K. *et al.* 1984. Phys. Chem. Min. **11:** 266.
11. STEBBINS, J. *et al.* 1985. Nature **314:** 250.
12. OHTANI, E. & C. A. ANGELL. 1985. Nature **314:** 78.

Empirical Requirements on Models of Relaxation[a]

A. K. RAJAGOPAL,[b] R. W. RENDELL,[c] K. L. NGAI, AND S. TEITLER

Naval Research Laboratory
Washington, DC 20375

We consider here the consistency with experimental observation of four prototypical models of relaxation in complex systems with the three empirically established features

TABLE 1. General Expressions for Relaxation

Model	Expression for Relaxation Function	Conditions
Free volume	$\phi(t) = \int_0^\infty e^{-t/\tau(\nu)} g(\nu)\, d\nu$	$\tau(\nu) = \tau_0 \nu^x, \quad x > 0$ $g(\nu) = A \exp(-C\nu^y), \quad y > 0$
Hierarchically constrained dynamics	$\phi(t) = \int_0^\infty e^{-t/\tau} G(\tau)\, d\tau$	$G(\tau) = \sum_n W_n \delta(\tau - \tau_n)$
Continuous time random walk	$\phi(t) = \exp\left(-c\int_0^t I(t')\, dt'\right)$	$I(t) = \int_{\gamma - i\infty}^{\gamma + i\infty} du e^{ut} \cdot \left[\frac{1}{(1 - \tilde{\psi}(u))P(0, \tilde{\psi}(u))} - 1\right]$ $\psi(t) \sim t^{-1-\alpha}, \quad 0 < \alpha < 1, \quad t \to \infty$ $\tilde{\psi}(u)$ is a Laplace transform of $\psi(t)$. $P(0, Z)$ depends on the geometry of the random walk. c is the concentration of random walkers.
Random energy model	$\phi(t) = \int_{-\infty}^\infty P(f_\alpha) e^{-t/\tau(f_\alpha)}\, d f_\alpha$	$\tau(f_\alpha) = \tau_0 e^{-\beta f_\gamma} \Big/ \left(\sum_\gamma e^{-\beta f_\gamma}\right)$ $P(f_\alpha) = \beta x \exp[\beta x(f_\alpha - f_c)]$, $0 < x < 1$ f_c is an upper cutoff.
Coupling model	$\phi(t) = \exp - \int_0^t W(t')\, dt'$	$W(t') = W_0, \quad \omega_c t' < 1$ $= W_0(\omega_c t')^{-n}, \quad \omega_c t' > 1,$ $0 < n < 1$ ω_c^{-1} separates the regimes for short and long times

[a]This work was supported in part by Contracts N00014-85-C-2315 and N00014-86-WR-24016 with the Office of Naval Research.

[b]On leave from the Department of Physics and Astronomy, Louisiana State University, Baton Rouge, Louisiana 70803.

[c]Sachs/Freeman Associates.

TABLE 2. Implications of Models

Model	Results or Predictions	Conditions
Free volume	$\phi(t) \simeq \dfrac{\exp - (t/\tau)^{y/(x+y)}}{(t/\tau)^{(y-2)/(x+y)}}$	$t/\tau_0 \gg 1$ $2/3 < y/(x+y) < 1$ Arguments for ½ instead of ⅔
	$\tau = \tau_0 \dfrac{Cy}{x}\left(\dfrac{x}{C(x+y)}\right)^{(x+y)/y}$	$1 - n = y/(x+y)$
Hierarchically constrained dynamics[a]		Special choice for relaxation sequence $\omega_{n+1} = \omega_n/\lambda$ $\tau_{n+1} = \tau_n 2^{\mu_0 n^{-P}}, \quad P \geq 1$
	$\phi(t) \sim (t/\tau)^{-\beta/2} \exp - (t/\tau)^{\beta}$	with $P = 1, \quad t/\tau \gg 1, \quad \beta = 1/(1 + \tilde{\mu}_0)$
	$\tau = \tau_0 \exp(\tilde{\mu}_0 \gamma_0)(1+\mu)^{-\gamma\beta}\left(\dfrac{\ln \lambda}{\mu}\right)^{-\tilde{\mu}_0}$	Special choice of matrix elements of a master equation in ultrametric space gives the same result $1 - n = \beta$
Continuous time random walk	$\phi(t) \simeq \exp -(t/\tau)^{\alpha}$	$0 < \alpha < 1$ $t \rightarrow \infty$
	$\tau = \left[\dfrac{0.659\, C}{A\,\Gamma(\alpha)\,\Gamma(1-\alpha)}\right]^{-1/\alpha}$	A is a constant arising from the calculation of $I(t)$ for $t \rightarrow \infty$ $1 - n = \alpha$
	$\tau = \left[\dfrac{0.659\, C}{\alpha\,\Gamma(\alpha)\Gamma(1-\alpha)\,\Gamma(1+\alpha)}\right]^{-1/\alpha} \dfrac{1}{\lambda_0}$	Use of the barrier model of $\psi(t)$ with positive energy barriers only
Random energy model	$\phi(t) = [2\Phi(\theta) - \Phi(2^x\theta)]/(1 - x)$	$\theta = v_0\,\Gamma(1-x)(t/\tau_0)^x, \quad 0 < x < 1$ $\Phi(\theta) = (1-x)\{e^{-x} - \theta^{1/x}\,\Gamma[1 - (1/x)\theta]\}$
	$\phi(t) \sim [\exp - A(t/\tau_0)^x]/(t/\tau_0)^x$	$1 - n = x$ $t/\tau \rightarrow \infty$

[a]Choices other than those shown in the right-hand column will give other $\phi(t)$'s.

described elsewhere in this volume.[1] As shown in reference 1, a fifth model, the coupling model, is consistent with these features.

Lack of space here has forced a general abridgement of presentation. Our procedure will be to list the model type, indicate the basic assumptions involved, and make brief comments on its character. The corresponding general expressions for the relaxation and specific results are tabulated, respectively, in TABLES 1 & 2. The expressions for the specific results provide a context for general comments. As a general observation, we note that the four prototypical models mentioned below include, at best, the first relaxation feature, namely, the Kohlrausch form. None include the other two concomitant features described in reference 1.

1) The "free volume" models[2] involve association of free volume with the relaxing species in a parallel process of relaxation in which the relaxation time and the probability distribution of free volume are adduced from percolation theory. The Kohlrausch form is obtained in the limit of asymptotically long times. Instead of having the range $0 < n < 1$, the exponent of the Kohlrausch form has the range $0 < n < 1/3$. Arguments are presented for extending the range to $0 < n < 1/2$.

2) The hierarchically constrained dynamics[3,4] scheme is a serial relaxation model where the relaxation time at any stage depends on its precedent in a constrained way. Only the choice for relaxation sequence[3] or matrix elements[4] given in TABLE 2 leads to the Kohlrausch form for the limit for asymptotically long times. The total relaxation function with this choice can be fit numerically to a Kohlrausch form over a wide range of time, but with a different value for n than that for the asymptotic evaluation.

3) In the continuous time random walk model,[5] relaxation is caused by the interaction of random walker defects with a dipole. The relaxation depends on the specific form of the random walker's waiting time distribution. With the choice of waiting time distribution given in references 5 and 6 (see TABLE 2), the continuous time random walk model at present only has the Kohlrausch form in the limit for asymptotically long times. The chosen waiting time distribution is not consistent with that observed in ionic conductivity measurements.

4) The "random energy" model has been developed to describe the slow relaxation exhibited in spin glasses.[7] This model also leads to the Kohlrausch form only in the limit for asymptotically long times. The exponent of the Kohlrausch form is temperature dependent and vanishes at and above the spin glass transition temperature.

REFERENCES

1. NGAI, K. L., R. W. KENDELL, A. K. RAJAGOPAL & S. TEITLER. 1986. Ann. N.Y. Acad. Sci. This volume.
2. COHEN, M. H. & G. S. GREST. 1981. Phys. Rev. B **24:** 4091.
3. PALMER, R. G., D. L. STEIN, E. ABRAHAMS & P. W. ANDERSON. 1984. Phys. Rev. Lett. **54:** 365.
4. OGIELSKI, A. T. & D. L. STEIN. Phys. Rev. Lett. **55:** 1634.
5. SHLESINGER, M. F. & E. W. MONTROLL. 1984. Proc. Natl. Acad. Sci. USA **81:** 1280.
6. BENDLER, J. T. & M. F. SHLESINGER. 1985. Macromolecules **18:** 59.
7. DEDOMINICIS, C., H. ORLAND & F. LAINEE. 1985. J. Phys. Lett. **46:** L463.

Pressure Variation of the Glass Transition Temperature in Atactic Polystyrene[a]

J. R. STEVENS, R. W. CROAKLEY, K. W. CHAU,
AND J. L. HUNT

Guelph–Waterloo Program
for Graduate Work in Physics
University of Guelph
Guelph, Ontario, Canada N1G 2W1

Values for the frequency shifts, F, obtained by Brillouin light scattering are used to measure the variation with pressure, P, of the zero-frequency glass transition temperature, T_g, in atactic polystyrene at values of P of up to 3 kbar. The values of T_g obtained by Brillouin scattering are no different from those obtained by dilatometric and other "zero-probe-frequency" techniques that use the same experimental time scale for the variation of T or P.[1] Only the local viscoelasticity of the polymer is probed; a procedure similar to this one could be used for any liquid—glycerol, for example.[2]

We chose to use Brillouin scattering to obtain T_g because every aspect of our experiment can be very carefully computer monitored and controlled. One of the challenges is to see if the variation of T_g with P reflects the variation of T_2 (defined below) with P as predicted by extant theories.[3,4]

Values of F are related to hypersonic longitudinal velocities and are obtained from light scattered at 90° using a Fabry–Perot interferometer. Measurements of F for each run were begun at $T > T_g + 50°$ C and were continued as T was slowly (~2° C/hr) decreased. Atmospheric pressure was maintained throughout the first run; after this and each subsequent isobaric run (except the last), the sample was returned to the liquid-like region and P increased by about 0.25 kbar to prepare for the next isobaric run. The sample was very clean, chemically and physically, and had a weight-average molecular weight, M_w, of 8.0×10^5 and a number-average molecular weight, M_n, of 3.6×10^5 ($M_w/M_n = 2.2$). Other experimental details may be found in Stevens *et al.*[1]

For a particular pressure the variation of F with T in the solid (glassy) region can be described empirically by

$$F_s = a_{s(T-0)}(P) + a_{1s}(P)\,T \qquad (1)$$

Similarly, for the liquid-like region

$$F_\ell = a_{\ell(T-0)}(P) + a_{1\ell}(P)\,T \qquad (2)$$

The liquid-like and solid regions are separated by the glass transition region, which is characterized by $T_g(P)$. Values for $T_g(P)$ are determined by setting $F_s = F_\ell$. Values of $a_{s(T-0)}$ and $a_{\ell(T-0)}$ were obtained at $T = 0°$ C. These values and the values of a_{1s} and $a_{1\ell}$ are plotted in FIGURE 1 as functions of P.

[a]This work was supported by the Natural Sciences and Engineering Research Council of Canada.

The intercepts $a_{\ell(T-0)}$ (P) and $a_{s(T-0)}$ (P) were obtained for each of 93 isobaric runs—in which P ranged from 0.056 (in the first run) to 3.03 kbar (in the last run)—and were seen to increase with P in approximately the same way; they were almost linear and parallel over the whole pressure range. The slopes $a_{1\ell}$ (P) and a_{1s} (P) were demonstrated to be functions of P and T, the latter with a much weaker dependence than the former.

In the present experiment as exacting analysis indicates that the functions $a(P)$ can always be represented as negative exponentials of the form

$$a(P) = r - s\exp(-tP) \tag{3}$$

The parameters were obtained using a nonlinear least-squares regression analysis. The

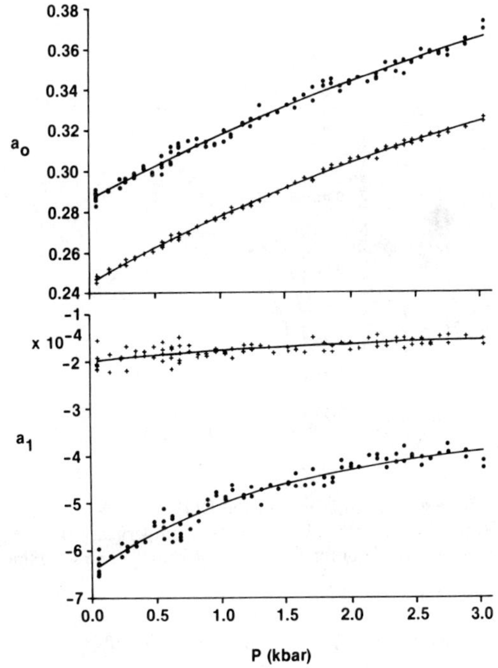

FIGURE 1. Plot of the fitting coefficients $a_{\ell(T-0)}$, $a_{s(T-0)}$, $a_{1\ell}$, and a_{1s} versus P (kbar). The solid curves are obtained by a nonlinear regression analysis fitting to exponential functions. Each of the curves drawn through dots is for the liquid (ℓ); each of the curves drawn through plus signs is for the solid (s). For a_o the upper curve is for the liquid; for a_1 the upper curve is for the solid.

values of r, s, and t are different for each of the parameters $a(P)$ although those used for $a_{\ell(T-0)}$ and $a_{s(T-0)}$ are within 20% of each other; these values may be found in Stevens *et al.*[1]

Our results for T_g are extrapolated to high pressures (~40 kbar). We were encouraged to make this projection because of the behavior of the coefficient $a_{1\ell}$ (FIG. 1), which apparently reaches an asymptotic limit even at 3 kbar. Using equations 1, 2, and 3, we found that T_g at $P = 0$, $T_{g(P-0)}$, is 91° C (364 K); T_g at $P = \infty$, T_{gc}, is 313° C (586 K); and dT/dP at $P = 0$ is 38 K·bar^{-1}. The value of $T_{g(P-0)}$ obtained in this experiment agrees with those obtained by Chang[5] and Richards and Savill[6] for similar temperature rates. We found that T_g (extrapolated) as a function of P is approximately represented by $(T_{gc} - T_g)/(T_{gc} - T_{g(P=0)}) = \exp(-0.14P)$; the error is about 5% but is not random. The experimental and extrapolated results are shown in FIGURE 2.

It is recognized that the glasses formed under isobaric conditions at different formation pressures have slightly different volumes in the glass. These slight differences, however, have a negligible effect on the Brillouin phonon frequences.[7]

DiMarzio, Gibbs, Fleming, and Sanchez[3] have used a lattice model as the basis of a theory (the DGFS theory) of the glass transition. A second-order thermodynamic (equilibrium) transition is postulated to occur at a temperature T_2 when the configurational entropy decreases to a critically small value. According to the theory, T_g increases nonlinearly with pressure and asymptotically approaches a constant value at very high pressures (~10 kbar).

Purely for the sake of comparing our experimental results with the DGFS theory,

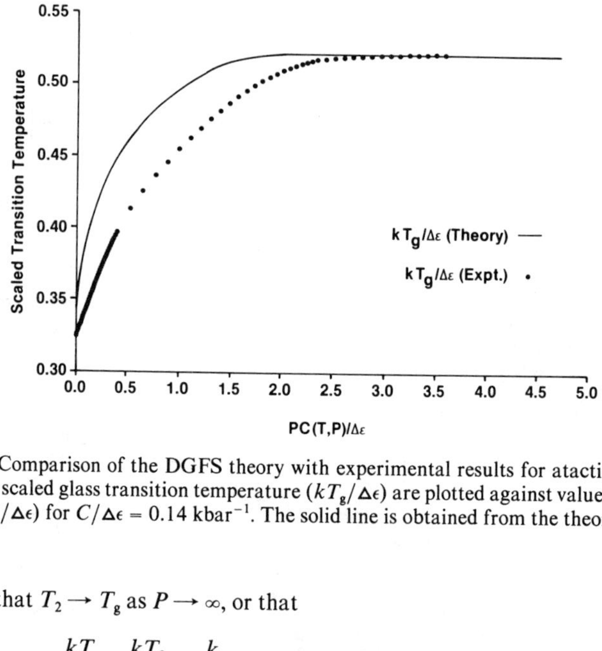

FIGURE 2. Comparison of the DGFS theory with experimental results for atactic polystyrene. Values of the scaled glass transition temperature $(kT_g/\Delta\epsilon)$ are plotted against values of the scaled pressure $(PC/\Delta\epsilon)$ for $C/\Delta\epsilon = 0.14$ kbar^{-1}. The solid line is obtained from the theory.

we assume that $T_2 \rightarrow T_g$ as $P \rightarrow \infty$, or that

$$\frac{kT_g}{\Delta\epsilon} = \frac{kT_2}{\Delta\epsilon} + \frac{k}{\Delta\epsilon}(T_{g(P=0)} - \mathrm{T}_{2(P=0)}) \exp(-0.14P) \quad (4)$$

where k is Boltzmann's constant, and $\Delta\epsilon$ is the energy difference between bonds flexed out of their lowest energy state and bonds in the low-energy state. Each cell in the lattice model has a volume C, and we take $C/\Delta\epsilon = 0.14$ kbar^{-1}. We were guided in our choice of $T_{2(P=0)}$ by the proposals of Adam and Gibbs,[8] which are based upon the DGFS theory.[3] FIGURE 2 compares our extrapolated values of T_g with the theory.[3]

We believe that we have demonstrated that in atactic polystyrene glass there is a limiting critical value (T_{gc}) at high pressures for the glass transition. At T_{gc}, $dT/dP = 0$. This implies that for temperatures above T_{gc} the "liquid" can never be compressed to a glass. The DGFS equilibrium entropy theory for polymers qualitatively predicts this result, but quantitatively has some shortcomings. The molecular theory of Bengtzelius *et al.*[4] for a classical monatomic liquid indirectly predicts this result.

It would be interesting to check the finding with other glass-forming materials. Glycerol has been studied using dielectric relaxation techniques up to pressures of 53 kbar,[9] but the authors acknowledge that the results are preliminary. Too few data points, technical difficulties, and the imposition of linear extrapolations, especially at very low frequencies and high pressures, make meaningful conclusions difficult. They acknowledge, however, that at constant frequency of maximum loss ω_{max}, dT/dP is not linear, and a rough examination of their data indicates that it decreases with increasing pressure at low ω_{max}.

Questad *et al.*[10] have also measured dielectric relaxation in a polyurethane elastomer up to 7 kbar. Dilatometric measurements were also made. Their results show very clearly the difference between the glass transition temperature as defined here and the α-relaxation, the former being independent of probe frequency. They observe that dT_g/dP decreased with increasing pressure up to about 3 kbar but then claim that a constant value is reached, a result with which we disagree. Atake and Angell[11] report on pressure measurements up to 2.5 kbar on various organic liquids that form glasses. These include polyphenyl hydrocarbon *o*-terphenyl and polyphenyl ether 5P4E. They also observe that dT_g/dP decreased with increasing pressure, but indicate that their differential thermal analysis data above 1 kbar are generally of poorer quality.

In conclusion there is support in the literature for the trend in dT_g/dP observed in our study. We believe that the predictions of the two theories cited are confirmed and that $dT_g/dP \rightarrow 0$ at high pressures.

REFERENCES

1. STEVENS, J. R., R. W. COAKLEY, K. W. CHAU & J. L. HUNT. 1986. J. Chem. Phys. **84:** 1006.
2. PINNOW, D. A., S. J. CANDAU, J. T. LAMACCHIA & T. A. LITOVITZ. 1968. J. Acoust. Soc. Am. **43:** 131.
3. DIMARZIO, E. A., J. H. GIBBS, P. D. FLEMING III & I. S. SANCHEZ. 1976. Macromolecules **9:** 763.
4. BENGTZELIUS, U. & L. SJÖGREN. 1986. J. Chem. Phys. **84:** 1744.
5. CHANG, S. S. 1984. J. Polym. Sci., Polym. Symp. **71:** 59.
6. RICHARDSON, M. J. & N. G. SAVILL. 1977. Polymer **18:** 3.
7. COAKLEY, R. W., J. L. HUNT & J. R. STEVENS. 1983. J. Appl. Phys. **54:** 5603.
8. ADAM, G. & J. H. GIBBS. 1965. J. Chem. Phys. **43:** 139.
9. JOHARI, G. P. & E. WHALLEY. 1972. Faraday Symp. Chem. Soc. **6:** 23.
10. QUESTAD, D. L., K. D. PAE, B. A. NEWMAN & J. I. SCHEINBEIM. 1980. J. Appl. Phys. **51:** 5100.
11. ATAKE, T. & C. A. ANGELL. 1979. J. Phys. Chem. **83:** 3218.

Existence of "Critical" Frequencies in the Dielectric Loss Properties of Glasses

U. STROM

Naval Research Laboratory
Washington, DC 20375

The dielectric loss in glasses exhibits certain universally observed dependencies on frequency. These include a quadratic dependence of the conductivity $\sigma(w) = w\epsilon'' \sim w^2$ just below the dominant phonon modes,[1,2] and a dependence of $\sigma(w) \sim w^n$ $(n < 1)$ in the region below $\sim 10^8$ Hz.[3,4] These regions delineate the disorder-induced coupling to phonons and dielectric loss, respectively. Each of these dependencies can be observed to change relatively abruptly within a narrow frequency range.

The higher frequency w^2 loss can be expressed in terms of the product of a vibrational density of states and a coupling coefficient. Recent neutron-scattering measurements in vitreous SiO_2[5] have determined the low-frequency vibrational density of states; such measurements allow the separate determination of the far-infrared coupling coefficient. It has been found[1] that the coupling coefficient to an w^2 phonon density of states can be expressed as $c(w) = 1 - 1/(1 + q^2l^2)^2$, where $q = w/v$ (v is the Debye sound velocity) and l is a structural length of glass. Fits of $c(w)$ to the far-infrared dielectric loss yield $l = 18$ Å for glassy SiO_2.[1,6] The condition $ql \simeq 1$ defines a "critical" frequency $f_c = v/2\pi l \sim 5 \times 10^{11}$ Hz for SiO_2. This frequency can also be viewed as marking the onset of phonon localization.[7,8]

The lower frequency w^n loss is observed in many glasses and polymers. This loss is often due the thermally activated motion of molecules or ions that cannot be described in terms of the Debye theory. Deviation from w^n power law behavior is theoretically expected for sufficiently high frequencies, which generally fall into the microwave range (10^9 to 10^{11} Hz). Dielectric loss data in this frequency region is very limited. Where data is available it is evident that other loss mechanisms, such as the direct coupling to phonons, often obscure the w^n loss. The phonon loss mechanism, however, has been shown to be considerably less important for the disordered ionic conductor sodium beta alumina, which exhibits many of the characteristic thermal and dielectric properties of glasses, including a non-Debye-like low-frequency dielectric loss.[4] This loss, which extends well into the microwave regime, is characterized by a well-defined frequency, $w_o \sim 10^9$ Hz, above which the ionic conductivity no longer follows w^n behavior.[9] Although this observation has been described in terms of the energy dependence of the density of states of certain low-energy excitations,[4] the relationship of w_0 to w_c and the relationship of w_o to the glass structure are not understood.

REFERENCES

1. STROM, U. & P. C. TAYLOR. 1977. Phys. Rev. B **16:** 5512.
2. STROM, U. 1985. *In* Relaxations in Complex Systems. K. L. Ngai & G. B. Wright, Eds.: 229. U. S. Government Printing Office. Washington, DC.
3. STROM, U. 1983. Solid State Ionics **8:** 255.

4. NGAI, K. L. & U. STROM. 1983. Phys. Rev. B **27:** 6031.
5. BUCHENAU, U., N. NÜCKER & A. J. DIANOUX. 1984. Phys. Rev. Lett. **53:** 2316.
6. STROM, U. 1986. Phys. Rev. Lett. **56:** 2110.
7. JOHN, S., H. SOMPOLINSKY & M. J. STEPHEN. 1983. Phys. Rev. B **27:** 5592.
8. KIRKPATRICK, T. R. 1985. Phys. Rev. B **31:** 5746.
9. STROM, U. & K. L. NGAI. 1983. Solid State Ionics **9** & **10:** 283.

Glass Transition Kinetics in Polystyrene Using the Pressure-jump Volume Relaxation Method

J. J. TRIBONE, J. M. O'REILLY, AND J. GREENER

Research Laboratories
Eastman Kodak Company
Rochester, New York 14650

An apparatus has been constructed that permits the measurement of time-dependent changes in volume following a sudden change in pressure near the point of vitrification. The same instrument is used for measuring steady state pressure, volume, and temperature properties that are necessary for a proper analysis of the dynamic measurements. The former experiments are referred to as pressure-jump volume relaxation (PJVR) measurements and are considered a direct probe of the structural relaxation process that occurs in all glasses.[1] This type of experiment relies on the nonequilibrium nature of the glassy state and the relatively slow approach toward equilibrium near the vitrification temperature and pressure.[2]

Experimentally, the measurements are performed as shown in FIGURE 1. Following equilibration on the liquidus line (point A), the sample is subjected to a rapid change in pressure to enter the nonequilibrium glassy state (point B). The pressure is then held constant during the relaxation of the volume toward a new equilibrium state (point C). The cycle is completed by executing another pressure step of equal magnitude, but opposite in sign, resulting in another nonequilibrium state (point D), followed by recovery toward the initial volume (point A). Experiments are then repeated at different starting pressures and temperatures.

The present experiments have been performed on polystyrene (T_g at zero pressure is 100° C) at temperatures from 110 to 150° C and at pressures of up to 2 kbar using pressure steps of ~500 bars. The qualitative observations are analogous to those obtained at atmospheric pressure by rapid changes in temperature,[3] namely, 1) nonlinearity with respect to the magnitude of the departure from equilibrium, 2) asymmetry with respect to the sign of the departure from equilibrium, and 3) memory effects associated with complicated temperature or pressure histories. Each of these effects can be accounted for by a phenomenological order parameter model that has been extended to include the effect of pressure.[4] This model assumes that a distribution of relaxation times exists and that these relaxation times depend on the external variables (temperature and pressure) as well as on the instantaneous departure of the system from equilibrium. This can be represented as a product of independent shift factors

$$\tau = \tau_0 \cdot a_T \cdot a_\delta \cdot a_P \tag{1}$$

where τ_0 is a reference relaxation time, $a_T = \exp[-\theta(T - T_r)]$, $a_\delta = \exp[-(1 - x)\theta\delta/\Delta\alpha]$, and $a_p = \exp(b_P P)$. Note that T_r is a reference temperature, θ is proportional to an activation energy, x is an adjustable parameter that reflects the nonlinearity,[5] $\delta = (V/V_\infty) - 1$, V_∞ is the equilibrium volume, $\Delta\alpha$ is the change in the

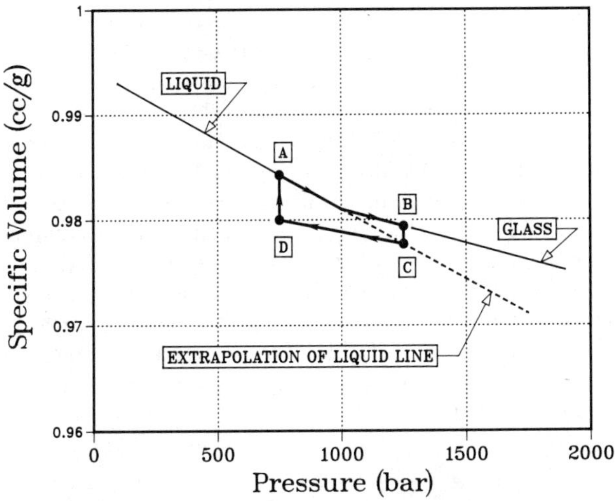

FIGURE 1. Schematic diagram of a PJVR experiment. Numbers shown correspond approximately to polystyrene at 130° C.

thermal expansivity at the glass transition, and b_P is another adjustable parameter. The distribution is represented by the Williams–Watts function,[6] whose exponent β is a measure of the breadth of the distribution. The instantaneous structure dependence of the relaxation times is governed by a parameter x that was empirically postulated by Narayanaswamy.[5]

Each contraction-expansion cycle described above is analyzed by nonlinear optimization. At low temperatures and pressures the fit is excellent. Deviations between

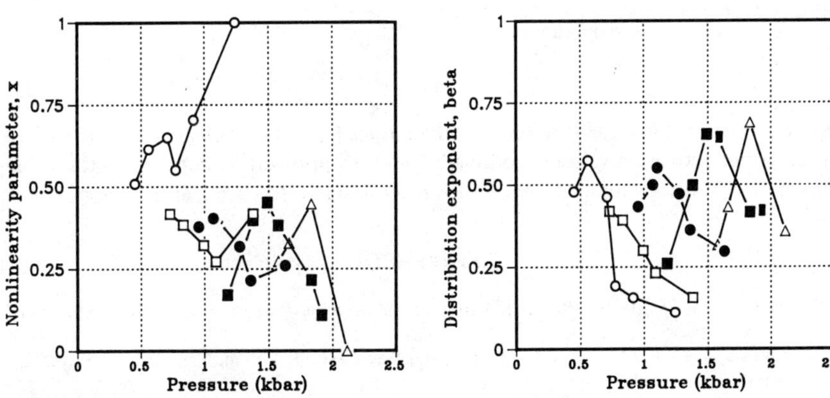

FIGURE 2. Plot of the adjustable parameters x and β versus the pressure during the contraction isobar at various temperatures: ○: 110; □: 120; ●: 130; ■: 140; and △: 150° C. Each point has been determined by nonlinear optimization over one contraction-expansion cycle. The points at $x = 0$ and the point at $x = 1$ are probably artifacts associated with the instability of the calculation at these points.

theory and experiment, however, increase as temperature and pressure increase, this being manifest mostly in a predicted recovery curve (expansion isobar) that recovers the equilibrium volume more quickly than the experimental data; the contraction isobars are usually predicted within experimental error. Two of the adjustable parameters of the model (x and β) are found to depend on the pressure and temperature history in a way that does not lend itself to a clear physical explanation (FIG. 2). The parameter b_P is found to depend on temperature and pressure in a systematic way (FIG. 3). From this one calculates an activation volume, $\Delta V^{\ddagger} = RT(\partial \ln \tau/\partial P)_{\tau} = RTb_P$, which ranges from 100 to 600 cc/mol. An intriguing trend is for b_P (and therefore for $\Delta V^{\ddagger}$) to increase with increasing temperature; this suggests that at higher temperatures (and therefore at higher pressures) more segments are involved in the relaxation and recovery process, the calculated number ranging from 5 to 60 segments, assuming that the activation volume itself represents only a fraction of the dynamic unit (as observed in molecular glasses).[7]

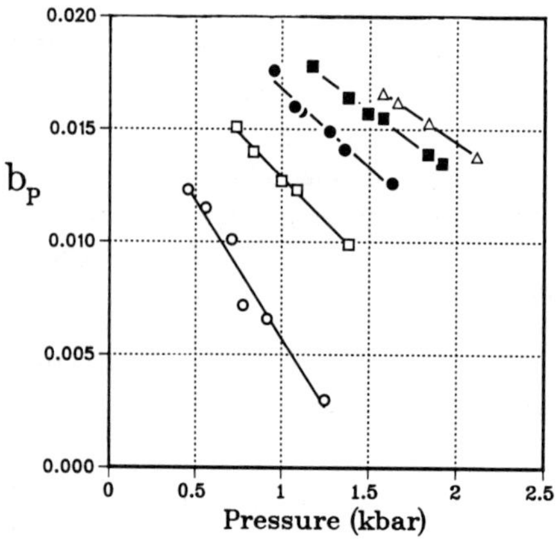

FIGURE 3. Plot of the adjustable parameter b_P versus the pressure during the contraction isobar. The symbols have the same meaning as in FIGURE 2.

[NOTE ADDED IN PROOF: More recent calculations have shown that b_P, and therefore the apparent activation volume, obtained from the optimization depends on the value chosen for θ, the activation energy. This is discussed in reference 2 of this paper.]

REFERENCES

1. STRUIK, L. C. E. 1978. Physical Aging of Amorphous Polymers and Other Materials. Elsevier. New York, NY.
2. TRIBONE, J. J., J. M. O'REILLY & J. GREENER. 1986. J. Polym. Sci.: Polym. Phys. Ed. Submitted for publication.
3. KOVACS, A. J. 1963. Fortschr. Hochpolym.-Forsch. **3:** 394.
4. KOVACS, A. J., J. M. O'REILLY, J. J. TRIBONE, J. GREENER & A. R. RAMOS. 1986. J. Polym. Sci.: Polym. Phys. Ed. Submitted for publication.
5. NARAYANASWAMY, O. S. 1971. J. Am. Ceram. Soc. **54:** 491.
6. WILLIAMS, G. & D. C. WATTS. 1970. Trans. Faraday Soc. **66:** 80.
7. ANDERSON, J. E. & W. P. SLICHTER. 1966. J. Chem. Phys. **44:** 1797.

Index of Contributors

(Italic page numbers refer to comments made in discussion.)